Keeping to the Marketplace

Keeping to the Marketplace

The Evolution of Canadian Housing Policy

JOHN C. BACHER

McGill-Queen's University Press
Montreal & Kingston • London • Buffalo

ISBN 0-7735-0984-4

Legal deposit second quarter 1993
Bibliothèque nationale du Québec

Printed in Canada on acid-free paper

This book has been published with the help of a grant from the Social Science Federation of Canada, using funds provided by the Social Sciences and Humanities Research Council of Canada.

Canadian Cataloguing in Publication Data

Bacher, John C. (John Christopher), 1954–
Keeping to the marketplace: the evolution of Canadian housing policy
Includes index.
ISBN 0-7735-0984-4
1. Housing policy – Canada I. Title.
HD7305.A3B33 1993 363.5'0971 C93-090167-3

This book was typeset by Typo Litho composition inc. in 10/12 Palatino.

Contents

Contents

Preface

The roots of my long odyssey into Canadian housing policy lie in early childhood memories of the destruction of cherished valleys, orchards, vineyards, and historic buildings. The scenes I witnessed were part of a significant Canadian environmental controversy, the loss of the lands of the unique Niagara fruit belt. To understand what drove this process, which appeared to me as a destructive force of bulldozers unleashed on the land with the acceleration of urban sprawl after the Second World War, I delved into the St Catharines public library. Here I discovered even more disturbing urban-renewal and transportation studies, which called for the levelling of most of the urban core and interminable road extensions. But discovering a method behind the apparent madness of ecological and architectural degradation helped to spark an interest in urban history.

As a graduate student in history at McMaster University in 1979, I planned to examine more fully the evolution of environmental land-use problems in the "Golden Horseshoe" arc of southern Ontario, from Oshawa to Niagara. Aware – from the writings of James Lorimer and Humphrey Carver – of the connection between federal housing policy and the problem of urban sprawl, I originally hoped to combine a study of both problems. Fortunately, however, I was steered by the three wise men of my MA thesis advisory committee, John Weaver, Lou Gentilcore, and Harry Turner, to the more manageable topic of housing. My research was supported by generous scholarship assistance from the Canada Mortgage and Housing Corporation, and guided by my adviser John Weaver. Housing policy remained the focus of both my master's thesis and my doctoral dissertation. After graduation I had an academic's dream come true: a chance to combine study with direct government experience by researching a history of the Metropolitan Toronto Housing Company. This was helped immeasurably by the frank recollections of its president, George Barker, and notable social-housing pioneers Albert Rose and Clare Clark.

My experience of service in the voluntary sector, with the aim of protecting the rural landscape, has continued to reinforce my academic and professional research. The importance of combining both can be seen in the common tendency to disparage any reform in the land-use planning process as an infamous booster of housing costs. A partner in both academic research and social activism whose work proved critical to mine was the late Anne Mason Apps, whose brave research into the dark areas of organized crime and politically connected land speculation should provide inspiration for generations of Canadian intellectuals.

Although this book frequently seemed an impossible task, and despite the pessimistic view suggested by the title, the history I retraced in its writing was frequently inspirational. For while Canadian housing policy has been characterized by a zeal to uphold the private marketplace, a cause championed mainly by the powerful federal Department of Finance, there is a brighter side to the story. Its history has been full of heroes who challenged the ideas of the anti-hero of much of this book, W.C. Clark. While social housing was long delayed by Clark's ingenious machinations, the final arrival of subsidized shelter helped countless Canadians to escape the poverty trap. Although originally social housing in Canada was visually designed to proclaim that it was inferior accommodation intended to serve a low-income group, the vision of the generation of "flower power" helped to spark a vibrant third sector that aimed to make high-quality shelter affordable for all.

Writing the history of Canadian housing policy has made me more conscious than ever of the importance of the "Red Tory" tradition to Canadian social history. The unhappy contrast between the dazzling new public-housing projects of the New Deal and the relief camps where Canada's homeless were wasted was one of the conditions that led to the continentalism of that critical shaper of national consciousness, Frank Underhill. Later events, such as the burning of American ghettos and the demolition of poorly maintained public-housing projects, served to stimulate the distinctive ideals of Canadian Tories. The American early-warning system sounded an alarm that halted the urban-renewal bulldozer in its tracks and replaced it with small-scale, income-integrated housing projects built by municipalities, native housing corporations, non-profit associations, and co-operatives.

The Red Tories I have met and read about in the process of drafting this book have been quite refreshing. It seems appropriate that many of the critical ideas shaping heretical social-housing policies should have been formulated in the magnificent chateau Chorley Park, which housed the lieutenant-governor of Ontario, Herbert Bruce. This Kremlin of Red Toryism was, with appropriate symbolism, closed down by the most Whiggish of Ontario premiers, Mitch Hepburn. The delight that this foe of tradition took in evicting the king's representative from a palace where radicals gathered seems to have been surpassed only by his pleasure in organizing strikebreakers to assault industrial unionism in the infamous Oshawa General Motors–UAW conflict.

The ideals Bruce nurtured in Chorley Park first sprouted in the public-housing innovation of Regent Park and then blossomed more luxuriantly in the co-operative Alexandria Park. Along the path of these three parks trod Harold Clark, a Bay Street financial wizard who was the antithesis of W.C. Clark. French Canada had its own Red Tory champion in Paul Dozois, a businessman and respected social reformer who brought public housing to Quebec over enormous opposition. In western Canada similarly socially responsible business figures brought public housing to hostile environments by uniting with labour and social activists over a fierce chorus of protest from real estate interests; the same alliances triumphed in the Atlantic provinces. An ethic of community united diverse and even conflicting social groups.

Creative class coalitions searching for alternatives to the housing market eventually won out in the most adverse circumstances. Polarized conflicts sparked protests over evictions and rent increases, but led to only short-term victories for working people. Social solidarity set the conditions for longer-term gains through the establishment of innovative measures for social housing. While Communists in Vancouver and Montreal in the midst of the post–Second World War housing shortage organized dramatic squatting occupations of vacant properties, their comrades in Toronto more quietly and discreetly worked with the conscience of big business to bring about the nation's first public-housing project.

Although W.C. Clark waged a determined battle to prevent the construction of a single unit of social housing, he did so out of a deep, even fanatical personal conviction, not in service to the interests of any identifiable sector of capital. Indeed, his lordly crusade resembled the ascetic craft of a high priest or the ruling ideologue of a Marxist regime. For the small-scale building, retail lumber, and real estate interests that benefited from his opposition to social housing Clark had supreme contempt; he hoped to replace them with more socially responsible large-scale developers who would plan entire communities. Much of Clark's success came from demonstrating a concern for urban form similar to that of his opponents. Later critics of federal policy in the 1960s would wrongly seize upon the big developers he fostered as the key villains on the urban scene – a mistake they would soon realize when residents concerned with real estate values challenged even the most sensitively designed mixed-income projects.

While Red Toryism has had many visible successes, its survival, as George Grant noted, is always uncertain in the shadow of a powerful United States with a very different philosophy. Much of the often weak Red Tory legacy is now being assailed by the Reform Party, emerging out of Alberta, the province that resisted the introduction of any social-housing programs longest. Indeed, the party is headed by the son of the Social Credit premier who blocked the construction of a single unit of public housing in the province during the long period from 1949 to 1966, contemptuous of federal legislation used in other jurisdictions. But even the most clever alliance of narrow-minded interests, focused on the primacy of pecuniary land values over human suffering, will have to contend with the enduring strength of a social-housing movement that continually seeks new forms to realize the values of community.

Critical Stages in Canadian Housing Policy

1912	First limited-dividend housing projects in Toronto
1919	First federal housing measure
1935	Passage of Dominion Housing Act establishing joint loans
1948	Amendments to National Housing Act (NHA) for limited-dividend housing
1949	NHA amendments for publicly subsidized low-rental housing and land banking
1954	New NHA replacing joint-loan program with mortgage insurance
1956	NHA amendments for urban renewal
1964	NHA amendments to speed public-housing construction
1969	Urban renewal terminated
1973	NHA amendments to encourage third-sector housing
1978	End of federal land banking, transfer of much program administration to provinces
1986	Ontario begins Homes Now program, first major, exclusively provincial social-housing program
1992	Federal government ends direct delivery of co-operative housing program

Far left, W.C. Clark, deputy minister of finance 1933–53, who masterminded the marketplace thrust of Canadian housing policy. The others in the picture, reading from left to right, are R. Bryce, Sir Wilfred Rady, Norman Roberston, the Honourable Malcom MacDonald, Gordon Munro, the Honourable J.L. Isley, and William Lyon Mackenzie King.
Courtesy the National Archives of Canada, PA 150450.

Above right: Toronto businessman and social reformer Harold (W.H.) Clark, who played a critical role from the 1940s through the 1970s as a promoter of public housing and of the "third-sector" approach adopted in the NHA amendments of 1973.
Courtesy the Clark family.

Above left: Humphrey Carver, a pioneer who campaigned for social housing in Canada through his work both in government and in the voluntary sector.
Courtesy the National Archives of Canada, PA 150446, photo by Paul Orzdale.

Below left: David Mansur, first president of CMHC (1946–54), who modified but maintained the policies of his mentor, W.C. Clark, by making them more flexible.
CMHC, *Habitat* 8–9 (1966): 4. Courtesy the Metropolitan Toronto Library Board.

Slum housing in Toronto at the time of the Bruce Report, 1935.
Courtesy the City of Toronto Archives, Globe and Mail Collection, 33–226.

In its elegance, the first home financed under the Dominion Housing Act (1935) typified the failure of both the early joint-loan program and subsequent mortgage insurance to reach any but the affluent top 20 per cent in income among Canadian families. The housing was built by Wilfred Cude, famed National Hockey League player, and is in the Town of Mount Royal, Montreal.
CMHC, *Habitat* 2, no. 6 (1959): 5.
Courtesy the Metropolitan Toronto Reference Library.

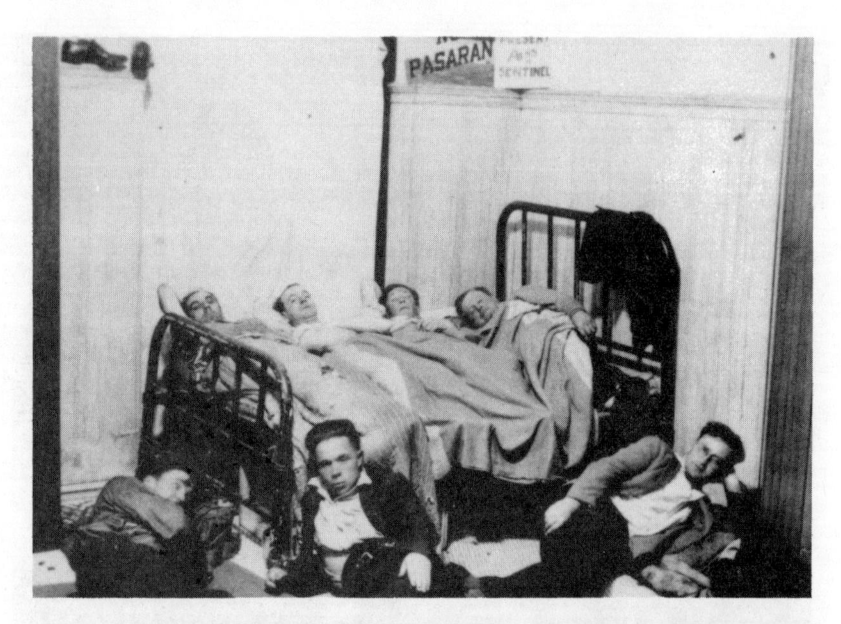

These photographs of Depression-era hostels for homeless single men illustrate the harshly punitive conditions they had to endure.
Courtesy the National Archives of Canada, C 13236.
Courtesy the Archives of Ontario, S 15754.

This line-up in front of the Toronto Emergency Housing Registry in 1945 was for very bleak accommodation, generally in former army barracks.
Courtesy York University Archives, Toronto Telegram Collection, box 278, file 1955.

Severe housing conditions in Montreal during the Second World War. Courtesy the National Archives of Canada, PA 129012.

A veterans' Wartime Housing project, converted into a limited-dividend seniors' project by a service club in Vancouver. Few such projects were successfully converted into long-term social housing, and seldom for low-income families. Courtesy the National Archives of Canada, PA 170064.

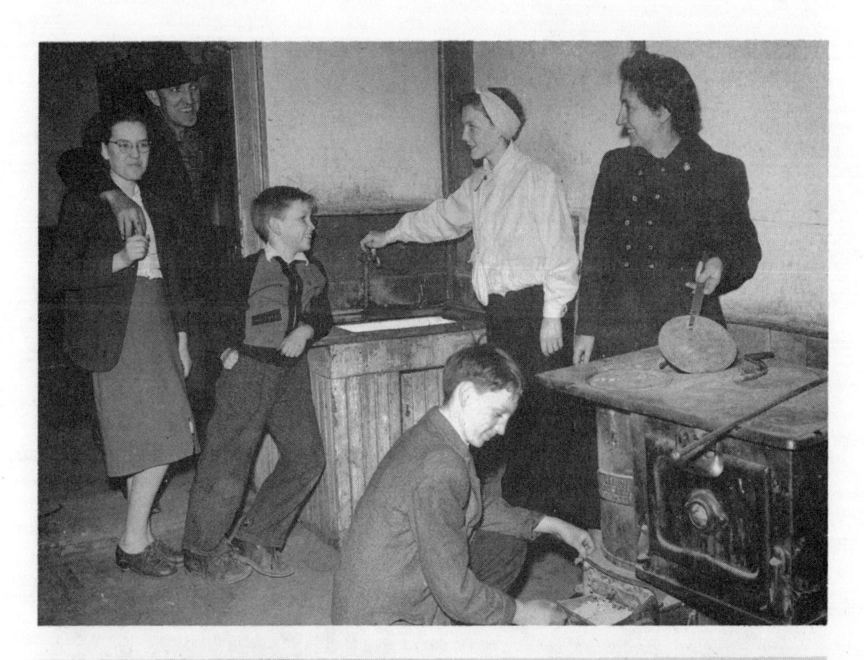

The opening of Toronto's pioneering Regent Park project (1949) was attended with a sense of high idealism, especially evident in the concern for the welfare of children. The before and after photographs illustrate the dramatic improvement in tenants' living conditions after their move to the project.
Courtesy the City of Toronto Archives, Globe and Mail Collection, 132432.
Courtesy the National Archives of Canada, PA 128760.

The reliance on the financial support of a business association for the Regent Park project is a reminder that in the early days of public housing it was difficult to obtain funding for recreational programs.
Courtesy the City of Toronto Archives, RG 28–7, photo by Canada Wide Photo; RG 28–94, photo by McFadden.

The Parkchester Apartments, New York, built by the Metropolitan Life Insurance Corporation and used by W.C. Clark as a model to encourage private capital investment in rental housing.
Architectural Forum, June 1939.

The Toronto Housing Company's Bain Avenue project. The nation's first limited-dividend project, built in 1914, became one of the first acquisitions of Toronto's Cityhome Corporation after passage of the 1973 NHA amendments. It was later acquired by its tenants and converted into a co-operative. Courtesy the City of Toronto Archives, SC 18–8.

Keeping to the Marketplace

1 Canadian Housing Policy in Perspective

> The fundamental premise about housing has undergone a tremendous change. It has become a Public Utility, in accepted theory at least, if not as yet in any complete sense of accomplishment. The right to live in a decent dwelling has taken its place among the "national minima" – the right to good and abundant water, to sanitation, to adequate fire and police protection, to the use of paved and lighted roads, to education, to a certain amount of medical care, and, in most European countries, to various forms of social insurance.
>
> Catherine Bauer, 1934

CANADIAN HOUSING POLICY IN INTERNATIONAL PERSPECTIVE

When drafting the lines above in depression-ravaged America, Catherine Bauer was attempting to instil in her fellow citizens some of the admiration she felt for the social-housing achievements of post-war Europe. Her book *Modern Housing* was part of a campaign she undertook to achieve "an active demand on the part of workers and consumers ... strong enough to over-balance the weight of real estate and allied interests." Through her writing and her efforts to interest the American labour movement in housing issues, Bauer was able to see her ideals realized in the Wagner Housing Act of 1937, passed at the apex of the "social-democratic" phase of the later New Deal.[1] While Canadian reformers had similar visions and pressed for an approach to housing based on public utility and human rights, similar to the direction taken in democratic European states, their ideals would not begin to be realized until the building of Regent Park, largely through the initiative of the city of Toronto, in 1949. Although provided for in the 1949 National Housing Act (NHA) amendments, public housing would flow at a trickle of an average of 873 units per year until the moribund program was revitalized in 1964. Dominated by urban redevelopment considerations, public housing fostered a backlash as a result of its association with high-rise towers and the brutal clearance of low-income families from their cherished neighbourhoods. Not until the NHA was amended in 1973 could anything approximating the well-planned, subsidized low-income projects praised by Catherine Bauer in 1934 – designed, as she had urged, by co-operatives and non-profit associations – be found in Canada on a significant scale. Moreover, this legislation was introduced under the unusual pressure of an NDP-influenced minority government and was quickly weakened, upon the return to a Liberal majority government, by the removal of all its provisions to support land banking.[2]

Although Canada has been slower to adopt the "comprehensive" housing policies of other Western industrialized democracies, particularly the predominately social-democratic states of Scandinavia, this conservatism has not been rooted in a peculiar national exemption from housing problems. Indeed, the same "crises" that subsidized rental housing programs were designed to address were widespread in Canada during the rapid urbanization of the "Laurier prosperity" boom before the First World War.[3] Compared to problems in Europe, however, the housing crisis in Canada, characterized by the overcrowding of a limited supply of low-rental dwellings, came much later. Sir Herbert Ames, pioneer Canadian social investigator, in the 1890s documented and described the marked superiority of housing conditions in Montreal over those of Europe, although this situation would end in little over a decade. Ames understood that Canada in the nineteenth century lacked the long debate and early modest efforts at government-sponsored housing already tested in Europe. This was largely because housing conditions were better in a nation where urban concentration and the industrial revolution were less advanced. In the twentieth century increasingly similar housing conditions caused subjective factors, such as ideology, rather than the dry world of housing statistics to determine the marked divergence in approach to housing questions in Canada and in Europe. Ideas that, on the eve of the First War, had long been tried and found wanting in Europe, such as limited-dividend housing, were just being attempted in Canada on an experimental scale. This lack of experience contributed to the modest nature of the federal government's first housing scheme, in 1919. But the strong role of ideology in shaping housing efforts is underscored by this scheme's abandonment as soon as the widespread labour unrest of the immediate post-war years had dissipated.

While the achievements of subsidized rental housing in Europe and Great Britain began after the First War concluded, they had long been in the minds of the visionaries of the European and British labour movements. Typically, at a 1912 meeting of Amsterdam architects, the controversial housing reformer Ary Keppler, after berating his colleagues for neglecting workers' housing, urged them to become involved in "the struggle of the working class" and to create "beautiful workers' dwellings, the monuments to that struggle." A few years after these remarks Keppler would be employed by the social-democratic Amsterdam municipal council to realize their shared ideal – that the workingman have the right to live in the beautiful, carefully designed surroundings that had once been the exclusive preserve of the propertied classes. Such housing, which memorialized the class struggle while meeting physical needs, fostered worker solidarity and served as advertising for the Dutch social-democratic party by proclaiming its success in obtaining housing for its constituents.[4]

While the armed conflicts that took place in the fortified working-class apartments in Vienna, named for such socialist heros as Karl Marx and Frederick Engels, illustrate the early divisions over social-housing policy in Europe, in the Scandinavian countries these policies ceased to be viewed in class terms and became part of national consensus. In part this was a function of the unusual longevity of local social democrats in government. In Sweden, for example, social democrats dominated every ministry from 1932 to 1976, except for a few months in the summer of 1936.[5]

In the United States, socialist parties have counted themselves fortunate to elect one congressman. The contrast in the degree of labour organization is also striking. Sweden's labour force has been at least 75 per cent unionized throughout the twentieth century, in contrast to the current American low of 27 per cent, which represents a decline from the upsurge in union membership during the later stages of the New Deal. In housing policies the contrasts are similar. In the United States the division between a residual "social" sector for the poor and a "market" sector for those who can pay their way is pronounced. Sweden is the strongest example of a society where the housing and property industry has been transformed into a social utility controlled by the users of the service and the workers in the industry.[6]

In Sweden between 85 and 90 per cent of housing is built on publicly owned land. Over time, land costs have fallen as a component in the price of housing and now represent only 5 to 10 per cent of the cost of a dwelling. Approximately half of Sweden's housing finance is taken from government revenue or the National Pension Fund. Various types of non-profit groups develop 58 per cent of the nation's annual housing production, greatly reducing opportunities for speculation in housing property.[7]

The strong co-operative sector in Sweden has further reduced opportunities for speculation in housing by entering the building industry. Firms owned by building-trades unions produce 10 per cent of new residential construction, and 6 per cent of housing is still built by the direct labour of the home owner. Municipalities are legally obligated to acquire and bank land sufficient for ten years of future development; stringent sales regulations against speculation in residential properties are enforced, and all profits in transactions are taxed in order to finance municipal land acquisition.[8] In order to retain the housing sector as an outlet for private investment, increased emphasis has been placed on obtaining greater productivity in construction, resulting in a 300 per cent increase in building productivity from 1950 to 1970.[9]

Comprehensive housing policies in Scandinavia have both minimized speculation and profiteering and maximized the extent to which housing occupiers, either individually or corporately, own their own homes or have a voice in their management. Most housing is either built for home ownership (without speculation-inflated land prices) or constructed by non-profit housing associations and co-operatives. Co-operatives undertake every aspect of building, including the manufacture of prefabricated kitchens and bathrooms, furniture, and raw-material extraction. Student housing has also evolved along co-operative lines. Unlike their British and North American counterparts, who conventionally live in university-administered residences, Swedish students are responsible for the design and construction of their housing and the management of the completed accommodation. By law municipalities are required to provide day-care facilities, kindergartens, and housing for the aged. Special attention has been focused on developing housing facilities adjusted to the varied needs and abilities of the elderly. As a result subsidized pensioners' hotels, shopping assistants, home-care attendants, and daily telephone checks have been developed. Rural communities have been relatively quickly provided with piped water, drainage, modern heating systems, electricity, kitchen equipment, and insulation.[10]

Although the strength of the non-profit housing sector is often viewed as the unique feature of the comprehensive housing policies of Scandinavian states, programs that encourage the provision of affordable and high-quality housing for owner occupation are also integral to them, including the socialization of housing finance as well as land. Since the mid-1930s Sweden has provided low-interest "homestead loans" to agricultural workers for the cost of purchasing or building their own homes.[11]

Comprehensive policies are carefully adjusted for individual needs. Homes of elderly pensioners with physical disabilities are refitted with new kitchen and toilet facilities, or home extensions, lifts, cranes, and ramps, to avoid premature institutionalization. Shelter subsidies are provided for 70 to 90 per cent of the population and take varied forms, from tax-exempt rent rebates per child to special low-interest loans that allow families with many children to buy their own homes.[12]

The political will to develop comprehensive housing policies emerged because of the strength and values of the Swedish trade-union movement. It inherited the adult-education ideals of the folk schools of the Scandinavian farm and co-operative movements. Currently, some 650,000 students take its classes annually, absorbing strong social and egalitarian values. In vivid contrast, a contemporary study of British workers found that only 2 per cent felt such social goals should be a priority for their unions. Also unlike Swedish unionists, the majority of British trade-union members felt that unions had too much power.[13]

The marked contrast in political outlook of Swedish and British workers goes a long way towards explaining the consensual nature of Swedish housing policies and the sharp divisions characteristic of British ones. Whereas in Sweden a long and progressive socialization of the housing market has occurred – largely unbroken by changes in government – in Britain attempts to introduce similar policies have been the occasion of fierce conflict between the Labour and Conservative parties. This has been evident in bitter battles over council-housing standards and the recent moves of the Conservative national government to force Labour municipal councils to sell off much of their stock of public housing to individual home owners.[14]

The same reflection of social trends in housing policy characteristic of Great Britain and Europe can be seen in the United States. In part the evolution of American policy was shaped by its slower acceptance of municipal building-codes standards. In Europe and Great Britain application of such codes discouraged workers' self-built housing and drove away private investment in housing. The impact of rent controls, imposed throughout Europe during the First World War, had a similar tendency to make the production of unsubsidized working-class housing more difficult. The American worker who built a home on inexpensive land on the urban fringe often faced other hardships, but such practices did provide a safety valve that reduced class-conscious political pressures on housing issues, and permitted low-paid workers to escape exploitation by real estate managers and mortgage lenders. As late as 1944 American public-housing advocate and administrator Nathan Straus observed the last forms of this self-built housing by low-income workers. He noted that the only new homes being built for the lower third in income of urban families were "shacks" erected in communities that were still "without adequate building codes."[15]

Building codes did not become significant in the United States until the beginning of the twentieth century. Their implementation repeated the European experience of such legislation by making the new housing constructed under their provisions too expensive for the poor, increasing overcrowding in the remaining "old by-law" flats. The advocates of code enforcement, led by Lawrence Veiller, opposed European ideas of promoting low-cost housing by building new communities through public land assembly. Daniel Burnham's ambitious 1909 plan for Chicago confined its remarks on housing to the observation that persons removed for street improvements would not have to be rehoused at public expense, as in Europe; in America, he believed, the unaided private market would painlessly provide new homes. After another Chicago reformer, Jane Addams, had acquired a block of property from a slum landlord, she soon concluded that the homes were beyond improvement and proceeded to demolish them. The tenants without subsidies had to be evicted since their homes could not be repaired within the economics of the private marketplace. The one European idea eagerly adopted by the Americans was zoning, which was primarily used to keep the poor out of affluent districts. Veiller attempted to characterize the 1924 housing legislation of the British Labour government as "nothing more nor less than public charity on a gigantic scale ... even more unsound than that of the Socialist government of Vienna."[16] Such attitudes prevailed through the booming 1920s; not until the Depression would circumstances favour Bauer and her supporters in their desire to bring "un-American" ideals to their country.

The evolution of Canadian housing policy fits into the broad contours of the contrasting experiences of Sweden and the United States. Canadian and American working-class movements shared a common inability to organize most unskilled workers until the coming of the Great Depression. The slower development of labour organization in Canada during the Depression reduced pressures for subsidized housing, accounting for its absence from the agendas of King and Bennett. An upsurge in labour organization during the Second World War, coupled with increased support for the Cooperative Commonwealth Federation (CCF), encouraged the adoption of a program of subsidized shelter through the temporary activities of Wartime Housing Limited. The momentum of labour militancy increased in the immediate post-war years and was accompanied by the veterans' rental-housing program and the continuation of national rent controls. Federal support for subsidized housing was recognized as necessary in 1949 in view of plans to secure the termination of programs fostered by heightened working-class organization and consciousness during the 1940s. Unfortunately, this commitment was a tokenist measure and served to wind down in quieter years an activity that had been forced on the government during the turbulent time of war and post-war adjustment. This sleight of hand was typical of the sound and fury, coupled with minor advances, that would prove to be so characteristic of the evolution of Canadian housing policy.

HOUSING AND CANADIAN FEDERALISM

Canada, unlike the unitary states of Great Britain and the Scandinavian nations, has a federal political system. But federalism has not proved anywhere to be a significant barrier to any level of government determined to pursue a bold housing strategy. The ambitious housing achievements of Vienna in the 1920s were possible only because that municipality acted on powers comparable to those of a provincial government, often with little encouragement at the national level. A more co-operative pattern between national and state governments was the rule in the Weimar Republic of the 1920s and the post-war Federal German Republic, and helped to sustain high levels of social-housing construction.

Although federalism in Canada poses some additional challenges to social-housing advocates by blurring the lines of political responsibility, the structure itself has not impeded any government from taking action. The absence of constitutional roadblocks is most striking when we consider that social housing in Canada actually emerged from the weakest link in the constitutional chain, municipal government.

The better-funded and more powerful federal and provincial governments did not bring social housing to Canada because their relative remoteness from activist groups made them more insulated from political pressure. The major breakthrough took place instead in the city of Toronto, where reformers had carefully organized to make public housing a major political issue that candidates for office were under considerable pressure to support. Consequently, Toronto accepted first what federal governments in Canada had been resisting for fifteen years: the principle that there were low-income families who needed subsidized rental housing. Reformers were so successful that they could persuade city property owners to approve such subsidies from their own pockets in a municipal referendum, taken at a time of considerable social idealism about a better world at the end of the Second World War.[17]

The victory of Toronto reformers was all the more notable because it was achieved on the toughest fiscal battleground. They had to persuade the more affluent property owners of Toronto to accept higher taxes so that their fellow citizens would enjoy a better life. The reformers' historic success in winning approval for the pioneering Regent Park project on this basis underscores the close connection between the campaigns of a relatively small group of Toronto reformers and the evolution of innovations in social housing in Canada.

The Regent Park breakthrough in the 1940s was the culmination of a series of campaigns for social housing in Toronto that can be traced to turn-of-the-century dialogues between enlightened manufacturers and socially aware trade-union leaders. In the "Progressive" era before the First World War, only Toronto reformers succeeded significantly in implementing two modest limited-dividend housing projects. During the Depression of the 1930s Toronto reformers wrote a new bible for social housing, the Bruce Report, named after a major establishment ally of their cause, Ontario lieutenant-governor Herbert Bruce. This report helped to inspire the back-bench revolt of a typical Toronto Red Tory Conservative MP and former mayor, T.L. Church, which prodded the Dominion Housing Act (DHA) out of the government of R.B. Bennett. In the ensuing decade the banner of establishment-led reform would be passed on from Bruce to Harold Clark, who bravely combined the worlds of social activism and high finance.

Under Harold Clark's leadership Toronto reformers cultivated enough support in the city to win the major breakthrough represented by approval of Regent Park. Clark gathered together an array of distinguished luminaries in the Toronto Citizen Housing Advisory Committee. These included the then adult educator and future philosopher George Grant, social-work professors Albert Rose and Stuart Jaffary, and future CMHC policy planner Humphrey Carver. The group's imaginative agitations made it difficult for candidates for municipal office to oppose public housing and be elected. This produced a unanimous vote of the city council in favour of Regent Park.[18]

The boldness of Toronto's unilateral action caused a conversion of Prime Minister Louis St Laurent as dramatic as St Paul's revelation on the road to Damascus. The new prime minister, who had vowed while campaigning for the Liberal leadership not to be a part of any Cabinet that sponsored public housing, suddenly reversed his position in time to open the new Regent Park project. His government then passed the 1949 amendments to the National Housing Act, which finally accepted the reformers' principle of federally funded, rent-geared-to-income shelter.[19]

Due to the continuing vigour of Toronto reformers, most of the public housing built in Canada during the 1950s went up in the Toronto area, but the reformers remained unhappy with the limited volume constructed, blaming this on the complicated federal-provincial-municipal structure of the 1949 NHA amendments. Aided by Humphrey Carver, who was now policy planner for CMHC, they continued to press for what had been sought in the 1940s, a federally funded, municipally administered public-housing program. Their efforts resulted in the NHA amendments of 1964, which allowed for a direct relationship between the federal government and provinces or municipalities. The result was the creation of the Ontario Housing Corporation (OHC), which rapidly expanded public-housing production in the province, some years consuming all of the federal government's social-housing allocation.

While meeting reformers' targets of needed social housing, OHC's less encouraging management policies and project designs sparked a determined move to "get the City back in the housing business." This promoted a third change in social-housing policy, an emphasis on "third sector" delivery by municipalities, non-profit associations, and co-operatives, in contrast to provincial bureaucracies such as OHC. This was achieved with the passage of the 1973 NHA amendments, long campaigned for by the remarkable team of Harold Clark and Humphrey Carver. Toronto's success in this new era was epitomized by the large St Lawrence project, which carefully mixed both diverse housing tenures and income groups. Such success, however, encouraged opposition from entrepreneurial housing developers and a reduction of city-sponsored and -assisted activity with the sudden end of federal land banking in 1978.[20]

While the commitment of the city of Toronto to social housing has been remarkably constant, this zeal has been balanced by dimmer enthusiasm from senior levels of government, generally more sympathetic to entrepreneurial developers. In the 1950s and 1960s social-housing supporters were undercut by federal policies, generally supported by provincial governments, that encouraged spartan, severe standards for public housing. These were intended to avoid competition with entrepreneurs, who, it was assumed, would build a better-quality product for those who could afford it. It was this very principle of non-competition that the second generation of reformers emerging in the 1960s was determined to reverse by having income-integrated social-housing projects, with a mix of market and subsidized tenants. These ideals were spelled out by federal housing-policy critic Michael Dennis and then implemented by him as president of Toronto's Cityhome Corporation in a manner that would outrage entrepreneurial housing developers.[21]

Generally unsympathetic to social housing, federal and provincial governments have tended not to promote its construction even after passing legislation permitting it. This pattern was set by the 1949 legislation, under which local activists had to push projects through a formidable obstacle course. After 1964, starting with Ontario, provincial housing corporations began to spark a trend reversed by the 1973 NHA amendments, which placed responsibility for initiative on the third sector. In this the federal government again played a generally passive role, giving more encouragement to builders of condominiums than to the emerging co-operative sector.[22]

While large urban centres have been supportive of social housing, their surrounding suburbs have not. This has benefited federal and provincial governments unsympathetic to social housing, since local opposition encourages delays and prevents more requests for additional allocations.

The contrast between city and suburban approaches to social housing is epitomized in the experience of Metropolitan Toronto. Originally, the existence of Metro encouraged the provincial government, supported by Toronto reformers, to impose social housing on conservative municipalities. This pattern, which had begun with the Lawrence Heights project in 1956 under the determined leadership of Frederick Gardiner, the first Metro chairman, with the support of Premier Leslie Frost, continued in the 1960s in OHC's practice of disguising its projects by having private developers design them. With the end of OHC's development role in 1973, only the city of Toronto picked up the challenge of social housing. No suburban municipality launched its own program, and Metro did not take up the controversial family-housing field until federal housing allocations had been seriously reduced at the end of the decade. With the exception of Peel Region, new social housing for low-income families became concentrated in the city of Toronto. Low-income families in the suburbs were largely confined to the income-segregated public-housing projects built by OHC in the 1960s.

The definition of municipal boundaries and the creation of regional government entities is entirely a provincial responsibility in Canada, and consequently is one of the major ways in which provinces have shaped social-housing provision. The Red Toryism of Ontario premiers Leslie Frost and John Robarts in the 1950s and 1960s would create regional instruments to blunt suburban housing opposition. In the 1970s such boldness in the face of conservative suburban areas was on the wane. Where Frost had used Metropolitan Toronto to impose social housing on suburban governments dominated by hostile real estate lobbies opposed to it, his successors surrounded it with regional governments that effectively blocked new projects in most developing suburban areas.

Ontario's sometimes bold interventions in public housing have proved to be an anomaly, however. Generally the pattern across Canada has been to sit back and see municipalities become the forum for often sharply polarized debates on housing issues.

Typical of the pattern of provincial passivity was British Columbia's response to the 1949 National Housing Act. While it duly passed the necessary complementary provincial legislation, the province cleverly ducked for cover in the ensuing heated battle over its implementation. This zone of conflict emerged instantly, as forces that had argued with each other over the virtue of the principle of social housing since the Great Depression took up their positions. The reformers' general and key strategist was none other than Leonard Marsh, who had written the blueprint for national reform in the 1943 report of the Advisory Committee on Post-War Reconstruction. Arrayed against Marsh's social workers and trade unionists were determined real estate industry activists. The reformers won after four years of debate, but secured only the building of a single project before the onset of urban renewal.[23]

Despite the intensity of housing battles, constitutional roadblocks never proved to be crucial in preventing action desired by critical personalities at any level of government. The city of Toronto showed that the weakest level of government could go it alone to achieve its social-housing objectives. The system was also flexible enough to leave room for the ideological opponents of social housing to do as much damage as their combined wills and imaginations could create. Usually this was done with some degree of subtlety. Premiers would generally not openly forbid municipalities to take part in social housing, preferring to impose a portion of provincial housing costs on them to make it less attractive. Suburban mayors would usually complain of costs rather than admit openly that they wished their communities to bar the entry of poor people. Like them, premiers could conspire to show loyalty to allies in the property industry by blocking social housing, but their intrigues required sympathetic federal politicians who would not devise policies without a provincial or municipal cost contribution, for without such fiscal excuses social housing would be politically too difficult to refuse.

The crucial role of ideology is evident in the otherwise astonishing position of Saskatchewan as an island of social housing in the prairie provinces in the 1950s. During this period it was the only province to take part in the 1949 federal legislation for public housing, although it was far poorer than oil-rich Alberta and the more economically diverse and affluent Manitoba. Its housing conditions were not notably worse than those of the neighbouring provinces either. Although the wheat fields and boreal forests across the borders were identical, these boundaries did delineate very different political principles. Consequently, the CCF-governed Saskatchewan did not pass on housing subsidies to municipalities and urged more vigorous social-housing policies. In contrast, a Social Credit government in Alberta and a Liberal regime in Manitoba passed all their costs on to municipal governments and did everything to hold back social housing.[24]

The role of ideological factors in shaping housing policy is also illustrated in the case of Quebec. Although nationalist posturing in the province has frequently been used to obscure these debates, their content has mattered more than mere jockeying for advantage between the levels of government. Since it suited his outlook to have an essentially tokenist federal public-housing program, the nationalist Quebec premier Maurice Duplessis co-operated closely with CMHC president David Mansur in framing the 1949 NHA amendments.[25] In keeping with his desire to foster home ownership and to maintain his alliances with related small-business and real estate groups, Duplessis kept public housing out of Quebec for five years. Large business interests supportive of municipal reform organizations, such as the Civic Improvement League and trade unions, were all ignored as alien to the premier's political constituency.

A peculiar set of political circumstances elicited a virtual deathbed repentance from Duplessis on the issue of public housing. This coincided with the election to the provincial legislature and subsequent elevation to the post of minister of municipal affairs of Paul Dozois. Dozois was one of the centrist, business-linked, reform-minded supporters of public housing in Montreal. His Union Nationale credentials made it impossible for realtors to continue their rhetorical equation of public housing with atheism and communism, and Dozois was able to persuade Duplessis of the merits of approving the first public-housing project in the province, which was named after that seventeenth-century hero of Canadiens, Jeanne Mance.

Even with the support of the powerful Duplessis, bringing public housing to Montreal remained a difficult battle. The premier's political opponents in Montreal, who had previously used the public-housing issue to distinguish themselves from his administration, began to reverse themselves. No longer held back by partisan tactical considerations, real-estate-minded members of the party assailed public housing without restraint. Their leader, Jean Drapeau, even had any discussion of Duplessis's favoured scheme removed from the agenda of Montreal's executive committee. This forced Duplessis to resort to the provincial legislature and to pass a Montreal city charter amendment to rescue the project.[26]

Ideological warfare over public housing burned with even more fierce intensity during the 1950s in Winnipeg. Here it became a symbol of the class and cultural conflicts that divided the city. Discussions of the issue became a ritual re-enactment of the Winnipeg General Strike.[27]

Polarization over public housing began in Winnipeg when the issue first arose in the Depression. The pragmatic David Mansur, then chief inspector of mortgages for Sun Life, attempted unsuccessfully to tone down the intense opposition of his company's Winnipeg branch office to the concept. His later 1949 NHA public-housing legislation was not received any more warmly by the city's real estate fraternity.[28]

The polarization in Winnipeg of the pro-public housing CCF municipal party and their business opponents was so intense that many centrist, business-linked municipal reformers who supported it would not openly break ranks with their political allies. The right-wing Liberal government of Premier Douglas Campbell encouraged such intransigence by passing all the province's share of costs on to municipal governments. This contributed to the defeat of public housing in a 1953 ratepayer referendum. It was rescued only by the fortuitous combination of Campbell's electoral defeat in 1958 (at the hands of the Progressive Conservatives, headed by Duff Roblin) and the 1956 NHA amendments, which linked federal funds to urban renewal to provide public housing for the displaced. Public housing only came to Manitoba when Winnipeg mayor Stephen Juba informed council, after a ritual pilgrimage to Ottawa, that urban renewal could not occur without relocation housing. In response, centrist aldermen, in an unrecorded vote, exempted the necessary public housing from a referendum vote, saving it from a likely repeat of the 1954 battle.[29]

Housing policies provide a remarkable litmus test for the values of politicians at every level of office and of the varied communities that influence them. Often this test measures simply the warmth or coldness of heart of the more affluent and secure towards families of a lower socio-economic status. Such values-based policy differences are most evident in the attitudes of municipalities of comparable financial resources towards co-operative housing, as in the contrasting cases of St Catharines and Welland, two Ontario cities in the Niagara region. Reacting to the concerns of residents wishing to exclude low-income people from their neighbourhoods, St Catharines has had a consistent record of opposing rezonings needed to facilitate co-operative projects. This contrasts remarkably with Welland, which has little resident opposition and a welcoming attitude towards co-ops. It not only routinely approves necessary rezonings but will sell land to co-operatives below market prices. Its mayor, Roland Hardy, ascribes the difference to his city's understanding of what it is like to be poor.[30]

The enormous controversies unleashed by the 1949 NHA amendments underscore the difficulties faced by social-housing advocates resisting well-organized and influential real estate interests. Despite their strength, however, social housing eventually became accepted everywhere across the country. Even when, with a return to harsh right-wing politics in the late 1970s, Manitoba, British Columbia, and Saskatchewan engaged in attacks on the housing innovations of their NDP predecessors, new social-housing production could not be eradicated because of the funding for co-operatives provided by the federal government.

While the constitutional framework has been flexible enough to permit determined opponents of social housing to block it, it has similarly been elastic enough to provide room for the initiatives of its supporters. In British Columbia, social-housing activists unable to secure support from municipalities have turned to labour unions, notably the carpenter's union, to support co-op housing. In Ontario activists unhappy with the reduced social-housing allocations of Brian Mulroney's Conservative federal government successfully persuaded the provincial Liberal government of David Peterson to launch its own Homes Now program. Toronto reformers unhappy with the quality of OHC projects could persuade both the city council and senior levels of government to initiate programs that met their expectations.

While the city of Toronto has played a pioneering role in social housing, such ground-breaking innovation cannot be ascribed to any provincial government. The province coming closest to such a role, Ontario, has a more chequered pattern of support for social housing. Premier Leslie Frost supported the 1949 NHA amendments in part as a way to curb the then independent interventions of the city of Toronto. He tended to tolerate suburban obstructionism to Metropolitan Toronto's housing program by accepting token levels of building activity in the city, thereby appeasing reformers while reducing burdens to the Ontario treasury.[31] A much bolder approach was taken by his successor John Robarts through the direct construction activities of the Ontario Housing Corporation, which did not require municipal partnerships. But when, under federal pressure, OHC slowly began to move towards a system of tenant self-management that irked entrepreneurs, Ontario stopped its development role, leaving new social housing to the third sector. While this approach worked in areas such as Toronto, Peel Region, Ottawa, and Hamilton, elsewhere in the province social housing was largely confined to the seniors' sector. Even with the revival of social housing by the province under the Homes Now program, the basic pattern remained: it went only where local activists were sufficiently organized to create their own projects or bring sufficient pressure to bear on municipal councils.

Ontario's experience in social housing, for all its twists and turns, has made it the exception that proves the rule governing the interaction of provincial and federal housing policies. The province, despite its frequent retreats, has accounted for 80 per cent of social-housing units built since 1964, with only 40 per cent of the population.[33] This ability to outperform other provinces on a two-to-one basis underscores the uncreative role played by most provincial governments.

Provincial governments have constrained social-housing production by passing on the costs of social housing to municipalities. Even in Ontario this 7.5 per cent cost to municipalities was a major barrier to social-housing construction. There it was not eliminated until 1980, and only after a high-pressure political poker game between Metropolitan Toronto and the province, involving the resolution of such issues as the creation of the Metropolitan Toronto Housing Authority (MTHA), approvals for new co-operative units, and the conversion of limited-dividend senior-citizens' apartments to seniors' public-housing units. Only after this barrier was removed would Metropolitan Toronto tackle the family-housing field, and new social housing for families return to the suburbs after being dormant since the end of OHC's development role in 1973.[34]

It seems clear that provinces have passed on the costs of subsidies for ideological reasons. While Liberal and Conservative provincial governments have resorted to municipal contributions to slow social housing down, their eliminaion has been a basic NDP strategy to remove bottlenecks to its delivery. New Democrats, like their social-democratic colleagues elsewhere in the world, believe strongly in reducing land speculation and in shelter subsidies to provide more social justice in the housing market. In Manitoba the share of housing costs borne by municipalities became a good barometer of provincial sympathy for social housing. Under the reactionary Liberal administration from 1949 to 1958, municipalities bore 25 per cent of the costs of housing subsidies – the full provincial share. Premier Duff Roblin's more progressive Conservatives lowered this to 12.5. Their total elimination was one of New Democrat Edward Schreyer's first acts as premier in 1969.[35]

The contrast between the enthusiasm of NDP provincial governments for social housing and the lethargy of others is striking and argues also that Ontario's concentration of social housing is an indicator of political will more than of financial resources. Very strong initiatives by New Democratic governments in Manitoba, Saskatchewan, and British Columbia have been followed by neo-conservative backlashes. Under Schreyer the Manitoba NDP banked over 3,500 acres of land in the Winnipeg area for housing and also purchased housing under its rural and native housing program. It assisted housing co-operatives by providing leased land. In 1975 the 2,037 units of public housing in Winnipeg constructed under the NDP government exceeded even the achievement of the zealous Toronto reformers under Michael Dennis. A still bolder approach was taken by the New Democratic government of British Columbia under David Barrett, which rapidly expanded the province's public-housing stock from 1,400 to 6,200 units in its brief three years, and provided substantial assistance as well to the co-operative sector. This success was achieved in part by the imaginative acquisition of a private development company, the Dunhill Development Corporation.[36]

The missionary zeal of NDP governments in western Canada to promote social housing was exceeded only by the determination of their successors to halt the innovations they had introduced, though they never dared, in even where most inclined to privatization, to sell the housing stock built up by their predecessors to the private market. The polarization of housing-policy debates in Manitoba, Saskatchewan, and British Columbia, however, demonstrates that ideology, rather than financial ability, has been the main reason other provinces have fallen behind Ontario's social-housing achievement. Red Toryism and social democracy have both proved too radical for electoral success in most provinces.

A surprising measure of the weakness of constitutional and fiscal explanations for provincial inaction is the province of Quebec. Maurice Duplessis's deathbed entry into public housing, with the enormous controversy it generated, scared off his Liberal successors. Consequently, the "Quiet Revolution" was particularly calm in the area of housing, and saw no new public-housing projects undertaken by the province.[37]

Although Lesage's Liberal administration embarked on preliminary discussions about creating a provincial housing corporation on the model of the Ontario Housing Corporation, they did not dare to tackle the controversial issue with actual legislation. This was done in 1967, when the Liberals were replaced by the Union Nationale, now headed by Daniel Johnson, and including in its Cabinet the crusading Paul Dozois. In response to Quebec's nationalist position, CMHC effectively delegated decision making to the province. Despite this concession, which eliminated the periodic delays and revisions that periodically stalled public housing in Ontario, Quebec could not meet Ontario's degree of success in producing social housing. It did become, however, the second producer in the country, matching its status in population.[38] Despite rhetorical attacks on federalism, the social-democratic outlook of the Parti Québécois government of René Lévesque in the late 1970s and early 1980s actually meshed well with the federal emphasis at the time on support for the third sector. Despite confrontations between the provincial PQ and federal Liberal governments over federalism, both parties' social-democratic-leaning housing programs during this period stimulated co-op housing, notably the remarkable Milton Park achievement.[39]

While the Atlantic provinces have not shown the ardour of Ontario or western NDP governments for social housing, neither have they been the scene of the fierce attacks of real estate interests on the concept characteristic of western Canada and Quebec. Despite the myth of Maritime conservatism, Newfoundland was the first province to take part in the 1949 federal-provincial public-housing program. It was shortly afterwards followed by Nova Scotia, which relatively soon launched a project in Halifax. Agitation for social housing by civic improvement leagues in St John and Halifax had begun before the First World War and continued during the Depression. When federal funds became available, real estate groups and related housing-market ideologues did not mount a determined effort to resist them.

The region was the source of a unique form of social housing pioneered by the Catholic co-operative enthusiasts of the Antigonish movement. In their "sweat-equity" co-operatives, a group of workers pooled their labour to build their own homes, which had a common mortgage. In 1941 the Nova Scotia Housing Commission was already encouraging such co-ops, and by 1953 CMHC agreed to participate jointly with the province in funding this program. Different versions of the program were adopted over time and were applied in all of the Maritime provinces. Modified schemes such as shell housing were developed, whereby a partially completed home would be sold to low-income people with a reduced loan and the substitution of sweat equity for loan payment. In Prince Edward Island the province used federal Regional Economic Development funds set aside for housing to provide grants to purchase for low-income buyers. In the 1950s versions of the Maritime co-operative scheme became popular across Canada, but were subsequently confined to rural areas because of escalating land costs. In the Maritimes land assembly by the federal-provincial partnerships helped to keep sweat-equity co-ops alive in metropolitan areas such as the suburban area of Halifax.[40]

The unfortunate tendency of provincial housing policies effectively to heat up rather than control the private housing market can be seen in the efforts to extend assisted home ownership, pioneered in depressed rural communities such as Cape Breton, into the major metropolitan centres of Canada. Provincial encouragement in this regard lead to CMHC's Assisted Home Ownership Plan (AHOP), which proved so disastrous as to plunge the Crown corporation into a state of technical bankruptcy.

Social-housing supporters within the CMHC such as Humphrey Carver were favourably impressed by the record of assisted home ownership in the Maritimes and recommended that similar approaches be taken in small towns and rural areas across the country. They balked, however, at such plans in growing urban centres, believing that they would fuel housing-price inflation as greater affordability was cancelled out by higher prices. This concern caused CMHC to reject a 1962 request by developer Robert Campeau for an expanded home-ownership program.[41]

While getting a chilly response initially from CMHC, developers favouring assisted home ownership got a more welcoming reception from provinces amenable to pumping more cash into the real estate market. Quebec was first to adopt the assisted home-ownership approach unconnected with sweat-equity schemes, which could not benefit the sales of speculative builders. In 1947, under Duplessis, Quebec passed its own Family Housing Act to subsidize home ownership.[42]

As land prices increased in the 1960s, developers lobbied successfully for provinces to develop assisted home-ownership schemes. The Social Credit government of British Columbia led the way in 1967, with a $500 grant for the purchase of an existing home and $1,000 for the acquisition of a new unit. After a request by the Regina Home Builders Association, a similar scheme was approved in Saskatchewan. In 1970 Alberta adopted a 2 per cent interest subsidy. Ontario, in typical dissenting Red Tory fashion, avoided interest-rate subsidies and capital grants, with their potential for housing-price inflation, and instead provided land subsidies from lots made available in banked land through its Home Ownership Made Easy scheme.[43]

CMHC eventually accepted the grants and interest-rate subsidies favoured by British Columbia, Alberta, Quebec, and Saskatchewan. This trend began under the $200-million Special Innovations Program launched by CMHC in 1970, which, despite complaints about the poor quality and site location of units built under the scheme, was expanded under the Assisted Home Ownership Program (AHOP) created by the 1973 NHA amendments.[44]

The pattern under AHOP, amendable to the private real estate market, was the opposite of that in the more controversial field of subsidized rental housing. In the case of AHOP, the federal government followed the lead of the provinces. In the case of subsidized rental housing the provinces were generally dragged along by the federal government, although the city of Toronto continued to lead any senior level of government. Only Newfoundland and Saskatchewan followed the federal lead quickly. Ontario premier Leslie Frost was slowly converted, and by 1952 British Columbia faced a political battle with Vancouver city council. The lure of federal funds for urban renewals had to be used to prod Alberta, Manitoba, and Quebec into acceptance.

The provinces generally have been more supportive to entrepreneurial developers than has the federal government. Despite its strong support for social housing, even Ontario, while launching its own Homes Now program, has not moved to create a complementary land-banking scheme in response to the federal exit from this field in 1978. Although co-operatives struggled for funding until the

1973 NHA amendments, it was always earlier provided by CMHC and not by provincial jurisdictions, which had full constitutional power to do so. Provinces directly involved in decisions affecting land use appear, in the absence of compelling ideological social-democratic or Red Tory imperatives, more susceptible to the influence of land-development interests, with whom they are in more regular contact. For this reason the federal government has remained the source of key innovations that socialize the real estate market, including land banking, shelter subsidies, and co-operative tenure.

The patterns of housing-policy debate vary as they flow from trade union halls to chambers of commerce, from social-work associations to real estate boards to co-operative groups, public-housing tenants' associations, and all levels of government. In these often heated discussions, which would, through the century, penetrate remarkably varied quarters of Canadian society, reformers and their opponents would cross many lines of class and consciousness. Reformers would usually be most effective when led by well-connected establishment figures, such as Herbert Bruce and Harold Clark. Battles over housing policy would produce unexpected alliances: large-scale construction companies would support social housing against retail lumber dealers. The reformers' sympathy for Henry George's notions of the unearned increment from real estate speculation might win allies among business unconcerned with real estate, while making enemies of millions of working-class families who had achieved home ownership.

The clash of principles involved in Canadian housing policy is best understood if personified by the two key shapers of Canadian housing policy who happened to have the same last name, W.C. (Clifford) and W.H. (Harold) Clark. Although both had extensive experience in corporate boardrooms, they could not have been further apart on the political spectrum of housing policy in Canada.

W.C. Clark served as a real estate investment broker in the United States during the boom years of the 1920s; W.H. Clark was head of the Toronto branch of Canada Trust Huron and Erie. This similarity of background, however, was overshadowed by a profound philosophical divergence. W.C. Clark was an intellectual protégé of O.D. Skelton, whose doctoral dissertation was a critique of the interna tional socialist movement. W.H. Clark took his Christian socialism from his English grandfather, a close associate of the philanthropic Cadbury family, with whom he collaborated in the creation of a model greenbelt town.

Each would dream visions of the future far different from those of the other. W.C. Clark's were of cities of skyscrapers; W.H. Clark's sense of beauty was closer to that of his brother, Spencer Clark, who built the Guildwood Inn near Toronto as a medieval-style artists' retreat. W.C. Clark's aesthetic sense, tailored to the tastes of real estate investors, prompted his rise to the top of the Canadian civil service in the then pivotal role of deputy minister of finance. W.H. Clark's principles led to his being investigated as a Communist. Cleared of this charge, he still faced dismissal from employment after he had declined a promotion out of Toronto that would have ended his voluntary service. While the two men's efforts in housing policies were always at cross purposes, their paths directly crossed when W.H. Clark risked his position at Canada Trust by drafting a public letter to Prime Minister Mackenzie King that criticized the proudest achievement of W.C. Clark's career, the National Housing Act of 1944.[45]

That so much of the evolution of Canadian housing policy in the twentieth century can be understood as a clash of principles between the two Clarks illustrates the validity of political scientist Keith Banting's thesis that public policy making is an elite process primarily involving Cabinet ministers and senior civil servants. While the two Clarks had competing constituencies for their respective housing agendas, both camps were small in comparison to the total Canadian electorate. The advocates of public housing never had a large, nation-wide membership capable of deluging governments with phone calls and letters, as home builders did in 1942 to rescue the NHA and curtail wartime housing, and as realtors were able to do to curb rent controls. They even lacked a national organization clearly committed to organizing public support for social housing. Reformers made two attempts to achieve this. The first, the National Housing and Planning Association (NHPA), did not survive the advent of the Second World War and dissolved before it would endorse the basic principle of reformers, support for subsidized low-rental housing. The second attempt was made at the founding of the Community Planning Association of Canada (CPAC) in 1947, where a reform agenda was scuttled and the organization tamed by the careful intrigues of CMHC representatives.[47]

The person the CMHC feared would assume the CPAC presidency, George Mooney, in many ways epitomized the strength and dedication of Canadian housing reformers. A long-time executive director of the Canadian Federation of Mayors and Municipalities, a former CCF candidate, and a close friend of Norman Bethune, Mooney was an effective orator. He could sway many otherwise conservative mayors and municipal governments to the public-housing cause. Mooney's career and the fear it aroused in CMHC shows that the reformers' message could be made acceptable to a broader Canadian public. For all his considerable talent and connections, however, Mooney would never be able to lead a strong advocacy organization primarily dedicated to the social-housing cause.

Mooney's fate at the founding convention of CPAC was symptomatic of the failure of early social-housing advocates. Housing reformers lacked strong organizational support, particularly crucial when business sectors were organized against them and in view of the hostility to their views held by the Department of Finance. The Canadian Construction Association (CCA) was essentially a fair-weather friend, supporting public housing in depression but keeping silent during periods of prosperity. Unions were similarly lukewarm in support, taking no part in the NHPA and CPAC and, as Frank Underhill complained in 1936, ignoring the social-housing issue in the labour press.[48] Occasional outbursts of public protest about housing from tenants could be carefully controlled in their impact on housing policy through temporary concessions, such as an expansion of veterans' housing in response to the squatters' movement.

Placed in an excellent position by his role as deputy minister of finance from 1933 to 1952, W.C. Clark was able to outwit the small group of well-intentioned and dedicated reformers who wanted a bold social-housing program for Canada. He accomplished this through a number of ruses intended to obscure the reality of conflict over housing policy, and he was often rescued by the intervention of less doctrinaire figures in the federal civil service who were able to create temporary programs, such as Wartime Housing, that modified public discontent. Clark wisely confined his attack on the reformers' key principle – the need for subsidized rent geared to income – to confidential memoranda for Cabinet ministers. In his address to the founding NHPA convention he avoided any allusion to disagreements and focused on the virtues of his recently passed National Housing Act of 1938, thus also avoiding any need to refute the detailed statistical arguments the reformers had assembled in support of rental-housing subsidies. Clark's astute public evasion of issues was even carried over into the texts of his 1935, 1938, and 1944 housing acts, all of which appeared to accept the principle of subsidized rental housing but made it impossible to deliver, for reasons best known to W.C. Clark himself.

Clark's ability to manipulate events was helped by the relative indifference of the public to housing issues. Popular farm and labour movements agitated for better housing conditions by urging higher prices for farm products and higher wages, and did not raise much of a cry to tax the affluent to pay for improving the shelter conditions of the less fortunate. Unions had ignored housing in the boom years of the 1920s and only took up the cause in response to unemployment. Desperate situations creating potentially explosive situations were carefully defused before serious political damage could be done to the legitimacy of the real estate market. Homeowners facing the loss of their homes in the Depression were rescued by provincial debt-moratoria legislation, and rent controls and wartime housing similarly took some of the sting out of the housing crisis of the 1940s.

Housing reformers in many ways were better organized in the second phase of housing-policy debate in Canada, after the death of W.C. Clark in 1952. The labour and co-operative movements, encouraged by such visionary figures as Eugene Forsey and Alexander Laidlaw, did begin to work closely together, starting with their efforts in 1954 to devise an attractive program of self-built co-operative housing for low-income families. This was later extended through efforts to achieve the first continuing housing co-operatives in Canada in the 1960s. Social-housing supporters also broadened their organizational base. A major milestone in this regard was the Canadian Welfare Council's 1968 national housing conference, which, unlike previous reform gatherings, drew the active involvement of low-income tenants. It went beyond the public-housing focus of the past and had a more comprehensive program, calling for land banking and the encouragement of subsidized and nonprofit housing for a broad band of income groups.[49] Nevertheless, although better organized and with a broader message than before, reformers in the turbulent years of the late 1960s and early 1970s were to be outmanoeuvred by the Department of Finance in essentially the same fashion they had been earlier by W.C. Clark.

The broad acceptance of the formal resolutions of the 1968 conference papered over a number of potential divisions among reformers, which their more united opponents eventually exploited to defeat reform ideas. Although urban renewal's alliance of real estate interests with social-housing supporters finally brought public housing to such bastions of conservatism as Edmonton, Calgary, Winnipeg, and Montreal and helped to expand the program even in cities where it had already enjoyed support, such as Toronto and Halifax, it roused a hornet's nest of new enemies. These ranged from historical preservationists angry at urban renewal's brutal razing of the urban past, such as the seventeen-acre clearance for the massive Scotia Square complex in Halifax, to homeowners who did not receive adequate compensation for their expropriated homes to purchase comparable new ones. Long-time reformers like Humphrey Carver were understandably dismayed to find that, because of the failures of urban renewal, profit-motivated real estate developers, supporters of public housing, and CMHC all came to be lumped together in the public's mind as related parts of a sinister, self-serving establishment.[50]

As Carver could understand from his unusual perspective as both a social-housing activist and a senior civil servant, much of the hostility directed against the public-housing supporters and CMHC was misplaced. Both the critics of urban renewal and public-housing supporters were concerned about ending income-segregated, high-rise public-housing ghettos and desired an active program in federal land assembly. But the cross-fire between the two groups would permit the Department of Finance to continue to guide policy by pumping public money into the private market.

The first effort to move federal housing policy in a more socially sensitive direction was the report of Humphrey Carver's CMHC Advisory Group in 1965. Its recommendations favouring widespread public land assembly, assisted home ownership for small towns and rural areas and the creation of a vigorous third-sector rental-housing industry were similar in spirit to the later recommendations of the Hellyer and Dennis and Fish task forces. However, this similarity of purpose would often be lost in the smoke of battles that obscured the critical role of the Department of Finance.

Rather than directly opposing the Advisory Group's recommendations on their own merits, federal co-ordinating bodies such as the Department of Finance and Privy Council Office created a maze of consultations with provinces designed to maximize opposition. Rather than being discussed directly in a White Paper statement, CMHC's proposals for amendments to the National Housing Act were enmeshed in the thorny issue of an intergovernmental urban council. Acceptance of these reforms became even more difficult when the Department of Finance delayed the mailing of the proposals for the 1967 federal-provincial conference on housing and urban affairs. This sabotage set the stage for a full-scale disaster when the federal minister responsible for CMHC, John Nicholson, denied any intention to proceed with land banking except for public housing.[51]

If CMHC's proposals for reform were bungled by Nicholson's ineptness, confusion would hamstring his successor under Pierre Trudeau, Paul Hellyer. While Nicholson's fascination with personal press clippings, along with his lack of interest in policy, freed the Department of Finance to pursue its traditional purposes, his successor's disdain for potential allies would have the same result, despite his remarkable vision and determination on housing matters.

Hellyer's vision of building new cities, part of a romantic and prophetic imagination, was similar in spirit to the CMHC legislative proposals, especially the New Communities Program, which involved public land assembly and related infrastructure improvements, with an emphasis on regional planning. Hellyer, however, had conceived as much distrust for CMHC as had the demonstrators against its urban-renewal projects, and it was only reinforced by the protests of affected residents while he was in office. His suspicions caused him to ignore CMHC reform proposals, encouraged in this course by the success of his earlier anti-bureaucratic battles against senior armed-services commanders who opposed unification of the armed forces.

The Hellyer task force recommendations were not that different from those adopted by the Canadian Welfare Council's National Housing Conference, which had recommended alternatives to public housing such as co-operatives. Both reports urged that future public-housing projects be small scale and more closely integrated into their communities. But the inflammatory tone of the task force's commentary on public housing, spread over ten pages of its eighty-five page report, caused a firestorm of criticism from reformers such as Carver and Rose, who had long campaigned for public housing. Predictably, organizations that took similar positions, such as the Canadian Welfare Council and the National Department of Health and Welfare, joined the chorus of protest against the report. The deputy leader of the NDP initially denounced it as "reactionary," a position repeated by the Quebec unions, the only segment of the labour movement that bothered to comment.[52]

All this tumult meant that two generations of reformers, sharing many ideals, had been separated essentially by differences of age and experience. Hellyer's supporters, a younger generation full of fire against the urban sprawl promoted by developers at the fringe of the city and the urban renewal that bulldozed away neighbourhoods in the historic core of cities, lacked the experience of the difficult struggle the elder generation had waged to gain acceptance for the principle of shelter subsidies for low-income people. To older reformers the criticism of their younger colleagues had some of the flavour of the arguments of their earlier, self-interested opponents.

More diplomatic behaviour on both sides could have brought an alliance. For instance, as long-time public-housing advocate Albert Rose pointed out in defence of his positions, public-housing agencies had already decided against large-scale concentrations of public housing.[53] But this lukewarm support did not prevent the roasting of precisely the controversial land-banking recommendations by Robert Bryce, deputy minister of finance.[54]

Hellyer's successor as minister responsible for CMHC, Robert Andras, was sympathetic to his predecessor's ideals; he had been a close ally in Hellyer's unsuccessful bid for the leadership of the Liberal Party. He consequently continued reform in the spirit Hellyer had initiated, encouraging tenant participation in the management of public housing, for instance, funding tenant associations, curbing urban renewal, and creating the short-lived federal ministry of state for urban affairs. As part of his general reformist drive Andras commissioned the CMHC low-income housing task force, whose report became popularly known as the Dennis-Fish Report after two of its principal authors.

The Dennis-Fish Report repeated many of the criticisms of public housing made by the Hellyer task force, but did so in a way that clearly could not be interpreted as reactionary by the social-welfare establishment, trade unions, and social democrats. The reform appeal of its recommendations was enhanced, especially after unsuccessful attempts to suppress the report led to its being leaked to the press by NDP federal leader David Lewis. Consequently, the NDP became very supportive of its recommendations, helping to incorporate major features, particularly improved support for public land banking and third-sector housing, into the 1973 NHA amendments.

The reformers' triumph over the ideologues in the Department of Finance was partial, however. The NDP, even in a minority-government situation, could not block the adoption of AHOP without being placed in the politically embarrassing position of seeming to oppose programs that assisted home ownership by working-class people. Yet AHOP diverted CMHC energies into such areas as promoting condominiums in a vain effort to make AHOP affordable in metropolitan areas. The tide of reformist energies was again set back remarkably in 1978 when the federal Liberals eliminated funds for land banking. The sharp ideological nature of this move is evident in the fact that the cut was made shortly after land banking was instituted by the city of Toronto for co-operatives and its own non-profit housing, which competed with entrepreneurial developers for the middle-class housing market. After this blow, the third sector would face annual battles to continue to expand at modest rates.

The division between reform generations was to have tragic consequences. The wisdom of the common position taken in the 1967 national housing conference was to be revealed by later events, after unity collapsed in bitter infighting. Carver and Harold Clark, key organizers of the 1967 conference, had developed a framework whereby the third sector would provide competition with public-housing agencies, which had previously had a monopoly on providing low-income families with subsidized shelter. The 1973 NHA amendments did provide for this framework; subsequent actions by provincial governments, notably Ontario, abruptly halted the future production of public housing for low-income families and confined it to seniors. This limited the benefits of the transition to a form of market socialism in social housing.

The wisdom of the social-housing elders was revealed in time in that most politically prickly of housing situations, Metropolitan Toronto. After 1973, the Metropolitan Toronto Housing Company (MTHC), an agency of the metropolitan government, continued to receive funding for building senior citizens' apartments under the public-housing formula of NHA. This was for 100 per cent rent-geared-to-income subsidized projects for low-income people. At the same time this public housing was being built at an accelerated rate, the MTHC encountered intensified competition from third-sector producers operating under the 1973 NHA amendments. Since as many as half the units in the MTHC projects were reserved for market tenants, amenity standards had to be increased to attract the more affluent, who could pay their way and were not captive to the projects' lower, subsidized rents. This created a number of problems for the MTHC. For the first time it had vacant units, as its tenants, in some instances, began to vote with their feet and move to new third-sector housing projects that happened to be located close to their former homes. Vacancies became to be especially severe in spartan, small bachelorette units. To eliminate these embarrassing vacancies, the MTHC was forced to upgrade its projects. Millions were invested in new facilities such as recreational centres in older projects, and maintenance standards improved. Bachelorette units began to be phased out: some were converted into apartments; others were leased to students and charities. Later, bachelorettes provided shelter for homeless people between fifty-five and sixty-five.[55]

The contrast between housing for families and for seniors in Metropolitan Toronto became vivid after 1973. Public housing for families was brought to an abrupt halt by a combination of provincial hostility to higher standards being imposed by the federal government and more vigorous suburban resistance. Consequently, the impact of the new third-sector projects for family housing was considerably weakened. As a result, social housing for families became overwhelmingly concentrated in the city of Toronto. Rather than third-sector competition and ensuing vacancies, OHC's biggest problem became a mounting waiting list, intensified by the 1980s as a tight private rental market encouraged of a lower participation rate among private landlords in provincial rent-supplement programs.[56] Under such circumstances, incentives for improvements in project quality were small indeed, as crusading reformer John Sewell would discover in his brief tenure as chairman of the Metropolitan Toronto Housing Authority.

While the younger generation of reformers felt that competition from the third sector would improve the quality of public housing for families as it had for seniors, their predictions proved to be overly optimistic because the volume of social housing for families built after 1973 was not sufficient to have such an impact. In this only partial prophecy, the younger generation had been shaped by the nature of their perspective. Because of their age or experience, instead of the difficulties of winning acceptance for the principle of subsidized rents for low-income people, they saw only the problems of some of the worst projects, often built as components of urban-renewal schemes that damaged the historic continuity of communities and imposed great hardships on the relocated. Consequently, while not responsible for the abrupt termination of public housing, they tended to be uncritical or even sympathetic to its sudden demise. The Dennis-Fish Report went further than the Hellyer task force in urging an end to all further public-housing construction, irrespective of design or site location. Michael Dennis, in his influential *Living Room* report, rejoiced that the city of Toronto's new housing program would mean the end of OHC projects "imposed from above."[57] This was fine for Toronto, where under his leadership the city was taking bold new measures to encourage social housing, but only such imposition could provide subsidized shelter in communities lacking a municipal council determined enough to form its own housing company, or without the social activists needed to create a housing co-operative.

In many ways Toronto's unusual social-housing success can be seen as a tribute to a reform ideology that combined technical expertise with a zeal for goals such as neighbourhood preservation, community participation, and a scorn for profit-motivated land developers. This was indeed a complex ideology that combined elements of the managerial efficiency beloved of centrist reformers with left-wing populism. Its success can be traced through the careful fusion of elements in the eclectic career of John Sewell. Appropriately enough, it began in opposition to the urban-renewal scheme of Trefann Court, where his diplomatic skills helped to forge a common alliance between low-income tenants and more affluent homeowners. Such a concern for resident participation, integration of new housing projects into the fabric of older communities, curbing land speculation, and historic preservation would guide the housing efforts of the varied tendencies on Toronto's reform council.[58]

Toronto's relatively bold social-housing record illustrates most clearly the potential for positive achievements in the flexibility of the Canadian political system. Toronto pioneered in limited-dividend housing in 1912 and in the public-housing experiment of Regent Park of 1948. While it cannot claim credit for co-operative housing, nurtured earlier in distant Cape Breton and Winnipeg, Toronto also gave the biggest boost to co-ops and third-sector housing generally with the ambitious program of Cityhome, launched in 1973. Both Toronto's experiments and the early struggles of co-op housing projects demonstrate that dedicated social reformers can coax the Canadian political system to respond to their proposals.

The social-housing innovations of Toronto point also to the darker side of Canadian politics. While the federal system has the virtue of flexibility, it also has the vice of providing dark corners for political evasion. Toronto's experience reveals the enormous opportunities that exist at varied levels of government for mischief and camouflage by opponents of subsidized housing. Exclusionist suburbs outside the city have disguised their intent by attacking obscure fiscal arrangments provided by the province. The federal government in 1949 designed public housing according to a complicated three-government formula that frustrated even the Metropolitain Toronto authorities who were its principal users in the country. In typical Machiavellian fashion, this program was developed by CMHC president David Mansur to limit the growth of public housing. After his retirement from CMHC Mansur was appointed by the federal government to head up Metropolitan Toronto's public-housing authority. Given his past record, the government of Louis St Laurent, lukewarm towards social housing, could rest assured that the new president of the MTHA would not rock the boat by promoting social housing too vigorously. In 1973 the Ontario government would establish a framework for municipalities to assume the responsibility vacated by OHC, but it would prevent all but the most determined from achieving much because it maintained the cost-sharing agreement whereby they would have to pay 7.5 per cent of the subsidies. Obscured by these twists and turns, the major debates of public policy would often be invisible to all but the most zealous social-housing advocate.

HOUSING POLICY AND THE CANADIAN STATE

The varied nuances of housing policy have challenging implications for understanding the Canadian state as an agent of social change as well as for penetrating the complexities of the federal system. Both liberal and Marxist orthodoxies concerning the evolution of the social-welfare aspects of the Canadian government fall short when they attempt to account for the history of social-housing programs in Canada.

The remarkable success of Toronto's complex reform ideology in the area of social housing, for instance, poses fundamental questions about the approach of labour historians to understanding social change. Their studies of "working people" have tended to focus on the world of work and have ignored workers' relationship to the residential environment. This is of particular importance in view of the difficulties experienced by Toronto reformers in uniting tenants and homeowners and the reality of the opposition of working-class homeowners to social-housing proposals, inspired by the perceived threat to their property values and by racial discrimination.

The trends of Canadian housing policy also pose major challenges to many Marxist analysts of the Canadian state. In a standard collection of such analyses, editor Leo Panitch identifies social housing as part of the "legitimation" role of the Canadian state.[59] Similar Marxist analysis has argued that major steps in the creation of the Canadian welfare state, such as unemployment-insurance and family allowance legislation, were taken to appease working-class discontent, heated by the despair of the Depression and radicalism of the Second World War.[60]

Marxist authors writing on the state in Canada have tended to ignore the evolution of social housing. There are many peculiarities in its history that do not fit with their analysis of the roots of the legitimation aspects of the Canadian state. Unlike unemployment insurance or family allowances, the 1949 acceptance of subsidized housing cannot be tied to a surge of unrest, since post-war labour militancy had long since peaked by this time. Similarly, major events in expanding social housing, such as the passage of the 1964

NHA amendments or those of 1973, do not belong in such a context, for labour agitation did not play a major role in their achievement. Business groups that supported social housing were motivated more by new markets, industry stability, and visions of empire building than by fear of agitation from below. Workers in conservative construction trades that did take part in lobby groups often followed their employers' initiatives in these matters. In many communities, such as Winnipeg, reform-minded business leaders would make alliances with trade-union-based NDP politicians to support social housing in the face of right-wing opposition from real estate interests. All this points to social housing not as a conciliatory trade-off to legitimate the system but as the product of a coalition of reformers, from diverse class backgrounds, achieved in the face of opposition from relatively minor business interests and more powerful ideological opposition in certain branches of government.[61]

An essay in Panitch's collection by Rianne Mahon makes the characteristic claim that the Department of Finance constitutes the "seat of power" of the "hegemonic fraction" of the Canadian state, serving "to give coherence to government policy." Although the Department of Finance has been a major influence in Canadian housing policy, Mahon and other Marxists err in overstating their case, implying the imposition on the Canadian civil service of a rigid marketplace ideology as severe as the strait-jacket of Leninism in a communist state. Mahon suggests that W.C. Clark succeeded, through the training of senior civil servants, in indoctrinating the whole of the state apparatus. Such an oversimplification is too easy to counter with historical detail. Even Clark's close personal friend David Mansur was no ideological clone; the departures he began from his mentor's stern opposition to social housing began with limited-dividend housing in 1948.[62]

The Marxist analysis of the Canadian state is generally not so dogmatic as to deny the existence of internal conflict; it stresses, however, that lesser departments aim at "the containment of the subordinate classes." The Department of Labour is viewed as neutralizing working-class rebellion, while Indian Affairs is seen as seeking to integrate its marginal group into capitalist society. Another contributor to Panitch's anthology, Martin Loney, argues that "overall government funding and involvement in the voluntary sector must be seen as a conservatizing force," believing that "political activity which falls outside the conservative paradigm will not be funded." Loney notes that the Canadian Civil Liberties Association, the Canadian Organization of Public Housing Tenants, and the National Indian Brotherhood are all government financed. He cites the

Canadian Council for Social Development (CCSD) as a typical instrument of the state, serving to "sustain the illusion of meaningful debate" while actually reinforcing "the very narrow ideological space within which that debate occurs."[63]

Loney's targeting of the CCSD highlights the weakness in the Marxist analysis of the Canadian state. Rather than a creature of social-welfare disciples of W.C. Clark, meant to gain acceptance of the Department of Finance's philosophy among social-welfare activists, CCSD was formed by dedicated reformers such as Humphrey Carver who had long been chafing against narrow ideological constraints. When agencies that have difficulty being heard in Cabinet encourage activities in the voluntary sector, that encouragement is best understood as an attempt to build a political constituency for the agencies' programs, especially needed in the face of hostile scrutiny by the Privy Council, Department of Finance, or Treasury Board.

While federal departments such as Defence, Finance, Industry, Trade and Commerce, and Energy, Mines and Resources have a supportive business constituency, those of Labour, National Health and Welfare, Indian Affairs, and the Secretary of State have a different basis in Canadian civil society for political support. Funding of poor people's organizations, public-housing tenants, native organizations, and social-service agencies assists these branches of the Canadian government to counter the well-financed lobbies of business groups with a vested interest in opposing their programs. Polluting industries giving large sums of money to the coffers of political parties are thereby partially checked in their influence by an Environment Canada grant to a local environment group keeping a watch on them.

Humphrey Carver's career as combined activist and civil servant gave him an interesting vision of how he could subsidize social change by persuading government to provide free and ample office space to voluntary-sector organizations. His "utopian dream" of a university-style campus for the national offices of non-governmental organizations was intended to make government more responsive to their concerns rather than to discipline them to the will of the Department of Finance.[64]

The Canadian government's relatively progressive role in housing policy, when compared to the usual stance of the bastions of reaction in provincial governments, points to the generally reformist nature of its interventions in Canadian society, especially when taken by departments outside the seat of power. Consequently, many actions such as the funding of citizens' groups to fight federally funded urban-renewal schemes,[65] which have appeared to be a conspiratorial effort to tame radical agitation, can be seen in a more benign light as encouraging greater social justice in a business-dominated society. Organizing hippies to attempt to block cars in Yorkville, saving a Calgary neighbourhood from the bulldozers of expansion for the Stampede, or having anti-Drapeau clubs to protest spending on circuses rather than social housing, all had the impact of the king organizing against the unchecked powers of rapacious barons. In this sense the controversial Company of Young Canadians were king's agents, although the CYC's brief life after attacks by Drapeau, and incidents such as a priest and racist Saskatchewan whites driving its volunteers out of town for aiding Indians, were not surprising.[66]

That a federal agency should, if only briefly, take swipes at the local power structure, argues that the civil service is more complicated than the usual profile of a monotonous mandarinate would have it. In making simplistic assumptions, Marxists have built upon the flawed work of liberal commentators. While accepting a pluralistic interpretation of social process in Canadian society, such liberal historians as Robert Bothwell, Ian Drummond, and John English posit that the "history of the central government between 1945 and 1970 is the story of a single and coherent group." These authors are drawing upon historian J.L. Granatstein's *The Ottawa Men*, a volume on "civil-service mandarins" focused almost entirely on External Affairs and the Department of Finance. The tendency to view the ideologically entwined Clark and Skelton as founders of the Canadian professional civil service stems in part from Granatstein's sweeping generalization that before the dynamic duo's arrival in Ottawa, only in the "technical branches of government" were able civil servants found, doing highly specialized tasks such as astronomy or land surveying. These skilled specialists, however, like the social reformers attracted to service in the Labour and Health and Welfare departments, would push programs sharply at odds with the technocratic liberalism of Skelton and Clark. Even a Red Tory "dollar-a-year man," such as Wartime Housing president Joseph Pigott, would have similar clashes.[67]

The federal civil service would be home to both the most fervent supporters and the most passionate opponents of social housing. In their views they were closely linked to the clashes in the housing issue in Canadian civil society and would help to mobilize competing segments to have their views prevail. W.C. Clark would solicit the views of the Dominion Mortgage and Investment Association, while Humphrey Carver would turn to the National Welfare Council. But in this complicated game of thrust and counterthrust, after David Mansur's support for limited-dividend housing in 1948, housing reformers could at least find a sympathetic home and source of support for their efforts. A buffer against the generally harsher sentiments of provincial legislatures and similarly real-estate-obsessed municipal governments would eventually be found in Ottawa, a far different situation from that in the heyday of the Department of Finance.

2 Prelude to Policy: Government Inaction before the Great Depression

Although the Great Depression saw the origins of a federal government housing program, it was not a particularly intense time for the escalation of national housing problems. All the problems of conflict between what low-income families could afford to pay and what they needed for adequate shelter, so carefully detailed in the housing surveys of the 1930s, had been recognized as social ills since the turn of the century. The greatest shelter problem that did increase in the Depression, that of homeless single men, was not understood as a housing question and so was absent from the reform agenda except in terms of the adequacy of relief payments.

That housing became a subject of political interest in the Depression was more a consequence of the collapse of the construction industry than the condition of the ill-housed. The poor were suddenly discovered by unemployed architects, town planners, trade-union leaders, and enlightened segments of the construction industry when they realized that ensuring adequate shelter for low-income families would bring prosperity to their own damaged industry.

Since the construction industry flourished in the generally prosperous years from 1900 to 1930, it is not surprising that political calls for government intervention in the housing market were relatively rare. The two periods of greatest economic decline, the recession of 1913 and that of 1919–23, were the only occasions when the call for publicly assisted residential construction was clearly audible before the Great Depression. The recession of 1913–14 brought about the limited-dividend projects of the Toronto Housing Company and of

Pointe-aux-Trembles, near Montreal, which were guaranteed by the Ontario and Quebec legislatures. The post-war depression of 1919–23 saw a temporary federal-provincial program that resulted in the building of 6,244 homes for sale.[1] Both these programs ended when the construction industry entered a boom. Housing was not even mentioned in the index of Hansard until 1918. This first reference was sparked by a need to consider the impact of overcrowded conditions among northern settlers when designing legislation to protect minors from parental vice.[2] From 1923 to 1932 the only reference to housing in the House of Commons was a few words by H.H. Stevens in 1929, praising the wisdom of the post-war Unionist government's short-lived housing program.[3]

Although there was little political recognition of housing as an issue, there was an understanding that housing problems had increased significantly during the boom prior to 1913. To many in the nation's economic elite, this crisis appeared to be a golden business opportunity. One public-health reformer in 1912 complained that the "overcrowded tenement and slum" had been "long tolerated as the natural concomitant of wealth and prosperity" and as the "insignia of business and commercial activity."[4] Indeed, in 1906 the *Toronto Globe* argued that "mechanics in steady employment" should be prepared to "surrender" the amenities of homes they enjoyed in small towns when they moved to a large city.[5]

Despite the enthusiasm for overcrowding among the more narrow-minded civic boosters, social reformers came to recognize the growing severity of Canada's housing problems. They were shocked to find that the affordability and lack of overcrowding they had prized in Canadian housing were rapidly giving way to the contrasting conditions of Europe. Although Sir Herbert Ames was a determined critic of the unsanitary conditions of much of Montreal's late nineteenth-century housing stock, he also recognized that – apart from a few districts such as Griffintown – the working-class families of Montreal could generally afford to obtain a room for each member of their family. With considerable pride he observed that "not more than one home in fifty will have what is for Glasgow an average, that is two persons per occupied room."[6]

The lack of overcrowding that Ames rejoiced in during his housing investigations of the 1890s would vanish in the following decade, and Ames would cease to claim any especially favourable housing conditions for Montreal. Without the aid of government legislation, he would sponsor his own limited-dividend project, Diamond Court. Like early Toronto housing reformer Goldwin Smith, Ames was a critic of public housing at the turn of the century. He evolved in his views, however. By 1919 he was criticizing the post-war housing scheme for not doing enough to provide affordable housing for inner-city low-income families.[7]

Ames's evolution over three decades shows how an enlightened businessman could respond to the deteriorating housing conditions of Canadian workers with increasingly bold suggestions for government action. However, despite mounting evidence of need from social surveys, it was a path that few would journey. Although many eastern Canadian manufacturers did develop an interest in alleviating the housing problems of their workers, the whole movement for housing reform was derailed by the fear of the poor that so permeated the social-control remedies favoured by the public-health pioneers.

The economic boom characteristic of early twentieth-century Canada saw an increasing proportion of the work-force faced with the choice of accepting shelter that was overcrowded, poorly serviced, or below minimal building-code and sanitary standards, or sacrificing other necessities of life. From 1900 to 1913 rentals across the country increased by 62 per cent. Wages rose by only 44 per cent.[8]

The overall inflation of rents for the boom years actually underrepresents the severity of housing problems in communities that experienced rapid urban growth. Nationally, rent inflation was modified by the low increase in communities that the boom passed by. In relatively slow-growing Prince Edward Island, rentals increased from 1900 to 1913 by only 30.5 per cent, less than half the national average. Its capital, Charlottetown, viewed as being "stationary in population growth," experienced "little or no increase" in rents, according to the federal Labour Department. Department investigators found the same situation in Owen Sound, Chatham, and Woodstock. The latter's growth was said to be "so gradual that no housing problem exists."[9]

Most major Canadian cities experienced increases in shelter prices that exceeded the national average. This was particularly true of the largest Canadian municipality of the period, Montreal. Here, in working-class districts rents soared from $6–7 a month in 1889 (recorded by the Royal Commission on the Relations of Labour and Capital) to $12–14 in 1913. This 100 per cent increase came at a time when the wages of the city's most skilled workers rose by only 35 per cent.[10] Various documents provide useful illustrations of the impact of this decrease in housing affordability. By 1905 Montreal's health inspectors had added a new category to their reports, the "dark room" without natural light or means of direct ventilation. In 1908 Elzear Pelletier, secretary of the provincial board of health, noted the emergence of "white mice architecture" caused by the conversion of single-family homes into multiple units. He also observed residential structures placed on damp soil and even "upon land filled with garbage, without covering the ground with concrete."[11] Public-health reformer Bryce M. Stewart, in his examination for the federal Department of Labour of the five largest Canadian cities, demonstrated that Montreal was undergoing by far the worst changes in housing conditions by comparing data from the 1901 and 1911 censuses.[12] Montreal, which had a population of 267,730 in 1901, saw an increase to 470,480 in 1911. The number of dwellings, however, had actually declined in the period, from 36,530 in 1901 to 35,677 in 1911. This caused an increase in the number of persons per dwelling unit from 7.6 to 13.3. While the census figures of the period are marred by confusion between dwellings and houses, Labour Department studies also noted that the city's housing conditions had "degenerated" as a result of a 50 per cent increase in rents from 1909 to 1914. This resulted in the doubling up "of families in the same apartment or house, overcrowding and ill health."[13]

Toronto, the second-largest Canadian city of the era, like Montreal experienced a deterioration of housing conditions above the national average. These trends can be described with more precision because of the greater abundance of studies from the era dealing with housing in Toronto. One contemporary examination was conducted by Professor James Mavor of the University of Toronto's Department of Political Economy. He examined sixty-eight homes lived in by Toronto workers, none of which received more than ordinary repairs. Mavor found that rentals between 1897 and 1906 had increased by 99 per cent while other items in workers' budgets had risen by only 30 per cent.[14] This massive rent rise was sustained. Homes that in 1906 rented at $12 to $15 commanded $20 to $22 by 1914. These trends meant greater economic pressures on Toronto's workers. The $10 to $20 jump in the price of unsanitary dwellings in working-class districts from 1900 to 1913 helped produce an increase of rent as a share of a fully employed labourer's wage from 22.8 to 35 per cent. Although rents had risen by 100 per cent, wages for Toronto's labourers had increased by only 32 per cent.[15]

Labour Gazette correspondents reported that Toronto rents for six-room dwellings in working-class districts with sanitary conveniences rose from $12–14 a month in 1900 to $23–27 monthly in 1913. For homes without such facilities the rate of increase was slightly greater, from $10 to $20 for the same period.[16] Phillips Thompson reported in November 1905 that Toronto's housing problems were about to be intensified "with the approach of winter." He found that during the summer months "large numbers of families found temporary accommodation in sheds, tents and disused street-cars, but the cold weather will compel them to seek better housing."[17] A 1911 investigation of housing conditions in six Toronto districts by the city's board of health revealed that 2,137 houses were inhabited by more than one family, while 198 families dwelt in single rooms. Some 447 persons lived in basements, 22 of these in cellars. One basement was found to house 12 persons. In lodging-houses, of which 41 were found in Toronto, 10 to 30 men were found to be "crowded into a small house, three, four, six men or even more in one small room." One tenement housing 6 families and used partly as a factory had one tap with a pail under it to provide water for 40 persons.[18] By November 1913 Toronto's public-health officer, Charles Hastings, reported that since the 1911 study conditions "have become greatly aggravated." By 1914 the city's health department found that over 9,000 residences were overcrowded. In 714 houses dwelt 9,439 persons. Hastings observed that "eight to ten families are today living in an ordinary 10 or 12 roomed house. Families of five, six and as many as ten are living in single rooms, interior dark rooms, damp cellars and basements."[19]

Toronto provides the best evidence of a pattern to escape from doubled-up housing, crowded tenements, basements, and lodging-houses that many North American workers sought in the early decades of the twentieth century. Various American studies have illustrated how workingmen found home ownership on the suburban fringe through bypassing the "formal economy" of mortgage loans, high land prices, and real estate fees and building their own homes on cheap land. This process, while producing home ownership, exacted a price, paid in terms of long periods of dwelling in partially completed homes, enduring inadequate water and sanitary facilities, and, most significantly, accepting the presence of boarders to pay for shelter. Crowding as many as 28 boarders into a modest one-and-a-half-storey house with a full basement and two sheds was frequent.[20] In Toronto, James Mavor estimated that in 1901 and 1902 more than 10,000 wooden shacks were built by workers in response to the rapid rise in rents. Workers, by building outside of the Toronto municipal boundaries, were saving money not only on lower land prices but by building below the city's fire-code regulations.[21] Dr Charles Hodgetts, medical adviser to the public-health committee of the Commission of Conservation, observed that "in these conditions a worker or members of his family would die in the making of a home – victims of unsanitary housing."[22] The *Toronto Globe* saw such trends in a favourable light, reflecting the prevailing opinion of the period. The *Globe* admitted that such self-help home builders "may tent and sleep in the open air" and that these areas would create "a hundred problems of sanitation and fire regulation." Despite such problems, the system was still viewed as a success since it did give "hundreds of Canadian mechanics and labourers and many a immigrant a little space on which to live ... to call home."[23]

Small- and medium-sized Canadian cities participating in the massive growth of the Laurier boom experienced marked declines in the housing conditions of their working-class communities. Windsor's massive rent increase was accompanied by a trend towards poorer-quality construction. Frame houses with concrete-block foundations became the norm for working-class homes.[24] Heavy rent increases took place even in small towns such as Orillia and St Jean, Quebec, which were experiencing rapid urban growth. Rent increases in Peterborough forced families to move into boarding-houses.[25] Booms in real estate in Port Arthur, Fort William, and Sault Ste Marie caused rents to rise "all round."[26] In Trenton, Nova Scotia, after the Eastern Car Company opened its factory, rents rose by 40 per cent and "all kinds of buildings had to be used as dwellings."[27] In Moncton, after railway shops were constructed in 1907, rents increased at a rate of 50 per cent a year.[28] In Halifax a rapid rent increase in working-class homes took place after some 250 families were displaced for new grain terminals. The impact of this demolition on the local property industry was praised by the media, civic, and real estate elite on the grounds that "everything in the shape of acreage within the peninsula" now "traded at prices largely in excess of previous years."[29]

The nation's business elite was divided on the question of the consequences of the decline in working-class housing conditions experienced in the boom. Business leaders in western Canada were largely indifferent, taking few initiatives in either company housing or limited-dividend housing corporations. In eastern and central Canada manufacturers frequently intervened to provide better-quality and less expensive housing to moderate wage demands. This was done in Stratford and London, communities experiencing heavy rent increases.[30] In Port Hope a deputation of workers asked their employers to enter the housing business.[31] Of course, such interventions would be at the expense of other business groups, which sought to maximize returns on real property, sometimes through a degree of organization comparable to the arrangement of a cartel. In London, for instance, the Real Estate Protective Association controlled 60 per cent of the community's supply of rental shelter, and was able to require all tenants renting from its members to sign a standardized year-long lease.[32]

Business indifference to housing-price escalation in western Canada emerged in part because real estate was one of the leading businesses in the region. Manufacturers were fewer and failed to stake out a separate sense of values from the predominant "booster" spirit of the period, which equated civic well-being with rising real estate prices. Calgary, a city of 26,000 persons in 1912, had 2,00 real estate agents. These operated out of 443 real estate firms. Saskatoon, with a similar population, had 267 real estate firms.[33]

Rents in western Canada were the highest in the country. On the prairies these rents were not compensated for by higher wages, although some relief was experienced in British Columbia. Vancouver labourers' wages did rise from 25 to 37 cents an hour from 1891 to 1914. Although Winnipeg's rents were 25 per cent more than Vancouver's, wages there in this period only rose from 20 cents to 27 cents. Winnipeg's rents were 20 per cent above Toronto's in 1913. The imbalance in other prairie cities was more pronounced. Regina, Moose Jaw, and Saskatoon all had rents 30 per cent above Toronto's. In 1913 Edmonton had the most expensive housing in Canada, with rents averaging 40 per cent above those of Toronto.[34]

The severity of western housing conditions was increased by the failure of the "informal economy" of workers' self-built housing to function effectively, due to the lack of low-priced land that workers could purchase within a reasonable distance from their place of work. A great radius of expensive lots rendered this impossible. Calgary and Edmonton each had 250,000 vacant lots, enough to house the city of Toronto. In 1914 Calgary had 26,763 vacant lots that were fully serviced with sewers and watermains, enough to house all of its existing population at a density of two persons per acre.[35] One grazing ranch inside the city limits was assessed at $1,000 an acre. This forced its sale to an English investor at $1,500 an acre, and a few lots were later sold on it for $4,920 apiece. By 1921 the remaining property had become a ranch again, valued at $50 an acre.[36] In Saskatoon land on the fringe of the city limits that was sold at prices of $10,000 a lot in 1913 remained vacant until 1976.[37] In Vancouver typical housing lots in 1900 sold for $100 to $200; by 1913 lots on its most distant fringe could not be bought for less than $600.[38] Labour leaders estimated that 50 to 75 per cent of the city's workers lost their homes after the 1913 depression.[39]

High shelter costs forced hardships on western workers. Some Calgary workers briefly lived in tents and shacks. In 1912, 2,500 persons, approximately 10 per cent of Edmonton's population, lived in tents. Others were housed inside a curling club.[40] The *Labour Gazette* noted a typical Winnipeg response to rising rents and stable wages; tenants would sublet rooms. Often a vicious cycle began as landlords increased rents after discovering subletting. Overcrowding became so severe that boarders would sleep on fold-out couches in houses of ten rooms that would be sublet by ten families. In such conditions Winnipeg workers turned to more crowded lodging-houses, where twelve occupants often lived in 13 × 12 × 7-foot rooms.[41] In 1913 houses in Edmonton's working-class districts rented for $40 a month. Under such conditions even the budgets of skilled tradesmen with children who could not add to the family income would be strained.[42] In Victoria subletting became common after a massive increase in rents arising from a real estate boom in 1912 that saw homes change ownership two or three times a month at increasing prices.[43]

A detailed examination of the relation of prairie shelter costs and wages has been made for Saskatoon for 1913. Here, only skilled workers such as plumbers, painters, bricklayers, and masons could pay $35 rental for a six-room house. Such accommodation would be without water or sewer facilities, on the outskirts of the city. Any more luxurious housing would exceed the one-quarter income for shelter deemed desirable by the social-welfare statisticians of the era. This was beyond the means of unskilled workers. Those with wages from $15 to $20 a week would have to pay 50 per cent of their income for shelter. Similar predicaments faced women workers. The highest-paid female teachers in the city earned only $21 weekly, the public-health nurse $25, and stenographers from $14 to $20. Women employed in garment-making on the piece-work system averaged only $12 weekly. As rooming cost $6 weekly, this would often absorb 60 per cent of a single female worker's income. One Baptist minister complained that "in case after case high rents prevent marriage." Labourers lived in rooming-houses for $2 monthly, with twenty to thirty men in one room.[44]

The spending of vast sums on municipal public works in western cities in this period, specifically to service the speculative schemes of the property industry, was accompanied by failure to extend services to working-class areas. For example, 60 per cent of the houses in Regina's Germantown district, a foreign ghetto, lacked such basic facilities as access to sewers and water-mains. In Winnipeg fewer than half the dwellings in the working-class North End had water and sewer connections.[45]

In the central city the speculative activities of financial institutions and the real estate industry played havoc with working-class housing conditions, comparable to that which prevailed on the fringes of prairie boom towns. The contradictions of "prosperity" were most apparent in the city core, where skyscrapers soared above the most wretched housing conditions of urban Canada. Here enormous increases in land prices squeezed out both workers' homes and grocery stores in favour of commercial land uses.[46] In an atmosphere of boosterism, a dispassionate study of the transition from residential to commercial use would not be published until 1918, five years after the collapse of the real estate boom. This report was written for Toronto's Bureau of Municipal Research by the era's most eminent housing and planning authorities, among them Thomas Adams and Charles Hastings. It examined 147.2 acres of downtown Toronto undergoing conversion from residential to commercial use, and was ominously titled "What the Ward Is Doing To Toronto." In the "majority of cases," the report found, land values in the district had "tripled ... and in many cases, quadrupled." Land on the corner of College and Elizabeth streets, valued at only $95 a foot in 1909, was worth $1,000 per foot by 1917. The study described how new non-residential land use had reduced the number of residential dwellings in the district, consequently increasing overcrowding. Buildings in the area declined from 1,761 to 1,656. The proportion of these classed as residential dropped from 76.6 per cent to 63 per cent. This caused about 450 to 600 persons to be displaced, and resulted in an increase in the average number of persons per room from six to eight. This, the report admitted, was "an understatement."[47]

The former homes of workers were now used for "the erection of several large modern buildings including an extension to the Toronto General Hospital." Land speculators interested in converting the area to commercial use, "with the chance of at least doubling the money invested," chose not to "repair dwelling houses which do not pay."[48] Tenants likewise hesitated "to ask for repairs for fear of eviction." The report presented a grim description of ill-repaired doors, vermin infestation, falling plaster, boarded windows, and fire hazards.[49]

Although effectively analysing the cause of deteriorating housing conditions, the ward survey was less bold in proposing solutions. Indeed, the nature of its recommendations explains why no effective action was taken to counter such trends. Despite its catalogue of suffering, no call was made in the report to stabilize the residential character of the area, enhance tenants' security of tenure, or provide alternate housing. The tendency of the largely Jewish residents with "a superficial knowledge of our laws and standard of living" not to leave the district but to "manage to crowd into some other dwelling in the vicinity" was disparaged. A long chart of family histories focusing on the deficient character of the inhabitants – such as single women "frequenting dance halls" and "staying out late at night," and married women continuing to "entertain men in the house," – was appended to the report. A "plea for a Canadian Standard of Housing" complained that "the foreigner, with his low standards of living, is usually found" ignorant of "Canadian ideals of housing." The "foreign element" also was seen as composing most Torontonians diagnosed as "feeble-minded."[50]

The fearful image of the urban poor promoted in the ward survey was part of a widespread Protestant middle-class desire for social control over an intimidating immigrant population. Such attitudes actually set back the preliminary campaigns for limited-dividend and municipal housing that were sparked from 1904 to 1906 by manufacturers and trade-union leaders. Interest in the housing conditions of low-income persons began to mount around 1911 as sensational accounts of immigrant life aroused fears for middle-class health and safety. The *Toronto News* warned that fights involving "foreigners" flashing "razors, knives and revolvers" taught that "it is against the product of the slums" that "prisons and workhouses" were maintained.[51] S.W. Dean of the Toronto Methodist Union warned that "either the church must destroy the slum or the slum will destroy the church."[52] The Canadian Manufacturers' Association journal warned that "out of the slums stalk the Socialist with his red flag, the union agitator with the auctioneer's voice and the Anarchist with his torch."[53] Toronto public-health officer Charles Hastings, author of the influential 1911 Toronto housing survey, blamed "the foreign element" for the "exorbitant rents" they endured. He warned the city against immigrants whose "ideas of sanitation are not ours."[54] Dr Charles Hodgetts, head of the Commission of Conservation's public-health committee, viewed immigrants as "willing to live like swine." Hamilton's public-health officer, James Roberts, wrote that only the "drunken, lazy and improvident" poor experienced housing problems.[55]

Public-health reformers turned discussions on housing issues away from providing affordable housing. The new direction focused on regulating the lives of the poor. Roberts urged that a battalion of inspectors check the homes of poor families monthly so as to go about "looking for trouble" systematically.[56] Provincial legislatures amended public-health acts to permit entry into lodging-houses at any time to enforce sanitary regulations. In Winnipeg sanitary inspectors claimed they solved housing problems through "stern repression." In Vancouver and New Westminster, Chinese communities were destroyed on public-health grounds.[57] A public-health officer for Port Arthur complained that immigrants viewed "health officers as their natural enemies."[58]

The consequences of public-health reformers' attitudes to housing was witnessed in rapidly growing Toronto. The public-health department's response to the conditions described in Hasting's 1911 report actually served to exacerbate them. Of the 600 cellar dwellings discovered by the report, 500 were closed. Of the 2,000 unsanitary dwellings discovered, 390 were closed and 100 were torn down. These actions only caused a poor housing situation, as Hastings admitted in a November 1913 report, to "become greatly aggravated." On the basis of his studies of overcrowding Hastings estimated another 10,000 low-rental dwellings were needed in Toronto.[59] He and his department continued to work in the opposite direction. J.S. Schoales, chief of the division of housing and industrial hygiene, reported that after the shortage was identified, "we commenced a crusade against all overcrowded dwellings and lodging-houses. Families were notified under threat of prosecution to remove, and advised to double up in single houses rather than live as they were. Several were summoned to court." By 1915, Schoales noted, 1,007 houses were closed, of which 500 "have been pulled down and new buildings substituted; factories, apartments, stores and dwellings."[60]

Schoales' account illustrates how thoroughly the health department's intentions were grounded in considerations of a cosmetic nature. The demolitions of his department removed unsightly structures that depressed property values. The new factories, stores, and higher-priced residential dwellings built on the former slums did not add to Toronto's limited supply of low-rental housing. Hastings shared Schoales' attachment to the visual appearance of Toronto over the welfare of slum dwellers. His February 1914 report to the board of health gave no indication of whether the residents of the demolished homes had experienced any improvement in their housing conditions; he remarked only on the allegedly superior use that had been made of the land from which the poor were displaced: "For more than a year we have condemned a row of places on King Street occupied by foreigners and a disgrace to the thoroughfare. Only recently we succeeded in getting them closed up, after the owner and occupants were summoned to court on different occasions. They are now being torn down to be replaced by more sanitary and pretentious dwellings."[61]

Hastings viewed certain immigrant groups as inherently unsanitary. This view was held most strongly about gypsies and was apparently shared by public-health authorities across Canada. These people were viewed as prone to "sleeping and living like animals." Hastings told the Toronto board of health that with the aid of the police he had "succeeded in having the whole of this class deported."[62] He also feared the "influx of Bulgarians and Macedonians coming in from some place." These would "live on the cheap, work out all day, crowd into the houses at night, bring the muck of the streets into their rooms, drink beer, play cards, and sleep in clothing worn during the day in closed rooms." Hastings made use of a Macedonian priest in "an effort to teach these people better methods of living," and he printed instructions in the Macedonian press. Unlike Hamilton public-health officer James Roberts, who castigated the poor for wastefully spending their money in cinemas, Hastings denounced them for their thrift in securing low-cost shelter. He complained that the "Bulgarian and Macedonian always has money and you never find him in debt. He always pays his way."[63] Such respect for the work ethic did not spare these people from surveillance. At one point Hastings told the board of health that his attention was directed at "the Bulgarian and Macedonian quarters, 114 of which have been inspected and 72 at midnight. Fourteen lodging-houses and restaurants have been condemned, four closed and fines to the amount of $200 imposed on owners and occupants since the first of March for the ruinous, filthy conditions found."[64]

Hastings battled with overcrowded lodging-houses. Midnight raids were conducted to determine violations of the space regulations. One such raid, accompanied by a closure order, was held against "the so-called Workingman's Lodgings on Frederick Street." Here "the conditions were disgraceful – overcrowded, no ventilation, windows and doors closed, filthy – some fifty men were overcrowded, lying on floor in benches." At these lodgings "the lowliest type of humanity were to be found," taking advantage of the low prices of five to twenty-five cents a night. One of the lodgers, when questioned by health officials, replied that "it is better here than on the streets all night, or in the Police Station."[65] Hastings' proposals to help the immigrant poor got nowhere as a consequence of the opposition of higher authorities. He urged that the former General Hospital and Trinity College buildings be used as municipal lodging-houses. Nothing came of these recommendations.[66] This pattern – the approval of repression and the rejection of positive reforms suggested by public-health reformers to assist immigrant workers – gives some indication of immigrants' powerlessness in Canadian society.

Public-health departments also used gentler methods of education to deal with foreign immigrants. Terry Copp has described the

Montreal Child Welfare Exhibit of 1912 as epitomizing the "romanticism" of this approach to urban problems. The exhibit featured a model kitchen and living room "furnished with hand-made furniture," and noted that seeds were distributed to children so they could plant flowers in their yards.[67] The Toronto public-health department created a special municipal housekeeping branch, a subdivision of its housing division. Hastings believed that "it is questionable if the placing of these people in new, clean houses would materially decrease death rates, or improve social conditions unless their methods of living were changed, as I am afraid that the sow that was washed would soon return to her wallowing in the mire – hence the extreme importance of the education work of the women inspectors." He reported that, as a result of these efforts, between "seventy-five and a hundred backyards in these slums districts that were formerly filled with all sorts of filth and truck, have during the past summer, contained flowers and vegetables." Now one could see "Mrs Hogan's cabbage patch if you care to take a walk through the ward in the summer."[68]

Canadian public-health reformers, unlike their American counterparts, did not adhere dogmatically to a negative approach to housing reform. Hastings, after all, supported limited-dividend housing, whereby investors' returns were limited to 6 per cent. However, the thrust of "positive" reform in Canada would produce meagre results. Legislatures would enthusiastically back proposals for regulating the life of the immigrant poor, but the same impulse was less generally applied to the powerful interests that composed the Canadian property industry. The inability of limited-dividend housing projects to improve the housing conditions of low-income groups, based on the past performance of these schemes in Europe and North America, was noted by Montreal architect Percy Nobbs. But Nobbs and most of his colleagues, with the apparent exception of Vancouver architect MacKay Fripp, were unwilling to take the economics of sanitary housing to the logical conclusion of public subsidies for non-profit projects.[69]

Impetus towards action on limited-dividend proposals came from the national tour of a leading British housing reformer, Henry Vivian, MP. His visit had been arranged in part by the governor general, Earl Grey, who saw housing and co-operative innovations as means by which class conflicts could be lessened. The tour coincided with widespread changes in attitude towards housing problems as a result of the depression of 1913. Interest grew in a "positive" approach to shelter problems, that of building more affordable rental housing. Although greatly interested in controlling the lives of the poor, Canadian public-health reformers were too sophisticated to pursue this approach to the exclusion of all other reforms. Hastings himself noted that limited-dividend housing had won the support of such sound British "financial and commercial leaders" as "Lord Brassey, Sir John Brunner, Mr Rothschild." Consequently, there was no equivalent in Canada to the purely "negative" housing reformers of the American school of Lawrence Veiller, who stressed the need for better building regulations and opposed proposals to encourage new housing construction for low-income families.[70]

Only Quebec, Nova Scotia, and Ontario enacted legislation for limited-dividend housing. Elsewhere opposition, or indifference, from business elites prevented the passage of such enabling legislation. In Saint John proposals for limited-dividend housing did not get past the preliminary-drawing stage. Prominent citizens took the attitude that "intemperance was the main cause of the housing evil." In Winnipeg the first report of the city planning commission recommended that the city take positive action to promote housing construction. Such recommendations were so coldly received by the Winnipeg city council that it refused to pay for the printing of the report.[71]

Ontario passed limited-dividend housing legislation shortly after Nova Scotia, which was similarly inspired by the shock of reformers after Vivian's tour. The pioneer leader of limited-dividend housing legislation in the province, Frank Beer, a philanthropist and manufacturer, persuaded Ontario premier James Whitney to pass the legislation, primarily on the basis of the threat the continued housing shortage posed to immigration. Although the act was only implemented in Toronto, attempts were made to apply it elsewhere in the province, particularly in Hamilton. Here, however, the act's restriction on dividends to 6 per cent made it unattractive for potential investors. They wished the ceiling raised to 8 per cent. Also, the prospective backers, who included the city board of trade and the Hamilton branch of the Canadian Manufacturers' Association, were discouraged by the fact that the best potential sites for such a project were all held by speculators or "tainted by industrial gases."[72]

In Toronto, the only Canadian city that produced a significant limited-dividend housing project, such shelter was seen by enlightened businessmen and social reformers as an adroit compromise with the vigorous demands for municipal housing by the city's labour movement. In 1911 the Toronto Trades and Labour Council drew up a plan for the erection of 2,400 units of municipal housing in a garden-city suburban development. The council estimated that such a project would lower housing costs by 40 per cent. One council delegate, with keen insight into the business values that ruled the politics of the day, observed that the labour council had better elect some members to the provincial legislature if such a plan were to be adopted. The plan had the city buying land on which homes from six to eight rooms could be built for workingmen to sell or lease at cost plus interest charges.[73] Labour in Toronto campaigned against the limited-dividend Toronto Housing Company, primarily because, as Trades and Labour Council spokesman James Simpson noted, it was viewed as "clear evasion of municipal responsibility." The labour council accepted its municipal committee's recommendation to endorse "a municipal housing scheme independent of philanthropic schemes and as free as possible from control by individuals who are not in sympathy with organized labour in its fight for better economic conditions and shorter hours of labour." William Glocking, past president of the council, described the THC as a "pure real estate proposition." Joseph Gibbon, president of the Toronto Street Railway Workers, complained that the company's projects would not help those who really needed better housing accommodation.[74]

Both the achievement and the failure of the Toronto Housing Company can be seen in the testimony of Arnold M. Ivey, president and founding director of the firm, to the 1935 parliamentary committee on housing. Ivey recalled that in 1913 "we found there was a housing problem ... and we endeavoured to solve it, but we were unable to do it; we could not build for the lower wage earners." However, the company had been able to "attack the problem at a higher level" by surpassing the standards of the commercial developments of the day.[75] Indeed, the corporation was able to build to an innovative and attractive design, more characteristic of the manner of English garden cities than the grim English limited-dividend philanthropic developments of the era. Heat and continuous hot water was provided by a central steam plant to all flats, and each unit had its own door to the street, to avoid the objectionable shared stairway of tenements. The units were built in an English cottage fashion, with half-timbered gable ends, wooden verandas, and Georgian pine trim; all were placed in a U-shaped manner, which opened at the street to enclose a centralized grass court. Despite such achievements, the problem of the gap between low-incomes and decent housing was not solved; for although its rents of from $19 to $39 monthly were around the average rent of $25 paid by workingmen making $15 a week, they were above the means of low-income earners.[76]

In spite of the surge of interest in both municipal and limited-dividend housing, much of the public-health establishment remained committed to a negative approach based on the enforcement of building and sanitary regulations. This treated housing as an extension of the problems of unsanitary factories and workplaces. Such a view was particularly strong in the Commission of Conservation's public-health committee, headed by Dr Charles Hodgetts. With Hodgetts holding the view that Canadian housing problems were primarily caused by immigrants living like "swine" because of their greed, the commission's remedies were largely repressive. It worked on model laws to enable health authorities "to condemn, and, if need be destroy the house which is not a home."[77] Likewise, in its scholarly publication *Conservation of Life* it reprinted in all seriousness under the title "How To Deal with Slums" the recommendations of public-health officer Dr A.K. Chalmers of Glasgow. The health authorities of Canadian cities were urged to "be properly led" by Chalmer's recommendations that slums "have got to be destroyed. What happens to the inhabitants is not the question ... Those in authority have one thing, and one thing only to do – they must destroy. My advice is; Do it, and watch what happens. Don't be frightened by this consideration or that. Do it and watch the results."[78]

The hysteria over the danger of slums, which peaked in 1911, and interest in limited-dividend housing, which waned after 1913, were replaced by a new enthusiasm for town planning as a cure for housing ills. Unlike previous approaches, this one appeared unlikely to get bogged down in class conflicts. The pioneer of the Canadian town-planning profession, Thomas Adams, was sought out and brought from Scotland to Canada by the Canadian Manufacturers' Association, the National Council of Women, the Imperial Order Daughters of the Empire, the Hamilton board of trade, and the Canadian Public Health Association.[79]

The sudden interest in land-use planning, after the collapse of a real estate boom, denoted a growing sense of maturity in Canadian society, which had risen above both the turn-of-the-century view of a housing crisis as an economic asset and the later, somewhat hysterical efforts at social control over the immigrant poor. This new maturity proved sufficient to press the Canadian government to obtain Adams' services. However, it was not sufficient to guarantee the implementation of his ideas, as would be shown by the rejection of his advice for munitions workers' housing during the First World War.

The excitement generated by Adams' arrival contributed to expectations that the simple application of technical expertise would solve housing problems without requiring any transfer of national income to lower economic groups. Adams would encourage such thinking in the Canadian public and within the new planning profession he nurtured. But most significant for the future was the fact that the list of Adams' supporters did not include financial and real estate organizations, whose activities would be viewed as nationally damaging in Adams' research studies into Canadian land-use patterns. While the social tension of the years 1913 to 1921 would encourage such moderate reformers as Adams, whose planning proposals promised social peace and more efficient economic development, his skills would be viewed as a luxury the nation could afford to be without in the more conservative political climate of the later 1920s, which dovetailed with a return to prosperity for the real estate industry. Also, the planning profession Adams nourished would lose much of the critical sense he encouraged, in an effort to survive in an era in which the values of the private market had forcefully returned.

Thomas Adams gave both the strengths and the limitations of his outlook to the new Canadian planning profession he shaped. A life-long British Liberal, Adams was remarkably free from the religious, ethnic, and racial prejudices that characterized public-health and city-beautiful reformers. However, his liberalism was also to cause him to disregard the economic indicators that would in any case tie housing problems to the exploitation of one class by another. This would encourage the misreading of ratios between shelter costs and incomes that resulted in the flawed 1919 federal housing intervention, the consequences of which would do much to discredit government housing assistance.

Given the contrasting views of public-health reformers, Adams' freedom from ethnic prejudice is remarkable. He was scathing in his assessment of the social-control experiments of his day, observing ironically of slum dwellers that, "after we have permitted them to become degraded, after we have allowed dwellings to be erected in which their sense of decency cannot be kept, we organize educational campaigns and preach at them and expect them to respond."[80]

Instead of blaming immigrant workers, Adams traced the Canadian housing crisis to the tendency of Canadian cities to regulate urban development primarily to ensure maximum returns to land speculators. He believed that through careful design and the control of land speculation, suburban housing prices would decline to such an extent that slum owners would be forced to lower their rents and improve their properties in the face of this stiffened competition. He went so far as to propose that Canadian municipalities provide free land to persons undertaking to erect a house for their personal use.[81]

Adams' views on the need to restrict land prices to encourage home ownership were passed on to other Canadian planners, including his various assistants such as Horace Seymour, A.J. Dalzell, and Alfred Buckley. Their work focused on the enormous waste to Canadian society caused by land speculation. Typical of their findings was Adams' conclusion that if Ottawa were built at optimal densities, it would need only 15 square miles for the next fifty years of growth. However, 65 square miles of land had already been subdivided into vacant lots. This loss of valuable recreational and agricultural land made it impossible to install sewers and forced children to travel long distances to go to school.

Such waste and inefficiency had been created, in Adams' view, by the machinations of "absentee owners, whose sole interest is in securing the profits of speculation." They financed their speculation "by trying to artificially control distribution and cultivate a few acres of vacant lots." Adams estimated that three cities in western Canada had such a speculative radius of 48,000 acres each.[82] In addition to encouraging more compact urban developments, he urged more narrow streets, built to the natural contours of the land. He also sought to have "real estate operators ... carry out their own local improvements before local development," abandonment of the rectangular grid in street layout, and planning legislation for municipalities to be passed by provincial governments.[83]

Adams' task was made difficult by the fact that his own studies showed that the Canadian urban fringe already had enough vacant lots for fifty years. Consequently, he was placed in the proverbially unhappy position of locking the barn door after the animals had fled. Adams also evaded an aspect of the housing crisis that emerged from the findings of Department of Labour statisticians, the relationship between workers' incomes and the cost of decent accommodation in the rental market. The gap was dramatized in the Montreal Child Welfare Exhibit of 1912, which estimated that unskilled workers continuously employed could not support a family of five, even on its model family budget, except by accepting "unsanitary quarters, sometimes below street level."[84] This weakness in Adams' analysis was grasped by some contemporaries. Industrialist and urban reformer Louis Simpson commented in the 1918 edition of *Industrial Canada* that while the Commission of Conservation had drawn "useful" attention to "the suicidal speculation in building lots," it had ignored the more fundamental question of how Canadian workers were to be "housed in comfort, *under good sanitary conditions*, and at a cost of fuel that will not be too onerous, having regard to the wages paid them."[85]

Despite such limitations on his work, Adams' views were in advance of what Canadian governments were prepared to accept. This is most clearly seen in his isolation in urging an emergency wartime-housing program. Canada during the First World War had no program for constructing housing for workers in war-related industries, in contrast to both Great Britain and the United States. As early as January 1915 Adams urged the adoption of an emergency program.[86] To garner support he toured several provinces and cities. However, the limited moves towards government-assisted housing in most Canadian municipalities was actually terminated because of the war's inflation of housing costs. The sole exception to the rule of war-induced inaction was Halifax. Here, as a result of the intensified housing shortage caused by the war, renewed efforts were made to secure a limited-dividend housing project, spearheaded by the local civic improvement league. Its prospectus illustrated the gravity of Halifax's housing situation, asking, "If the reader can imagine the discomfort a family endured living under a common roof with several other families and being limited to the common use of a single convenience and drawing water from a common tap in the hallway, he has only a partial view of the problem. He needs to know something of the cold drafts, the foul odors, the dark rooms, the plumbing frozen in the winter time."

The housing company aimed at building fifty double cottages on three acres of land it had purchased on the slope of Fort Needham, to supply "a snug four-roomed house with running water, electric light, sanitary conveniences and something on the way of amenity for a rental of not more than about $160 per year." Justice J.P. Wallace recalled how its efforts did receive "support from some of the younger businessmen of the city," but the project became one of the many victims of the Halifax disaster.[87]

After the end of the 1913–15 recession, caused by the war's absorption of surplus economic capacity, residential construction costs soared and housing production plummeted. Private capital feared to invest in housing, in anticipation of the expected price deflation after the war. Adams estimated that "the building of houses to-day will probably involve a total cost of from 30 to 60 per cent above that of houses that had been built immediately before the war." He concluded that, when "erected under such adverse conditions, private residential construction would cease."[88] This observation was borne out by the fact that only 59,000 non-farm dwelling units were constructed from 1915 to 1918, some 500 less than the production for the single year 1913. This shortage was aggravated by demolition; even its supporter for public-health reasons, Charles Hastings, estimated that in Toronto from 1913 to 1918, of the 1,600 homes demolished, only 1 per cent had been replaced. The Ontario Housing Committee, formed in 1918, found that the number of habitable homes had declined by 5 per cent since 1914. By May of 1918 Hastings estimated that "there are over 5,000 families in this city requiring sanitary dwellings which cannot be had." In Winnipeg a 1918 health department survey found that in an eighty-two-acre district inhabited by a "a good class of English speaking Canadians," some 122 houses had been subdivided for multiple family use. Consequently some 125 rooms were "too dark for occupation," while the sharing of water-closets and sinks among eight families was common.[89]

Adams was candid in his castigation of the Canadian government. He reported favourably on British and American wartime government housing efforts and added, "It is therefore somewhat strange ... that so far no federal, provincial or municipal government has initiated any housing scheme."[90] The planner warned that, "if we are now short of houses to provide for newlyweds and industrial workers, what will be the situation when great numbers of soldiers return?"

The formation of the Ontario Housing Committee and the Winnipeg housing survey were in themselves indications of growing public demand for government-assisted housing construction. Thomas Roden had warned his colleagues in the Canadian Manufacturers' Association that "the indifference of the guiding classes" in Canada to housing problems was encouraging "that condition that brought about the downfall of Russia." Following Roden's advice, the Toronto chapter of the CMA resolved that Canadian housing problems posed a "menace" to the "industrial, social and political welfare of the whole country." This chapter of the CMA, together with representatives of organized labour, the Toronto Board of Trade, and the Great War Veterans' Association, formed a delegation that met with Conservative Premier Sir William Hearst. After this meeting Hearst agreed to the formation of an Ontario Housing Committee, which he appointed such veteran housing reformers as Roden, G. Frank Beer, and Sir John Willison.[91]

The Ontario Housing Committee found compelling evidence of housing shortages throughout Ontario. Its survey of Toronto, which excluded the notorious "ward," or central Jewish ghetto, found that 54.9 per cent of dwellings were occupied by more than one family. High rents had driven low-wage earners with dependent families to resort to subletting to pay for shelter.[92] In Windsor five hundred families were found to be doubled up. Severe housing pressures were also found in Brantford and Sault Ste Marie. Sixty municipalities called for provincial housing assistance. In Stratford three hundred houses were deemed to be "unfit for human habitation," while another four hundred were considered "out of condition." Employers complained that housing shortages made it difficult to obtain workers.[93]

Hearst quickly took action by announcing, on 17 July 1918, a $2-million housing scheme. In doing so he stressed the need for additional assistance from the federal government. The $2 million would be lent by the province to municipalities at 5 per cent, which would be repayable over twenty years. Municipalities were to lend this money for homes for sale or rent that did not exceed $2,500 in price.[94]

Given the wartime inflation of building costs, the $2,500 price ceiling was unrealistic. This figure emerged as a result of the housing committee's concern with social control. The committee viewed the provision of better housing for workers than they had enjoyed before the war as encouragement for labour militancy and class conflict. Beer told Hearst that his housing program "must be made a tonic." He cautioned that his government should "not encourage a standard of living that may not be justified by subsequent events." To do so would "add to labour unrest." Instead, the government housing program should actually be designed to teach workers to face "a lower level of living comforts."[95]

In September 1918 the Ontario Housing Committee told the federal government that, because the war had so disrupted the residential construction industry, housing should become a federal responsibility. The committee pointed out that Ontario's own 1913 legislation had become inoperative, since war-bond drives had dried up available sources of capital. It also stressed the increased price of an Ontario workingman's home from $500 to $800. As a minimal figure, the committee recommended that the Canadian government lend the provinces $10 million through a federal housing board. This board would also "distribute responsibility and ensure speedy action" by "provinces, municipalities and employers of labour."[96] Calls for federal action were also made by November 1918 by the newly formed Canadian Construction Association, representing a variety of industries involved in the building trades.[97]

Prime Minister Sir Robert Borden's Unionist Cabinet was divided on the housing question. Ontario premier Hearst obtained the support of federal Liberal Unionist ministers such as Privy Council president N.W. Rowell and immigration minister J.A. Calder.[98] Rowell warned that there was "less respect for law and authority" than ever in Canada and that without an adequate program to deal with the unemployment problem, no one could "foresee just what might happen." The Department of Labour warned that returning veterans would be far more militant than the pre-war unemployed.[99]

The principal opponent of any federal housing action was the powerful Conservative minister of finance, Sir Thomas White. White abandoned his opposition out of fear of the social situation in the first winter after the war. That such fears were reasonable is demonstrated not only by later events such as the Winnipeg General Strike but by the increasing incidence of industrial disputes in the later war years. White's reasoning was similar to that of the British and European governments that embarked on housing programs to avert revolution at this time. It was also embodied in other reforms of the Unionist government, such as Dominion aid to vocational education and the Employment Service Act, which created the first national network of labour exchanges.[100]

The federal government's commitment to a housing program was announced at the federal-provincial conference begun on 26 October 1918 to discuss post-war adjustments. On 3 November 1918, the final day of the conference, White was still holding to the view that housing was a municipal function, likely to test the extent of provincial support for federal participation. In response Premier Hearst argued that housing was a "war" and a "national" problem, especially in view of the additional accommodation that was a "necessity for the returned soldier." Hearst also favourably described the national munitions' workers housing projects in Britain and the United States. He stressed the need to encourage residential construction to prevent unemployment after the war's end. In response to White's acceptance of federal lending to provinces for home construction, Hearst obtained federal agreement to make these loans at 5 per cent, which was below prevailing market rates. Hearst pointed out that Ontario's plan had the province lending at this subsidized interest rate to compensate for abnormally high construction costs. He was surprised that White, serving as acting prime minister, agreed to his suggestion at the conference "without consultation with his colleagues other than those who surrounded him at the table."[101]

White kept his commitment, and eleven days after the conference, on 3 May 1918, a federal housing scheme was established by order-in-council. It followed the verbal understanding with Hearst that federal loans to provinces would be made at 5 per cent. The program provided $25 million for this purpose, to be repaid over twenty years. Funds would come from the war appropriation and would be spent under the terms of the War Measures Act to avoid immediate parliamentary approval. A second order-in-council, passed on 12 December 1918, gave responsibility for refining the federal scheme to a five-man Cabinet housing committee chaired by Newton Rowell. Thomas Adams served as the committee's secretary and played a critical role in the formulation of its "general scheme," which was announced by order-in-council on 20 February 1919.[102]

Adams had earlier sought a more ambitious approach to housing problems with the federal Cabinet. He outlined an integrated program of "housing, local transportation and land development." This would have been administered by a federal board that was intended to work with provinces and municipalities to build new towns, national highways, and veterans' farm settlements and to direct the production of materials for the reconstruction of post-war Europe. A minimum $40 million federal contribution to this scheme was recommended by Adams. This proposal clearly lacked the political wisdom of Hearst's plan for meeting immediate housing problems that threatened to cause social unrest.[103]

Despite the fact that his personal plans for post-war reconstruction were rebuffed, Adams was able to bring the imprint of his outlook to the federal housing program. It accepted his basic approach of encouraging home ownership among workers by lowering land costs. At this time Adams was deeply committed to such a solution; he wrote that since "the habit of home ownership has become so ingrained in Canada ... it is best to encourage it in preference to renting."[104] In the 20 February 1919 order-in-council, many advances on the earlier federal ouline show the influence of his thinking. A new objective, unmentioned previously by the Cabinet, was Adams' cherished goal of placing "within the reach of all workingmen, particularly returned soldiers, the opportunity of acquiring their own homes at actual cost of building and land acquired at fair value, thus eliminating the profits of the speculator." The provinces were urged to make "statutory provision ... for a cheap and speedy method of compulsory taking of land for housing purposes. "On such" acquiring of land at its fair value and at a cost that workingmen can afford to pay "was said to rest the" success of the housing movement."[105]

The federal order-in-council also sought to encourage workers to benefit from its scheme by imposing cost ceilings on homes to be financed with its funds. More realistic than Hearst's ceilings, these were $3,000 for four- or five-room frame houses, $3,500 for five- and six-room frame homes. Brick, stone, and concrete homes were to be limited to $4,000 and $4,500 for the same respective categories.[106] Adams shared the opposition of Hearst and the federal Cabinet to the heavily subsidized rental projects being undertaken in Britain and Europe. He favourably cited the Edinburgh city engineer's view "that the Canadian project is on sounder economic lines than the proposed housing schemes in Britain." Despite his praise of the government program, Adams admitted that it had been adopted only for such expedient reasons as the post-war housing shortage, pending unemployment, and "the desire to assist in avoiding industrial unrest."[107]

Enthusiasm for the housing scheme exceeded what it could achieve. Typical of the overly ambitious expectations was the view of C.B. Sissons, secretary of the Ontario Housing Commission, who exulted that there would now be "no reasons for denying to the humblest city workers of the future the charms of Rosedale without its inconveniences."[108]

The conservative approach of reform-minded manufacturers such as Roden, Beer, and Willison, combined with the town-planning remedies of Thomas Adams, produced a housing program that was particularly ill suited to a period when construction costs were still inflated by the disruption of the war. Their formulas, the product of an unsuccessful effort to make housing affordable for most industrial wage earners, encouraged the building of poorly constructed homes. After the depression of the post-war years ended, those homes would fall largely into the hands of municipalities, who would frequently have to pay heavy costs to upgrade them.

The federal-provincial scheme's reduction of interest rates by 1 per cent and elimination of the costs of land speculation did not go far enough to produce housing affordable to more than the top 20 per cent of the population in terms of income. The cost formulas for model homes produced by the Ontario Housing Committee were too optimistic. This was pointed out to the committee by the Toronto building contractors Ansell and Ansell. This firm estimated that homes listed for $2,500 would cost $3,200, and criticized the generally superficial nature of their designs. In response, the Ontario Housing Committee agreed to raise cost ceilings and indicated that its designs were "in no sense compulsory." However, it declined to issue a more realistic list of model-home designs and so remained in the dark about the likely true costs of its program.[109]

The gap between the costs of building and what workers could afford to pay was clearly evident in Ottawa. Here, the planning of the federal-provincial project was undertaken by Thomas Adams, who sought, in the Lindenlea subdivision, to achieve the dream of a working-class "garden city" Rosedale. He had the city housing commission purchase a large, twenty-two-acre site. This provided

169 building lots and had generous provisions for parks and streets.[110]

Adams' care for land use in planning the subdivision could not, however, compensate for the damage done when builders tried to cut corners to build homes within the costs of the federal-provincial price ceilings. A city of Ottawa investigator later discovered that builders frequently "had to leave off such things as verandas, front steps, double windows and various other things." Often, even more drastic savings were attempted, and the homes became a public danger. One city inspector recalled that it was found, by digging around a foundation, that a "house was simply standing on the bare ground," without a foundation. It was in such a poor state of repair that the home "became a menace, even to the public because it was tending to fall apart." To demolish such homes and pay the costs of major repairs, Ottawa had to issue $45,000 in debentures in 1922. Three years later the Ottawa Housing Commission secretary left town, stealing $82,000 in funds. By 1935 Ottawa had accumulated a $332,499 loss on the scheme. Lindenlea provided a striking example of how opposition to federally subsidized rental projects produced poor-quality homes for sale, subsidized by poorly financed municipalities.[111]

Although the outright graft associated with the Ottawa scheme was unusual, the experience of building unsound homes – later repossessed by the city, which was forced to make many repairs – was repeated throughout the country. Percy Nobbs, in reference to Quebec, observed how "the mess" from the 1919 scheme was "found all over the province."[112] When a surge in residential construction by 1923 ended the post-war housing shortage, those who purchased homes under the 1919 act found that they were paying more in principal and interest payments than they would to rent equivalent homes. As the home's purchase price would now be less than what they had paid, many owners let municipalities repossess their dwellings. These were frequently in a poor state of repair, required heavy investment by the municipality to bring them up to standard, and could not generally be resold until the favourable seller's market of the early years of the Second World War.[113]

In the wreckage of the 1919 scheme there was conceded to be one oasis, the project of the city of Winnipeg. This project succeeded because its creators had no illusions about building housing for industrial workers. In Winnipeg and other communities where the program proved to have some success and popularity, the scheme simply provided low-interest money to help someone who already owned a lot build a home for his personal use. Winnipeg's housing inspector, Alexander Officer, pointed out that the scheme worked because it gave "white-collar men ... who had standing salaries" money to "take this place and have a home built on it."[114]

The revival of the residential construction industry, which financially damaged municipalities that participated in the 1919 federal-provincial scheme, also discouraged interest in housing as a public issue. This did not mean that the problems of overcrowding and affordability that emerged in the boom before 1913 had vanished. Instead, those who suffered from wretched housing accommodation were ignored, or blamed for their own plight by the nation's tiny corps of social workers. It was anticipated that growth, demolition, transportation improvements, and planning measures would, in the long run, solve any temporary housing problems. No programs of shelter assistance involving the redistribution of income were ever recommended.

Although the total parliamentary silence on housing issues from 1924 to 1932 provides little by which to gauge the reasons for political indifference to housing as a public issue, the debates from 1920 to 1923 yield more insight. When the 1919 scheme was first debated in Parliament, the only opposing view was that it did not go far enough to meet the housing crisis of the nation. In subsequent debates, arguments of a ruralist and laissez-faire nature indicated opposition to the concept of government-assisted housing. Liberal Walter Nesbitt, a "general agent" from Oxford County, Ontario, saw housing problems as rooted in rural migration to the cities, where "there seems to be the greatest frivolity." Liberal F.S. Cahill, a broker from Pontiac, Quebec, argued that assisted housing was spoon-feeding. George Boyce, grand master of the Orange Lodge of Ontario, urged that housing problems be solved by closing "the picture shows and theatres" so that what "these young men earn during the day they may not spend at night."[115] In 1919 the threat of social upheaval was real enough to silence opposition to the principle of assistance for low-income housing. But by 1921 the political climate had changed so dramatically that housing had vanished from the political agenda altogether.

Although the years from 1923 to 1931 saw a great surge in housing construction, housing problems among working-class families remained, as is illustrated by the 1931 census. This was taken in June of 1931, just after the peak of the construction boom and prior to its precipitous collapse. In most Canadian cities, families paying fifteen dollars or less per month in rent tended to suffer from overcrowding, having on average more than one person per room. In the three prairie provinces, rural communities averaged more than one person per room. Some 40.8 per cent of Montreal's residents lived in overcrowded homes, and conditions were even more severe in Verdun, Regina, Quebec City, and Three Rivers. In Winnipeg and Toronto the figures were 35.74 per cent and 24.8 per cent respectively.[116]

In 1941, while reviewing the 1931 census figures on overcrowding, Dominion statistician Harold Greenway concluded that "a large proportion of the families concerned cannot afford even as much as $15 per month for rent." He concluded that the "only alternative" to "such unsatisfactory housing conditions" was "either a change in national income structure, or some sort of subsidized housing."[117] Such observations were made in Canada only after the residential construction industry had come to the verge of collapse.

Even the land-use planning solutions advocated by Adams went into decline in the prosperity and social peace that emerged after 1923. The Commission of Conservation was abolished, and Adams' own tenure as town-planning adviser to the Dominion government was not renewed. A.J. Dalzell, who most firmly championed his mentor's views on the need to control urban sprawl, lamented that there was "far too much loose talk about the necessity of dispersing population by providing for rapid transit."[118] By contrast, another Adams disciple, Noulan Cauchon, predicted that the automobile's ability to disperse population would solve the housing problem.

Dalzell's views would prove highly influential because of his appointment by the Social Service Council of Canada to conduct an investigation into Canadian housing problems. This commission was prompted by the council's surprise at the conclusion of the 1923 annual report of the Dominion Department of Health, which stated that it was impossible for private enterprise to have enough return to build homes for men who could not afford to pay twenty-five dollars a month in rent. Following Adams' rationale, Dalzell charged in his report for the Social Service Council that "the fundamental cause of most unsatisfactory housing in Canada is ... an essentially unsound land policy."[119] He provided ample examples of ruinous land speculation but avoided any examination of the relationship between income and the ability to obtain adequate housing, which the council had originally asked for. To obtain information on housing developments in Europe, the council hired Dr John W.S. McCullough of the Ontario Department of Health to prepare a monograph on the subject.[120]

The most radical housing-reform advocacy came in the May 1929 edition of *Social Welfare*, which reprinted an article by Edith Elmer Wood that aptly characterized the Canadian situation as being at the point where "Great Britain was 78 years ago, where Belgium and Germany were 40 years ago, France 35 years ago and Holland 28 years ago, debating whether or not nation, state and city should provide housing credits, on an at-cost basis, to cut down the price of wholesome housing to be within the reach of lower income groups that cannot otherwise attain it."[121]

More typical than Dalzell's zealous restrictions on suburban speculation and the Social Service Council of Canada's daring inquiries into non-profit European housing developments were prevailing attitudes towards the poor. As in the period of prosperity before the First World War, the spirit of regulation in housing was applied in a manner that reflected the class inequities in society. Zoning regulations preserved the amenities of middle-class neighbourhoods; working-class districts did not receive similar protection from the intrusions of commercial or industrial land uses. Public-health and social-work professionals did not call for greater curbs on land speculation and building regulations. Instead, they directed their reformist writings towards the need to exercise greater control over the lives of the poor.

The significant relaxation of the unprecedented housing shortages of the immediate post-war years caused a return of the tendency of middle-class reformers and professionals to blame poor housing conditions on their victims. Vancouver's public-health officer recommended the demolition of residences housing a total of two thousand persons, although the city was powerless to provide alternative accommodation.[122] Professional favour for social control was typified by a 1929 *Social Welfare* article entitled "The Social Welfare Workers' Attitude to Housing," by William McCloy. McCloy admitted to the "silence" and "scarcity of materials" on "actual housing conditions in Canada." Since few surveys had been made, "nearly every city in Canada" was ignorant of "the actual conditions that prevail." Despite social workers' urgings, poor people "refused to move to another section of the city." Even if such movements took place, they spent too much time visiting corrupting slum neighbourhoods because of the "lure of friends and familiar places." McCloy urged that this be solved by having social workers trained, in the paternalistic British "Octavia Hill" tradition, to manage tenants on "a firm business basis."[123] Dr Grant Fleming of the Montreal Anti-Tuberculosis League avoided the harsher methods of social control and instead urged the teaching of resignation. Medical workers could "teach people to make the best of what they have" so that most homes would be made "relatively" healthy.[124]

As long as the private residential construction market flourished, no demands would be made for public provision of housing at rents geared to ability to pay and reaching minimal standards to ensure the health of the poor and many labourers. A weakened labour movement, which in the past had advocated municipal housing, stopped making such demands. Its largest and most conservative section, the building-trades unions of the Trades and Labour Congress, would simply follow the initiatives of the business-dominated Canadian Construction Association. The CCA itself abandoned interest in federal housing legislation after the return of prosperity to its industry in 1924. The professions of architecture, social work, and planning harnessed their energies to the service of affluent elites that controlled government and could pay for their services. The era's experts would regard the nation's low-income earners as fit subjects for moral uplift to middle-class standards, not candidates for financial assistance.

The political evasion of housing problems can be understood partly as a sign of an unbalanced influence of realty and financial interests. Adams saw the nation's resistance to housing and planning legislation as arising from the "exceptionally strong" real estate interests that flourished in a period when Canada was "still being exploited as a new country."[125] The national emphasis on the export of resource materials and unfinished products encouraged a neglect of secondary industry. It also hindered development of a domestic market in consumer goods and stifled concern for the need for good housing conditions to ensure a stable and highly qualified labour force. Neglect was further fostered by the cyclical nature of many housing problems, which allowed for occasional improvements, such as the easing of the post-war housing shortage. All this enabled most political figures to treat the nation's housing problems as if they did not exist; or, if they were admitted to exist, the political economy of the day held that they could be left for resolution by the unaided private market. Meanwhile, experts placed the blame for the most serious interludes of crisis on the very persons who suffered from the worst housing conditions.

3 The Emergence of an Assisted Market Strategy, 1930–1935

The coming of the Great Depression finally saw the creation of a national housing policy and the adoption of federal housing legislation with the passage of the Dominion Housing Act of 1935. This usually overlooked aspect of Bennett's New Deal was, however, infused with profoundly conservative objectives. The sudden emergence of the Dominion in the housing field came in response to widespread demands for subsidized housing for low-income families from construction unions, social workers, architects, and important sectors of the building industry. In DHA, Prime Minister Bennett and deputy minister of finance W.C. Clark managed to defuse calls for social housing by passing legislation geared to strengthening the mortgage market for building homes that could only be afforded by the top 20 per cent of households in national income. Demands for subsidized housing were conveniently shelved by transfer to the Economic Council of Canada, which was never appointed. This would be typical of the tactics employed by the federal government, under Clark's leadership, to delay public-housing legislation until the passage of the National Housing Act amendments of 1949.

What led to the passage of even the minimal DHA of 1935 was the emergence of a new social approach to housing issues, which prodded a reluctant Bennett government to assume a responsibility for housing. An important conflict of values had emerged in Canadian society over housing issues. If the boom in the construction industry in the 1920s had allowed builders, architects, and even union leaders and social workers to ignore the poor or blame them for their fate, the collapse of the building business in the Depression helped to generate new attitudes. The same groups that, in prosperity, had ignored the slums or attributed them to their residents' failings, suddenly realized there was a severe problem of affordability for low-income families. A broad consensus quickly emerged to restore prosperity to the construction industries and to provide a better life for Canada's poor families. Many sectors of society believed that a national social-housing policy would provide jobs for the underemployed businessmen, tradesmen, architects, and planners who would build the warm, comfortable, and healthy homes for disadvantaged Canadians.

The reformers' solutions were straightforward and vigorous. They urged the building of homes for low-income families at standards that would not endanger the tenants' welfare because of uneven temperatures, lack of adequate utilities and sanitary facilities, and overcrowding. If such housing was rented at levels sufficient to pay the cost of its construction, it could not be afforded by the very income groups in need of assistance. Therefore the reformers invariably stressed the importance of public-housing subsidies. Anticipating attacks on the implications of such subsidies for public finances, the reformers argued that their costs would be less than the heavy expenses poor housing encouraged, such as those for fire and police protection and for hospitalization. They also maintained that their solutions to housing problems had been carried out in Great Britain and Europe since the First World War, and were now being applied vigorously by the United States Public Works Administration as part of the New Deal policies of Franklin Roosevelt.

The first significant seeds of the reformers' housing ideals were sown in the report of the Halifax Citizens' Committee on Housing, which was published in 1932. This proved to be the first of many housing surveys of Canadian communities undertaken during the Great Depression that would make bold recommendations for a public program of subsidized housing construction. The Halifax report expressed the new ideals characteristic of the ideals of the Depression-era social-housing advocates.[1] That there had been a shift in values is highlighted by the citizens' committee's rejection of the recommendations of A.J. Dalzell. Dalzell had been hired as a consultant in recognition of his status as the foremost expert on Canadian housing problems and the most prolific author in the field. Repeating his own and the planning profession's analysis of the past decade, Dalzell claimed that – with proper controls on land use – it would be possible to shelter all income groups in occupant-owned single-family dwellings. The only exceptions to this rule would be a class of "undesirable tenants such as the infirm, aged or handicapped" or such less-deserving poor as the "habitual drunkard" and the "feeble minded woman with a large family of children and not yet accepted as an institutional case." For such cases municipal housing would be appropriate, and would be run along the lines of a modernized version of the Elizabethan poorhouse. The "constant supervision" it would provide should ensure that "those that prove to be incurable because of their own bad habits or from complete mental inefficiency are put in their proper place."[2]

A different diagnosis was advanced by the secretary of the Halifax Board of Health, Arthur Pettipas, who argued that the incomes of the city's workers were too low for them to obtain healthy accommodation. As the latest available census returns (1921) had illustrated, many wage earners were in families earning less than $1,400 per annum and so could not afford to pay more than $20 per month in rentals. Since no houses were available at this rate, such families were glad to pay that amount simply for a few rooms, although the resulting overcrowding sometimes went as high as ten persons per room.[3]

An alternative approach furnished to the citizen's committee came from Major W.E. Tibbs, chairman of the Halifax Relief Commission, who supplied a "Report on Housing Developments in Europe" based on findings from a recent trip. It was one of the first accounts of European housing programs published in Canada since the early 1920s. Tibbs reported favourably on British municipal and limited-dividend housing projects and upon "the Vienna Housing Programme, a purely socialist scheme." He described this effort with evident enthusiasm, observing that:

> The development consists of enormous apartment houses with court yards, gardens, lawns, and playgrounds, although some of them have individual gardens. I visited one flat that was delightfully compact, and obviously appreciated by the tenants who were taking great care of their property. Rentals vary from about one ($1.00) dollar up, according to the size of the lot. There are central laundries with up-to-date equipment available for the tenants. There are also baths in central locations. Children are also well cared for as regards kindergartens, etc. The method of financing was by appropriation from the rates of about 60%, the remaining 40% being obtained by means of a special tax on rentals. All the people in these apartments appear very contented and happy. Beside providing houses the Socialistic Government assists them through general health programmes, educational opportunities and unemployment relief.[4]

Faced with contrasting European interventionist and American laissez-faire solutions to its housing problems, the citizens' committee pondered, "Upon which shall we decide? Shall we follow the Dalzell recommendation? In spite of world wide experience to the contrary is Mr Dalzell right in his recommendation to leave the matter to private enterprise?"[5]

In answering this fundamental question the committee concluded that even with the reduction of housing costs set forth in Dalzell's plan for "adjustment of street expenses, and civic services, and by a town-planning programme, the shelter needs of low-income tenants" would not adequately be "met by private enterprise." However, the committee then attempted to steer a middle course between European socialism and American capitalism, on the grounds of "the principle fundamental in social reform that the object aimed at should be secured with the least possible dislocating of the existing social arrangements." The costs of subsidized socialist housing would be "constantly added to the taxpayer's burden." Consequently, the citizen's committee recommended that low-interest loans be provided by the province to private companies in exchange for rents at levels that would "not be available through the normal channels of private enterprise." This recommendation, similar to New York State's housing laws of the previous decade, was passed by the Nova Scotia legislature in April 1932.[6]

The Halifax Citizens' Committee's compromise with the prevailing marketplace ideology indicates a stage in the transition to new values in a significant segment of informed public opinion, a process that would culminate in 1934 with the city of Toronto's housing survey, popularly known as the Bruce Report. It came out fully in favour of publicly owned and subsidized rental housing. In this process of radicalization the views and values of planners, social workers, architects, and even construction corporations were transformed. Percy Nobbs, in his 1932 presidential message to the Royal Architectural Institute of Canada, indicated that the organization's executive had been "much occupied" in "rather depressing subjects" such as "an effort to deal with speculative jerry builders and loan companies." Nobbs outlined patterns "of the wild exploitation of the land, materials and labour, which run counter to wholesome development in the building industries." In November 1932 Nobbs' successor as president, Gordon M. West, reported that "there has come to light a feeling that the financial background of the building trade in Canada has operated in favour of the speculative interests therein as against those of conservatism and sound ownership, and architects undoubtedly feel they are allied with the latter rather than the speculative interests." Nobbs had noted that "the speculative builder who does not employ an architect is usually a contractor, but seldom a member of the local Builders' Exchange or the Canadian

Construction Association." Consequently, architects and construction contractors with perceived conflicts of interest against loan-company-financed speculative builders formed a joint committee to explore means to recover their loss of employment. Their committee eventually expanded to include the Engineering Institute of Canada, the Canadian Manufacturers' Association, the Trades and Labour Congress, various construction-supply associations, and the Canadian Chamber of Commerce.[7]

At a three-day conference held in February 1933 at the Royal York Hotel in Toronto the unlikely allies sought a recovery remedy for the construction industry. The conference observed that in 1933 almost no capital expenditures had been made at any level of government and announced that "this policy of retrenchment we have so far never severely criticized, but today we are firmly convinced that its continuation will only lead to disaster." By May 1933 the alliance had taken on an institutional expression as the National Construction Council (NCC). One of its first steps was to send out a questionnaire to its members, boards of trade, and larger municipalities, inquiring into the desirability of certain construction projects. While not originally on the council's list of worthy public-works projects for government consideration, public housing rapidly became one of the most popular suggestions to combat the Depression. A program of "low-cost housing and slum clearance" was urged upon the federal government by the Royal Architectural Institute of Canada at its May 1934 convention. In its journal, architect James Craig wrote that although public housing in Canada and the United States was subject to bitter attacks by speculative builders and real estate firms, who believed their markets were being invaded, it had "become obvious that Canada can be classed with nations that are seriously battling against the depression only when she adopts national, provincial and civic measures to eradicate the slums and improve the housing conditions of the under-privileged classes."

By 1934 the NCC had come out in favour of public housing and announced that it was undertaking a survey to determine the extent of the need for a program of low-rental housing and rehabilitation.[8] The council's survey increased support for local movements for public housing that repeated the example of the Halifax Citizens' Committee. In Vancouver the British Columbia committee of the council found that "there is no recognized slum district or area in Vancouver as the term is generally understood, but there are hundreds of single buildings, cabin blocks and terrace blocks scattered throughout the city, which have within the last five or six years been allowed to develop through lack of attention, into a state menacing the health of the city at large."

In Kitchener, Guelph, and Galt, Ontario, the regional committee of the NCC estimated that $6,762,000 in rehabilitation and replacements were required. In Regina the local committee found that of two thousand substandard dwellings, half lacked sewers and water supply. In London the committee called for the demolition of slum areas and "the erection of a block of low-rental apartments, with proper playground facilities in connection with the same, properly supervised, along with the apartments, by the municipality."[9] In Ottawa the council, along with the Ottawa Welfare Board and the city's town-planning commission, conducted a survey that found, "as the depression years have deepened and family resources have run low, furnishings have become depleted and even beds and bedding are often lacking. Families have been doubling up seeking cheaper and cheaper accommodations. Sixty-five per cent of the families on relief at the time of the survey were doubling up to two or more families in the single family unit. Many families not on relief but on the borderline of relief are in identical conditions of housing."[10]

In Montreal housing conditions were investigated by a survey undertaken by a joint committee of the Montreal Board of Trade and the Civic Improvement League. The committee included Percy Nobbs, Leonard Marsh, representatives of the NCC and the Montreal Builders' Exchange, social agencies, and a sprinkling of industrialists and businessmen including George Elliott, president of the Montreal Real Estate Board. The committee found that some seventy thousand Montreal households were paying a disproportionate share of their income in rent. Low-income families who avoided this hardship did so only by living in deteriorating housing. This was especially true for families on relief, who were prohibited from paying more than $6 to $12 per month rental by the regulations of the Relief Commission. It was estimated that the cost of the welfare recipients' accommodation would have to range from $30 to $35 per month for the small proprietors who provided such housing to be able to afford the cost of adequate repairs and earn a reasonable rate of return. Consequently, since providing low-rental housing was so unprofitable, only a limited supply was available. This resulted in a great "competition for these dwellings, which accounts for the high mobility ... in poor districts."[11] It was concluded that "to spend more than $120.00 or so of an income of $600.00 on rent, means sooner or later some curtailment or deprivation of food and fuel, and certainly of clothing, minor luxuries, and recreational expenditures. The provision of low-rental housing tends to reduce undernourishment, tuberculosis, hospitalization, destitution, with their attendant social costs, and to release working class purchasing power for the other necessities, comforts, and conveniences of life."[12]

The Montreal study was greatly influenced by a Cleveland survey conducted by a young priest, Robert Bernard Navan, in 1932. The committee noted Navan's revelation that Cleveland's slums yielded only $10.12 in municipal tax revenues per capita, while requiring expenditures of $51.10 per capita as a result of heavy fire-prevention, police, prison, and health costs. Similar patterns were found in Montreal, where the highest death, tuberculosis, and juvenile-delinquency rates were found in the areas of the worst housing conditions. The committee concluded "that the cost to the community at large of leaving the low-income groups to find accommodation in deteriorated structures is not economically sound – that it is cheaper for the community at large to bear a substantial part of the cost of providing adequate accommodation for these wage groups."[13]

Consequently, the committee saw provision of a basic standard of shelter as a public utility: "Assuming certain minimum standards of accommodation and amenity, these standards can be secured only by regarding their provision as a matter of public responsibility – as we now do education and water supply."[14]

The Montreal study urged that ten thousand low-rental homes be built within twenty years. Fully 70 per cent would be located on fringe land, the rest divided equally between land in partially developed areas and in slum-clearance schemes. Projects involving slum clearance would involve four-storey apartments in the inner city, two- and three-storey flats on middle land, and two-storey grouped, self-contained cottages on the fringe. In the middle and fringe areas there was an abundance of land, as Montreal had 98,000 lots unoccupied compared to only 88,000 occupied. Many of these properties had come into public hands as a result of the Depression.[15]

The Montreal committee rejected the approach of socialist municipalities in Prague, Vienna, and Germany during the Weimar Republic as too "lavish." Their rooms were "over-generous" in size. A Prague apartment complex with "workmen's flats" of 234 square feet was described as "highly extravagant." Such projects were financed through taxes. The capital costs were written off, and tenants charged on the basis of the cost of maintenance. The committee favoured instead the British approach of loans at 3.5 per cent to municipal housing authorities and private building societies, with payment of subsidies to ensure low-income affordability. Such loans would be made by a national housing board directly to local housing corporations. The board would examine the merits of proposed schemes and set standards of accommodation for government-assisted dwellings.[16]

In Winnipeg similar proposals arose for low-rental housing, and much more controversy was generated than in Montreal. A survey conducted in 1934 by Alexander Officer of the city's health department found that of 1,890 houses surveyed in four districts, 1,300 had rooms used for both cooking and sleeping. On average there were 3.85 families to each wash-basin, 2.19 to each bath, 1.79 to each sink, and 1.93 to each closet.[17] To alleviate such conditions five members of the Royal Architectural Institute of Canada proposed a low-rental housing project to be built on municipally owned tax-default land.[18]

The city council voted to spend $1.5 million on the scheme, provided that the federal government would extend aid to the measure as an unemployment-relief project. The Winnipeg council called on the federal government in spite of substantial opposition from vested interests. Mayor R.H. Webb recalled that, after he urged the federal government to become involved in low-cost housing, he had "immediately received a flood of telegrams opposing his proposal." According to Webb this arose from "Real Estate organizations, Boards of Trade and others who don't have to live under these conditions."[19]

The report of Ontario lieutenant-governor Hebert Bruce's committee on housing conditions in Toronto became the key document cited by advocates of a social-housing policy until the release of the report of the federal subcommittee on housing and community planning (the Curtis Report) in 1944. The Bruce Report, largely written by Eric Arthur, architect, and Harry Cassidy, professor at the University of Toronto's School of Social Work, found that at least two thousand Toronto families were inadequately housed. This meant that they endured roofs that leaked, "uneven and often rotten floors," toilets and sinks that froze into a "a chunk of ice in winter," and an absence of hot water unless heated over stoves. Tenants lacked facilities for washing clothes and even storing food. Heating was frequently so poor that tenants dreaded "the winter because they cannot possibly keep themselves warm." Rats bit small children. In response, parents kept lights burning during the night to prevent "the bugs and the rats from eating them up."[20]

To provide improved but affordable housing, the Bruce Report urged that a slum-clearance and public-housing project be undertaken in Moss Park. The subsidized accommodation to be built there would charge rentals of $15 to $25, substantially lower than what tenants paid even for slum accommodation. This $500,000 project was only part of the $12,000 program outlined in the report. The larger sum included expenditures for rehabilitation of repairable substandard structures and the construction of at least one thousand new low-cost rental dwellings.[21] The report stressed the need for financial assistance from the federal government, concluding, "It should be urged on the Dominion Government particularly that no public-works grants are so urgently needed as those for the rehousing of the poorest member of the community; further, that in order to make such grants most effective, a National Housing Commission should be appointed to assist provincial and municipal authorities in the formulation of plans, in the choice of materials, and possibly, if a nation-wide housing scheme can be initiated, in securing economies by the large-scale purchase of such materials."[22]

The assumption of a housing responsibility by the federal government, despite growing pressures from varied interest groups, was not originally conceived as part of the Bennett New Deal. On 2 February 1935, approximately a month after his New Deal broadcasts, Bennett wrote to Herbert Bruce: "I am interested in the housing problem. The only difficulty is the financial one. Unfortunately everyone is turning to government for help. If we assist agriculture with low-priced money, I am afraid we will have to leave the cities to the private lenders."[23]

Bennett's views help to further our understanding of the shape of his New Deal, which began with the Farmers' Creditors Arrangements Act and virtually ended with the Dominion Housing Act – legislation passed two weeks prior to the end of the 1935 parliamentary session. Both acts dealt with mortgage lending and were arrived at only after negotiations with the Dominion Mortgage and Investment Association. In contrast to most New Deal acts, neither was overturned on questions of constitutionality.

In his broadcasts Bennett cited the Farmers' Creditors Arrangements Act as his first piece of New Deal legislation. Unlike the DHA, it was apparently initiated by Bennett himself, through a dramatic assembling of the representatives of the Dominion Mortgage and Investment Association in his office. T. D'Arcy Leonard, chairman of the investment association and later the virtual author of the DHA, handled negotiations for the DMIA. Leonard stressed that government agricultural loans should be second mortgages. Borrowers would deal with the private company holding a first mortgage – an approach somewhat similar to the joint-loan method established by the DHA.

While aspects of the Bennett New Deal developed by mortgage companies breezed through, more stormy sessions ensued with construction lobbyists. Prior to his New Deal speeches Bennett had attended extensive meetings with representatives of the Canadian Construction Association. They had already expressed hostility to private financial institutions and called for easy credit to create employment. CCA deputations seeking an expanded public-works program in September 1931 and April 1933 achieved few positive results. As a consequence, the CCA undertook comprehensive research into all Dominion public-works projects that had been held up over the past three years because of the Depression. A successful meeting on 17 November 1933 was held between CCA representative Joseph Pigott and Bennett to discuss a program based on activating these delayed projects. Bennett announced that he was prepared to spend $35 million, and the two then decided on the list of projects to be undertaken from the CCA's brief. Pigott later recalled amazement at Bennett's grasp of the details of government.[24]

Bennett's rejection of public-works spending on a massive scale to alleviate problems of unemployment increased the rift between himself and his minister of trade and commerce, H.H. Stevens. Stevens championed the progressive reformist notions of small-business groups during this period. Their support of soft money policy and of measures to restrict monopolistic pricing activity would be led by Stevens, first within Cabinet and later as leader of the Reconstruction Party. Stevens had called for a public-housing program as a remedy for unemployment as early as 11 August 1933. The Bennett New Deal's sidestepping of public housing, and its open legalization of cartel agreements by the new Dominion Trade and Industry Commission, meant an important defeat for the progressive reform notions of Stevens and his small-business supporters.[25]

It appears that Bennett was forced to move on the housing issue because of back-bench pressure from his caucus in addition to the pressure he had been withstanding from Stevens and the broadening social-housing movement. On 23 January 1935 two Conservative back-benchers, former Toronto mayor T.L. Church and James Arthurs from Parry Sound, called on the government to "inaugurate at once a national housing, building and construction policy." Church quoted extensively from the Bruce Report and observed that "a wave of emotion seems to be sweeping the nation on the subject of slums." Church's motion received support from several members of the CCF farm-labour group.[26]

Bennett seized upon a parliamentary committee on housing as an easy way to defuse a potential controversy. He pledged that whatever view "the house thinks desirable the government is only too anxious to accede to." Despite such magnanimity, Bennett's speech revealed the great extent to which he was committed to a private ethos regarding housing. In response to the bleak rural housing conditions described by Agnes MacPhail and the socialist Alberta MP G.S. Coote, Bennett observed that "there are many farmers in this country who, being thrifty souls, would not care to incur the financial obligations involved in connection with that work." Bennett told the house that his government's efforts at sprucing up public buildings had been successful since "people who had the means themselves began to make their places look a little better, to replace the broken windows, repair the sills and put the houses in better condition." In reference to a controversial proposal in Winnipeg for a municipal low-rental housing project, Bennett explained that the government had been presented with a proposal for shelter with centralized heat, water, gas, and light and with "streets laid out in an orderly fashion." However, his warm approval of the project had changed when "to our surprise petitions came to us asking that we should not go forward with the work, because it might interfere with the renting of properties in the city in question, and stating that the time was not ripe to carry forward a housing project of that character." Bennett also admitted that he was "appalled" at the Bruce Report's suggestion "that the sum of $12,000,000 would be required to provide proper dwelling houses for a certain class in that community who now live in dwellings that should be condemned, and that this amount would not be self-liquidating."[27]

Aside from T.L. Church and Winnipeg CCF member A.A. Heaps, the members of the housing committee had little previous experience with housing as a political issue. None the less, as they approached the topic with open minds, they were favourably impressed by the detailed findings presented by reform-minded housing experts. Typical of their transformation was the view expressed by the committee's chairman, New Brunswick Conservative MP and candy manufacturer Francis Gagnon, who observed: "As chairman of the committee I met a good many persons who might be called town planning and housing cranks. I worked with them for two months and at the end of that time was converted. I think rather than cranks they are torch bearers to something that is coming. This housing problem must be faced. We cannot continue to allow thousands of families in this country to live in one room under unsanitary conditions as they are and have been for the last few years."[28]

Until the final day of the hearings nearly all the witnesses who testified before the committee affirmed the need for a social-housing policy based on adapting the principles of British housing legislation to Canadian conditions. The crisis in low-rental housing accommodation was stressed by several experts. Percy Nobbs presented the findings of the Montreal Board of Trade and Civic Improvement League study. Dr E.J. Urwick, who headed the University of Toronto Department of Political Economy, gave the evidence of the Bruce Report. Support for social housing was also expressed by Dr J. Cudmore, chief of the General Statistics Branch and editor of the Canada Year Book. Alexander Officer, the housing inspector from the Winnipeg health department, offered a frank description of the behaviour of speculative builders and those who refashioned older homes for multiple family use. He noted:

In the case of speculative building we have seen in the past how frequently dwellings have been constructed without due regard to comfort and sanitation ... if only our dwellings were built in such a manner as to exclude cold in winter and heat in summer. Even when sufficient precautions have been taken to build a fairly sanitary structure, we are constantly coming across dwellings erected on flimsy foundations of wood, or rough stone, or what is worse, concrete of a doubtful character. In a year or two these so-called foundations get out of level and the buildings which rest on them get out of shape, with a consequent cracking of plaster and buckling of floors, walls, etc. If there is no cellar the plumbing and water pipes become subject to the action of frost. Even where a cellar is provided, unless a furnace is installed, freezing will take place.

The situation was even worse regarding the provision of low-cost rental accommodation. This was done by "the house farmer," who came along and obtained a former commodious residence from "some real estate man at some nominal sum, gets secondhand plumbing and gas stoves of questionable value and puts them in the rooms and puts in families." Officer told the committee that in such accommodation,

attic rooms only suitable for storage purposes are often found occupied by families. Many of these rooms, having low sloping ceilings and only small gable windows, are scarcely suitable for bedroom use; but when gas stoves are in use they are a menace to health. There is also the danger that if fire should occur in such premises, families living in attic rooms could be trapped. In addition, they are in danger of asphyxiation from gas fumes, due to defective and improper connections for gas stoves. Many of these stoves are fitted by incompetent persons, and as already stated, often the connections are of rubber tubing. Few attic rooms are fit for family use at any time. In winter time storm sashes are usually screwed on tight, and the inner sashes frozen, so that little daylight and practically no sanitation is available. In summer, these rooms being so close to the roof, are almost insufferably hot.

Officer indicated that nothing could be done to alleviate these conditions until a supply of public, low-rental housing was provided that would ease the housing shortage and give health department officials sufficient leverage to force private landlords to provide better accommodation. The parliamentary housing committee appears to have been impressed by this reasoning, for in its report it highlighted Officer's observation that "I have not the soul or the conscience to throw these people out, because there is no place for them to go, but as soon as our expectations are fulfilled, and the Dominion Government helps us out with cheap money and we can build places, we will apply pressure and get them out. But there is no use applying pressure now. There is no place for the people to go."

Officer told the committee that Winnipeg needed "a large number probably no less than 2,500 or 2,000 or more homes for the low paid workingmen." He pointed out that this should be accomplished through federal funding of municipal housing projects since, "if the money was given to them rather than to real estate men or anybody else they would do a good job, one that the Dominion could look back on and see how well they had accomplished what they had set out to do."[29]

The most comprehensive proposals for a national housing policy were presented by the National Construction Council. Its evidence was submitted in two parts by five witnesses. They included architects W.L. Sommerville and James J. Craig and the council's president, Gordon M. West, and its secretary, I. Markus. The council urged the establishment of a permanent national housing authority. It would set minimum standards for new and existing housing, make regulations for the rehabilitation of dwellings that fell below such standards, administer these rehabilitation loans, and make loans at rates of interest not exceeding 2.5 per cent for low-rental housing. The council also urged that private capital be encouraged to invest in home rehabilitation and the construction of houses, through the reduction of risk to private lenders. It was stressed that, for any of this federal aid to be given to a locality, comprehensive town planning – extending into suburban and rural districts – would be required.[30]

Craig filed a separate presentation on behalf of the council's finance committee. It recommended that the federal government sell $400 million in government securities to the Bank of Canada. Of this sum, $120 million (the proportion allocated to slum clearance) would pay interest of 2 per cent, while the remainder (to be used for repairs) would be invested at 2.5 per cent. As municipalities would pay 2.5 per cent for slum clearance and home owners 5.5 per cent for renovations, the entire project for the construction of forty thousand units of rental housing and the rehabilitation of ninety thousand existing houses would be "self-liquidating." This meant that no subsidies from general tax revenues would be required.[31] This proposal was typical of the soft-money policies advocated by the construction industry and other middle-class groups in the period, such as the rising western Social Credit Party. The plan attempted to reject socialist-tinted proposals for housing subsidies from general tax revenues in favour of a complicated bond scheme that might have entailed currency devaluation.

Even with such determination to stay within the bounds of market economics, the council's scheme drew some suspicion from committee members. Liberal MP Wilfred Hanbury, a lumberman and manufacturer, complained that the council had failed to produce "some practical and economically sound method by which funds could be provided for rehabilitations." To overcome such reservations, the council prepared more detailed proposals, which were presented in the form of a draft federal housing act. Its proposals for rehabilitation were now directed at buildings "basically fit for use and not in a declining or slum neighbourhood." When planner Noulan Cauchon, who served as an unpaid adviser to the parliamentary committee, suggested that some homes in a slum area might be rehabilitated, Sommerville replied: "Absolutely no; most emphatically no. The whole district has to be cleared out."[32]

Craig stressed that for an effective program of slum clearance, public authorities rather than limited-dividend companies should be relied upon. He told the committee that "in the United States they found that the federal projects were the ones that were possible to get under way, and that private projects or limited-dividend projects did not make an appeal, were not attractive to private enterprise." Detailed proposals for a mortgage-insurance fund to encourage new private construction were also presented by the council to the committee.[33]

The greatest surprise the parliamentary committee received from the council's suggestions was a proposal for the prevention of premature subdivision on suburban land. To explain this part of the council's draft legislation, Cauchon pointed out that at present municipalities were powerless to prevent scattered development unless it was located "down in a swamp that is undrainable or away up on a hill where the city can't furnish a water supply without extra cost." Two committee members admitted to having been badly burned themselves by their participation in land speculation on the urban fringe. Calgary Conservative MP George Stanley observed, with a touch of irony, that "our western cities have that beaten; all properties of that kind have reverted to the city for arrears in taxes."[34] The committee was evidently favourably impressed by the council's report, as it tendered a special "vote of thanks for the splendid evidence submitted" by the construction council.[35]

Noulan Cauchon had shared in the growing radicalization of the planning profession in its understanding of housing ills. Cauchon told the committee that "with workingmen if you exceed roughly 20 per cent of his income he is going to skimp on something else. He has only got so much money and if he takes off more for his rental than represents a reasonable proportion he is going to have to cut down on food and clothing for himself and his family – there are no two ways about it."[36]

The committee observed that every country in Europe and North America, with the exception of Canada, either had a housing policy or had initiated one. It recommended the creation of a national housing authority that would lend at the lowest possible rate of interest to provinces, municipalities, societies, and corporations for low-cost housing and repairs. It could also directly undertake its own housing projects.[37]

The committee's failure to spell out the interest rate at which money could be lent points to an unresolved conflict in its deliberations. Noulan Cauchon estimated that a 3.5 per cent interest rate (the amount at which the federal government borrowed) was low enough, when lent to municipal authorities and limited-dividend companies, to secure adequate and affordable accommodation for low-income families. In response to a question from A.A. Heaps, Cauchon admitted that rates of 1 or 2.5 per cent would produce housing that "would be so much better." He explained to the committee how, when the government of Vienna was faced after the war with a housing shortage, it had

> immediately executed a capital levy, and they built houses for those who had no capital. When I was last in Vienna in 1931, 11 per cent of the population of that country was living in municipally owned houses, and they were paying hardly any rent. The rental they were paying was barely sufficient for the taxes and upkeep. They were not paying rent in the ordinary sense, because the money had not been borrowed. They simply took it from the people who had it by that simple process. Now, you read in the magazines that their buildings are magnificent houses also, if you could take the money out of somebody else's pocket.[38]

The conflict at the committee's hearing over the issue of housing finance was reported by Collin McKay, who covered its testimony for the *Canadian Unionist*. McKay observed that when the public-housing question arose, "the inquiry halted, turned aside, or turned backward. Some witnesses balked at attempts to explore the possibilities of public financing of new housing. And even those who admitted the necessity of public action to deal with the worst features of the housing problem were troubled by that necessity. 'Public action, yes. But it will probably mean a heavier financial burden, more taxes.'"

McKay believed that the hearings demonstrated that the "fetish of finance" amounted to a taboo condemning civilized man to suffer from "poverty in the midst of plenty." McKay further noted that "the statesmanship which balks at the problem of financing a war on poverty nevertheless finds ways to finance preparation for a bigger, if not better war of destruction of human life. One might think the ruling classes were insane."[39]

However, the conflicts over financial subsidies expressed by Cauchon, Heaps, and the labour observers of the committee's proceedings were mild in comparison with the great gulf between all these parties and the conservative views of deputy minister of finance W.C. Clark, Prime Minister Bennett, and the nation's leading residential-mortgage lenders. The division was apparent in the contrast between the report of the parliamentary committee and the final Dominion Housing Act, which, quipped A.A. Heaps, was no more like the committee's report than "a pig resembled pig iron."[40] The views of Clark and the mortgage corporations were made known to the committee on the last day of its hearings.

D'Arcy Leonard was the first speaker, and, according to his testimony, his Dominion Mortgage and Investment Association represented "most of the mortgage companies in Canada, most of the trust companies – the Canadian Life Insurance Companies." Leonard estimated that these organizations had capital of $60 million. Of this they had at least $25 million "which they would like to get out now on new construction in Canada."[41]

When Calgary Conservative MP George Stanley asked Leonard if he had a concrete proposal in connection with desired government housing policies, Leonard replied, "I do not think that it could be said that we have a concrete proposal in connection with desired government housing policies." And then he repeated, "I do not think that it could be said that we have a concrete proposition. We can just put before you that we have this money." The companies' difficulty was the lack of suitable borrowers. Leonard noted that "the problem to-day, as a matter of fact, is to get our money out." He pointed out that his companies were prohibited by law from lending more than 60 per cent of the value of a house and that "there are not many people who can put up the difference between 60 per cent and 100 per cent." In response to questioning from the committee's chairman, Leonard forecast that if government put up another 20 per cent (as was later done in the 1935 legislation), the amount of building would undoubtedly increase.[42] He added that his companies would not want to go beyond 60 per cent – even if permitted to do so by law – unless economic conditions improved and "you had a strong covenant, you had proper supervision, you had good construction, economic management, a monthly payment plan, provision for taxes, and the legislation was cleaned up." Like Clark, Leonard expressed a desire for large housing corporations in order to lend on a higher percentage basis. He indicated that, "if the government can see any way towards the organization of companies which would be soundly managed, and arrange in some way for junior financing, certainly there is no question that we would, I think, be willing to assist."[43]

In reply to questioning by Noulan Cauchon, Leonard revealed his association's hostility to public, subsidized low-rental housing projects. Leonard believed that "the construction of a certain number of houses of that class if they did not remove other houses where the rent is being paid and which are overcrowded or uninhabitable, really – practically slums – would have the effect of bringing down rental values on the next class of houses and you might say above it; and thereby affect the rental values generally and affect them on the class of security on which we would be lending, which would be a workman's house where he was able to pay a rental based on the actual cost." Leonard predicted that low-rental housing, unless it was restricted to replacing demolished slums, would bring down the rental structure throughout the next several classes.[44]

W.C. Clark's presentation shared Leonard's private-enterprise approach. Clark told the committee that he started from "two general principles." The first was that "it would be wise to avoid any hasty commitments in regard to the most difficult and the most complicated aspects of housing: for instance the problem of slum clearance." Clark recommended that "some appropriate body" study these issues further and added, "I would not jump into it overnight." Secondly, Clark argued "that we should concentrate essentially on the immediate emergency problem of using housing as a stimulant to business recovery and as an absorber of unemployment; and I would suggest there that we should try to make the federal dollar go as far as possible in stimulating business recovery."

Clark urged federal aid for the formation of housing corporations to produce housing on both an ownership and a rental basis. The federal government would finance these corporations through the purchase of their preferred shares, while common stock would be purchased by any parties interested in receiving a dividend limited to 6 per cent. He emphasized the need for government to provide the junior money for such housing corporations since "as Mr Leonard told you the most difficult thing to get is the junior money, the junior mortgage money; it is practically impossible to obtain it from private sources." Officials of any private lending agency that purchased the company's first-mortgage debentures would be responsible for inspections to ensure that "the appraisal put on a specific property is a sound appraisal, that the construction costs are not unduly padded, and that the money was paid out in accordance with the usual safeguards." Clark stressed that this scheme would "make use of private lending agencies instead of driving them out of business." It would assure "the carrying out of the program on a sound and business-like basis." Clark also proposed, as an alternative scheme, that the federal government set up an insurance corporation to guarantee the last 20 per cent of an 80 per cent mortgage.[46]

Clark's views were unaltered by the parliamentary committee's conclusions, as can be seen in an exchange of correspondence he undertook with consulting architect F.W. Nicolls, whom he would later appoint to head the National Housing Administration branch of the Department of Finance. Nicolls viewed Clark's scheme for limited-dividend housing as unworkable. Investors were given too low a rate of return for their involvement to be attractive, and its interest rates of 5 to 5.5 per cent were too high to build quality low-rental housing.[47] In reply to such criticisms Clark stressed the burdens of an extensive low-rental housing scheme on the federal treasury. Moreover, Clark wished to leave mortgage lending to private business. He told Nicolls that it was important that loans be made by private financial companies at rates of 5 to 5.5 per cent because of the "undesirability of pushing government competition with private institutions beyond reasonable limits."[48]

The housing committee presented its recommendations to Parliament on 16 April 1935. It urged that national housing policies should be based on the principle "that the provision of a minimum standard house for every family in the country should be adopted as a national responsibility."[49] It recommended a national housing authority that would build its own projects and lend federal funds for municipal low-rental projects. This was simply a repeat of the Bruce proposals that had earlier "appalled" Bennett.

Between 16 April and 24 June 1935, when the Dominion Housing Act was introduced to the House, Bennett's government evolved pioneering Canadian housing legislation on the basis of the best deal that could be achieved between itself and the Dominion Mortgage and Investment Association. On 8 May 1935 DMIA president D'Arcy Leonard told Clark that his association had "been working steadily in the past three weeks endeavouring to work out a plan whereby the lending institutions might co-operate with the government in a housing scheme." He predicted that "within the next few days we may have something in the nature of a concrete proposition which would be very practical and fairly simple."[50] The secret meetings and unpublished exchanges of letters and cables between Leonard and Clark were in marked contrast to the open hearings of the parliamentary committee.

In even greater contrast were the views on housing, town planning, and urban development held by Clark, a former vice-president of the American real estate investment firm S.W. Straus and Company, and Noulan Cauchon, unpaid adviser to the housing committee. A founder of the Canadian planning profession, Cauchon's conflict with Clark is epitomized by their differing views on one noted symbol of American capitalism, the skyscraper. Cauchon cited Leonardo da Vinci's observation that buildings should be no higher than the width of the street facing them, which he considered "a very good approximation for access of sunshine and air, traffic adequacy and the freedom from foreshortening if one is to enjoy aesthetic appreciation of the building itself."[51] By contrast, in a work published by the American Institute of Steel Construction, Clark wrote that the best indicator of the optimal height of a building was the maximization of the structure's profitability to its owner. Clark calculated this to be sixty-three stories, given the state of existing technology, and argued that "the higher the land value, the higher is the economic height." He stated that height limitations from eight to twenty stories would result in disastrous consequences since "the whole economic fabric of society is built to an important degree upon current values of city property."[52] Clark saw even greater benefits from the scraper's future development on large plots covering an entire city block. He predicted that "as large development becomes typical, ownership will be in more responsible hands and, therefore, a more scientific determination of supply and demand conditions will be made before each new development is begun." Taking that logic to an extreme, he advocated the "self-contained city, accommodating many thousands of people, carrying out practically all their activities in a single structure ... the most profoundly efficient and adequate conception of gigantic size ever created by man."[53]

Unlike Cauchon's views, Clark's were in harmony with those powerful realty and financial interests concerned with maximizing the capitalization of land. Clark envisaged large-scale corporate ownership of urban land, giant skyscrapers linked to city arterials by special express runways and joined together by elevated sidewalk arcades. This technocratic vision – which virtually begs to be parodied by lurid artistic nightmares of a dehumanized, mechanistic megalopolis – was not eccentric but reflected the best work of other experts whose work did not seek to challenge the status quo of the property industry. Clark approved of Thomas Adams' call, in his *Regional Plan of New York*, for the expressways, bridges, and tunnels that Robert Moses would later transform from paper to concrete. He also favoured the work of Miller McClintock, who envisaged four-level streets and one-direction, seven-lane highways and expressways linking quarter-mile skyscrapers topped with airports.[54] Moses, Clark, and McClintock were all men of definite views with profound imagination and energy, but their dreams were realized because they dovetailed so neatly with those business interests that sought higher returns from urban land.

Clark's housing theories were also influenced by his hostility to deficit financing as a recession cure, despite his reputation as the father of Canadian Keynesianism. As bitterly predicted in 1935 by the All-Canadian Congress of Labour's *Canadian Unionist*, deficit financing would not be applied until dictated by the requirements of war.[55] Clark had formulated his views on the role of government spending in counteracting recessions in a paper given to the International Association of Public Employment Services. Clark advocated the "deliberate and more far-sighted planning" of public capital-works projects, which would be adjusted to the business cycle. Capital works would be postponed deliberately during prosperity so as to avoid competition with business; during recessions capital works would be expanded. However, Clark explained that he did not "wish to be quoted as advocating the expenditure of large sums of money in construction of public works for the relief of unemployment." Rather he was simply "advocating the expenditure of forethought."[56] Such views on the role of government spending were shared by the Canadian financial elite. In a 1930 speech Sir John Aird, president of the Bank of Commerce, called for both balanced budgets and the careful planning of public works to counter recessions. Similarly, in a December 1931 speech to the Professional Institute of the Civil Service in Canada, Clark urged that "citizens of local municipalities ... take to heart the warnings and good advice recently tendered by the Canadian Bankers' Association to exorcise administrative waste and reduce borrowings to the minimum."[57]

Clark's fiscal views played an important part in the making of the DHA. This is evident in an exchange of letters between F.W. Nicolls and Clark in April 1935. Nicolls urged Clark to prepare an ambitious low-rental housing effort. He believed that further study was unwarranted, as other nations had been testing such schemes since

1920 and that during this time "every conceivable question has arisen, and there is no reason to think that Canada would require different solutions to be found." Nicolls saw the Depression as an ideal time for a low-cost housing program; the required data was available and labour was plentiful. He told Clark that if something was not done before the return of prosperity, the nation would "be faced with such a shortage and a demand that will send rents soaring, thereby causing certain classes of people to dwell in worse slum conditions." Clark responded "In view of the very high financial burdens upon the Dominion exchequer I tried to develop a plan which would make the federal dollar or federal guarantee do as much as possible. It might be desirable from many points of view to develop some grandiose projects for the use of very large amounts of Dominion funds of credit, but a practical appreciation of the financial burdens we already bear make one pause in considering such schemes."[58]

The negotiations between Leonard and Clark regarding the DHA continued almost until the day the bill was introduced. In one memo written before the bill was drafted, T. D'Arcy Leonard wrote:

> We believe that the soundest housing scheme would be one that would enable the existing mortgage lending institutions to lend up to eighty per cent on approved new houses, in approved locations, to be built for home owners ... If approved, a loan up to an amount of eighty per cent of the value of the property would be made by a lending institution, of which sixty per cent would be supplied by the institution and twenty per cent by the government. The government contribution would be in the form of a cheque handled by the lending institution so that the borrower would have only the one organization to deal with himself.[59]

This recommendation was later incorporated into the actual DHA. The only subsequent modification came at the behest of Leonard in a letter of 21 June 1935, just three days prior to the introduction of the act. An earlier draft had provided the federal 20 per cent share of a loan only if lending institutions put up a full 60 per cent, so that the mortgage would amount to 80 per cent of the full cost of a house. Leonard argued that in such localities such as Montreal it was customary for lending companies to lend only 50 per cent. Consequently, line 6 of section 5 of the act was changed from "equal to" to "an amount up to" 80 per cent of "the cost of construction of a house or its appraised value, whichever is the lesser." Leonard also warned the government:

> In view of the fact that the only method in which the Dominion Government proposes to advance money for housing is through the lending insti-

tutions, I think I should point out that there are localities in Canada where our institutions do not operate and others where they would not recommend loaning. I would like to make this clear at the outset so that it will be well understood that the measure may not be capable of general application, and as the institutions will require to be satisfied before approving loans, there will be many individuals and localities that will not receive their approval.[60]

Despite such reservations, Leonard was so pleased with the DHA that he sent a cable praising it to be read out in the parliamentary debate on the measure. Since the cable noted that "the entire joint mortgage will be protected against provincial moratoria" (legislation against mortgage foreclosures), Clark decided it could not be used in debate since it would raise "a long discussion on a difficult technical subject." Telegrams of support were, however, read out by the minister introducing the act, Sir George Perley, from Mutual Life of Canada and the North American Life Insurance Company.[61]

The Dominion Housing Act did introduce some revolutionary innovations into the mortgage market, of the type recommended by the Canadian Construction Association in 1921. Previously, mortgages had been short-term loans of five years with no vested right of renewal and with annual or semi-annual payments, which meant that foreclosure often resulted from the inability of borrowers to come up with substantial sums. The twenty-year, 80 per cent mortgage at a 5 per cent interest rate did represent a major gain for home builders and buyers. These gains were achieved through great generosity to the lending institutions. To achieve a 5 per cent rate of interest the government contribution was made at 3 per cent, a subsidized rate, as the government itself paid 3.5 per cent to borrow money. Lending companies lent their portion at their usual rate of 5.67 per cent. In areas where interest rates were higher, the mortgage companies tended not to accept applications for loans.[62]

The National Construction Council complained bitterly about the inadequacy of the legislation. One member noted that 90 per cent of Canadians had incomes of less than fifteen hundred dollars per year and consequently could not afford even the minimum cost of houses that could be financed through the legislation. For this ninety per cent the DHA offered only the possibility of further study into the matter of low-income housing by a yet-to-be formed Economic Council of Canada.[63]

Bennett did consider the recommendations of his executive assistant, R.K. Finlayson, to appoint a housing subcommittee of the Economic Council of Canada. Finlayson's proposed list of names was representative of the interest groups concerned with low-income housing: the president of the All-Canadian Labour Congress, A.R. Mosher; Noulan Cauchon; Toronto alderwoman and housing reformer Mrs A. Plumptre; Ontario farm and co-operative movement leader W.C. Good; Dominion statistician H.F. Greenway; Percy Nobbs; J.C. Reilly; the Engineering Institute of Canada's H.H. Vaughan; president of the Royal Architectural Institute of Canada W.S. Maxwell; Toronto Housing Corporation pioneer G. Frank Beer; and W.C. Clark.[64] Clark's name headed the proposed list of appointees as the committee's financial expert. Clearly, his views on housing would have set him in diametrical opposition to every member of the committee except Beer. Beer's inclusion appears a throwback to the pre–First World War period, when limited-dividend housing was popular. The fear of deadlock may have influenced Bennett's decision to appoint neither an Economic Council nor a housing subcommittee. This was a significant decision, for the promise of an Economic Council for national planning was accented in the New Deal legislation.

As the *Canadian Unionist* pointed out, the strongest opposition to the DHA in the parliamentary debate on the measure "came from the Labourites and the United Farmers." Apart from A.A. Heaps, the most caustic CCF critic of the bill was William Irvine. He described it as "merely a shot in the arm of the capitalist system," giving only "ten grains of pep, not enough to raise the arm or furnish a single kick." Irvine called for a $300-million housing program, under which loans would be repaid to the government according to the rate at which the accommodation constructed with such funds deteriorated.[65] H.H. Stevens, now Reconstruction Party leader, argued that even after the government's Economic Council had decided what was the best solution for low-income groups, the DHA made no provision for acting on their recommendations.[66] The Liberals were divided in their position on the DHA. One Liberal member of the parliamentary committee, Wilfred Hanbury, repudiated his earlier position when the DHA received first reading on 18 June. In taking this position Hanbury stated that "the function of any government, as outlined in Dr Clark's evidence, is to avoid the socialization" promoted by the parliamentary committee's housing experts. Further pursuing Clark's reasoning, Hanbury added that if "insurance and loan companies" were "left in charge of this affair," there would be "no fear that our money would not be properly spent." Hanbury's concerns were shared by former Liberal revenue minister George Euler, who felt the DHA was not sufficiently generous to lending companies.[67] In contrast, R.W. Gray, Liberal backbencher and member of the parliamentary housing committee, decried the weakness of the legislation and charged that "this scheme is launched" because of "the fear of the lending companies that unless they come forward with some suggestion, the government will introduce state aid legislation that will materially affect such companies."[68]

The challenge of the Reconstruction Party in the 1935 election reflected not only populist discontent with the absence of trust-busting zeal in the Bennett New Deal but also the widespread conviction that Bennett's housing policies served big business and financial institutions. The Reconstruction housing platform called for the sale of self-amortized bonds guaranteed by the Dominion government in the event of the failure of lending companies to use their "reservoir of capital" for a national housing plan. This was similar in spirit to the housing program presented by the Canadian Construction Association to the special parliamentary committee. The party's platform also reveals an antagonism to finance capital by calling for mortgages to be restricted to a 6 per cent interest ceiling.[69] Both platform and CCA program illustrate that the increased social consciousness that created the League for Social Reconstruction among professionals was broadly shared among important middle-class groups during the Depression. The CCA's positions on social housing were similar to those advocated by the CCF. During the election campaign Stevens severely criticized the DHA, describing it as a "bill for the relief and protection of trust, loan, mortgage and insurance companies." With insight he noted deficiencies like the concentration of mortgage lending under the act in the major urban centres of central Canada and the generosity of the act's provisions for private lenders. Minister of finance Edgar Rhodes replied that the fact that only eleven firms had participated in the DHA was a sign of how the legislation had "carefully protected the interests of the public and government."[70]

Both the limited participation of lending institutions praised by Rhodes and the regional variations deplored by Stevens stemmed from the hostility of lending institutions to certain features of the act. Although larger mortgage companies like Sun Life saw the innovative features of the DHA as helpfully serving to stimulate depressed mortgage markets, many other firms did not. Lending institutions' support for the DHA tended to correspond to its impact on the interest-rate structure in the communities in which they did business. Even enthusiastic firms such as Sun Life would not apply the act in areas where it was felt the 5.5 per cent combined interest would lower their rate of return. In the July 1935 negotiations with lending institutions over the regulations set under the act, the companies pressed for a differentiation in the DHA interest rate in various parts of Canada. Areas considered to be of high risk, or costly to service, such as the prairies and northern Canada, would be charged higher rates of interest under the joint-loan scheme than would Toronto and Montreal. Clark replied that such a geographical variance "was politically impossible." In response, as David Mansur, inspector of mortgages for Sun Life, recalled, the lending institutions indicated that under the uniform 5.5 per cent rate, "very few loans would be made in areas which are residentially as hazardous as northern Ontario."[71]

With even lending institutions that drafted the DHA lukewarm, the political benefits to Bennett accruing from the act in the 1935 election were small indeed. The major supporters of the legislation were not the mortgage lenders who largely drafted it out of a desire to avoid less palatable measures. The most enthusiastic group were building suppliers, especially retail lumber dealers. Indeed, this group would prove to be one of the strongest supporters of the joint-loan package during its entire life. However, even among middle-class aspirants to home ownership, the legislation received a cool reception. Disappointment was the impression that the newly appointed DHA administrator, F.W. Nicolls, detected among persons visiting the DHA publicity booth during the Canadian National Exhibition in Toronto from 25 August to 12 September 1935. Some of this reaction stemmed from encouters with lending institutions. Nicolls told Clark that "many spoke of open hostility on the part of loan company officials, a hostility exemplified even in the manner with which they received applicants for Housing Act loans. I heard stories of officials who openly stated that they disliked the Housing Act, giving such reasons as 'the excessive amount of red-tape,' the fact that it meant tying up their money for ten or more years, the clerical work involved in the monthly payments, etc."[72]

As a result of the high-handedness of the lending companies administering it, the DHA became unpopular among prospective middle-class home buyers. Nicolls found that "considerable damage" had been done to the act by lending institutions' "not giving the applicant a satisfactory – or indeed any reason, for the refusal." Even more inflammatory was the refusal of these companies to lend in certain districts under any circumstances. Nicolls told Clark that while he understood reasons given for "black-balling certain areas," "considerable injustice" had been done where "the security seemed to be quite adequate, the only obstacle being that the applicant proposed to build in one of the rejected areas. I would like to stress the importance of this particular group. Respectable and sincere, and seemingly with financial standing, they are embittered at their failure to obtain a loan, and easily convince others that the Dominion Government, in the words of one of them, 'is humbugging the public with the Housing Act.'"[73]

The most insightful assessment of the origins and limitations of the DHA was provided in 1936 by David Mansur. An important person in the implementation of the DHA through his work at Sun Life, Mansur was also an influential adviser to W.C. Clark and would later be the first president of the Central Mortgage and Housing Corporation. In private correspondence Mansur wrote that the DHA had been undertaken primarily for two reasons, first to "answer the popular demand for governmental aid in housing" that was created by the parliamentary committee, and secondly to "prime the pump in connection with the building trade." Particular care was taken not to be "injurious to the business of the lending companies." On 6 August 1936 Mansur told Canadian Industries president and National Employment Commission chairman Arthur Purvis that at the time the DHA was drafted, "the Department of Finance through the agency of Dr Clarke [sic] showed every co-operation and reasonableness to meet the demands and wishes of the lending company. He was assured by the Dominion Mortgage and Investments Association that he would have the co-operation of their members and with this in view he made many concessions which in the original instance he did not wish to make." Despite such assurances, the "attitude of all the lending companies, with the exception of four or five, has been to throttle the Act but to keep it in operation so that no other measure of more disastrous character can be brought into force." Indeed, many companies had signed agreements to participate under DHA while having "no intention of making a loan under the Act, believing that acquiescence to the Government is necessary to keep the Government from direct lending." One Montreal Trust company registered under DHA to encourage more people to apply for loans with their firm so that these potential customers could be used for loans "directly by then without the aid of the Act." Prospective borrowers were warned about unfavourable features of the DHA, with the result that in Montreal the act received a reputation "of being very cumbersome." Sun Life was even attacked by the "propaganda" of companies hostile to the DHA. Despite the fact that such a move would mean less business for his company, Mansur felt the attitude of most mortgage companies would justify the government's entry into direct lending.[74]

Mansur also illustrated how the DHA benefited only an affluent minority, for whom he believed the act was "never intended." Of one hundred Sun Life DHA loans, only three were for homes valued from $3,000 to $4,000. It was "generally accepted" that in major cities, homes for the average family of four or five persons could not be built for amounts below $3,500. Yet, based on all mortgage companies' principles that "a man should not own a house which costs more than 2 1/2 times his salary," Mansur concluded that "practically the whole population earning less than $1,500 a year will be not considered for loans under the Act." They were simply viewed by lenders as too high a risk for the minimum-cost $3,500-value home. This conclusion was significant, considering that, according to the 1931 census, 80 cent of male Canadian employees earned less than $1,450. While family income would be higher, Mansur pointed out that it was the income of the home buyer that was considered when mortgage lending decisions were made. He also noted that the nation's industrial and wage workers were denied the benefits of DHA because entire working-class communities were regarded as "undesirable districts" for mortgage lending. In Montreal almost all of Sun Life's lending was in middle-class areas such as Westmount and Mount Royal. Only "one or two loans" were made in the entire east end of Montreal.[75]

Clark's response to Mansur's critique further confirm that the benefits of the DHA were meant to be reserved for the wealthy and that the legislation was devoid of social purpose. Clark did not dispute Mansur's contention that the DHA had been designed in response to the parliamentary committee's call for a social-housing program and that most lending institutions supported it only to preclude implementation of such radical measures. What he challenged was Mansur's belief that the DHA was not originally intended for a wealthy elite. To the contrary, he said: all the government wanted was "to encourage building." This objective could be met "more effectively," in Clark's view, by "the building of high cost houses ... than the building of low-cost houses." Clark also rejected Mansur's request to Purvis that more publicity be given to the DHA: "Frankly, the reason that we have not engaged in any widespread publicity campaign was the fact that it would have done more harm than good to have stirred up widespread public demand for Housing Act loans if the mortgage companies were not willing to co-operate. The 'neck of the bottle' has not been the public demand for Housing Act loans but the co-operation of the lending institutions."[76]

Clark's own words confirm that the government had become a virtual prisoner of lending institutions in the setting of housing policy. It could not engage in an advertising campaign promoting its legislation out of fear of alienating the institutions relied upon to implement its housing program. His efforts to co-operate with financial institutions brought meagre results, save for the defeat of more far-reaching measures. Thus Canada's initial housing legislation was almost a caricature of the inequities of Canadian society, its benefits going mainly to the rich and middle class even when supported by public funds and wrapped in the rhetoric of concern for the underprivileged. The supreme irony in this pattern is that the DHA emerged in response to demands to improve the housing situation of the poor. The call for federal action came from a broad coalition including large construction corporations and allied industries, a variety of middle-class reformers, and both industrial and craft unions. Their efforts to create a social-housing policy in Canada only served to create a pattern of government intervention geared to the interests of the private market.

4 Housing and Public Relations, 1935–1940

The return of the Liberals and W.L.M. King in the general election of 1935 made little difference in terms of housing policy. The main immediate change was that the body assigned to "study" the low-income housing problem was transformed from a subcommittee of the Economic Council of Canada to King's newly created National Employment Commission (NEC). This move, however, simply highlights the cynical manner in which the housing problems of low-income Canadians, and the vision of reformers, were to be shuffled into oblivion by the skilful management of King and Clark.

Prime Minister King interpreted his victory as a triumph of the virtues of honest economy over the wickedness of reckless spending. In reply to John Rockefeller Jr's congratulations upon his return to power, King characterized the four parties that opposed him in the 1935 election as seeking "to outrival each other in the promises they made at the expense of the public treasury." He considered that the Liberal campaign had exposed this spending formula "as immoral and corrupt."[1] King enthusiastically embarked on a policy of retrenchment, to the smallest details of which he devoted remarkable attention. He believed that minister of finance Charles Dunning had "done herculean work in cutting down the federal budget" and that his own innovations had helped to save "a good many millions." This had been done by restricting federal grants to provinces, which he felt was the only way to cut such costs without provoking riots by the unemployed.[2] Clearly, the prime minister's conservative fiscal views were in harmony with the thrust of the DHA and the philosophy of W.C. Clark.

King believed he could further reduce the federal budget by appointing an efficiency-minded National Employment Commission that would also serve to harmonize competing regional and economic interest groups in Canadian society. In his third radio address of the 1935 election campaign, the future prime minister pledged that the NEC would "mobilize and co-ordinate the activities of agencies prepared to lend their good offices in providing work."[3] To achieve this the NEC established a national advisory committee intended to be "representative of industrial, occupational, philanthropic and social services organization."[4] King's appointment of the committee's chairman and vice-chairman on 13 May 1936 personified his ideal of a corporatist partnership of labour and capital. Arthur Purvis, president of Canadian Industries Limited, had been involved in a campaign to support the DHA. NEC vice-chairman Tom Moore had long held corporatist views as president of the Trades and Labour Congress, which was dominated by construction unions. Liberal Party stalwart Mary Sutherland served as King's watchdog on the commission. Economics professor Dr W.A. Mackintosh was from Queen's University, like W.C. Clark. Alfred Marois of Quebec, A.N. McLean of New Brunswick, and E.J. Young of Saskatchewan, appointed primarily to represent regional interests, played relatively minor roles. King was confident that the commissioners shared his concepts of voluntary corporatism and fiscal conservatism. After an early meeting with them he confided in his diary, "Purvis made an especially fine presentation – all tending in direction of getting back to voluntary effort. He admitted most of our problems, was undoing of mistakes of the last few years; Bennett's reckless spending etc."[5]

The commissioners saw as their first task the creation of a housing-rehabilitation loan scheme that would be universally accepted as sound by the Canadian financial community. The scheme was the penultimate expression of King's voluntary-corporatist approach to the Depression. In contrast to Bennett's New Deal, it avoided coercive legislation subject to legal challenge and used more sophisticated media manipulations than the melodrama of startling radio broadcasts.

The Home Improvement Plan (HIP) that the commissioners came up with would generate considerable enthusiasm but would do little to help those Canadians most in need of home repairs. It continued the principle of Bennett's DHA, namely the public underwriting of the private sector, but did this with considerable support from the companies involved. While the plan got great numbers of employed Canadians working together, Purvis himself recognized that it made slight impact on the overall unemployment situation. More far-reaching measures would be buried for two years.[6]

Purvis was informed by the National Construction Council of one success of his earlier DHA promotional work shortly after his NEC appointment. The NCC had devised "an intelligent general policy" of advertising that would sell "the industry as a whole." At the same time, the NCC told Purvis of their difficulty in reaching agreement on what government policies would be best for "credit facilities for rehabilitation work."[7] It was to this thorny question that the NEC would apply its first efforts.

W.A. Mackintosh made a "preliminary Report of Housing" at an early NEC meeting. He noted the great volume of employment potential in the rehabilitation of existing houses. However, he also saw barriers to its realization. Homeowners had difficulty financing rehabilitation because of collapsing real estate values. This caused owners who had paid off most of a mortgage to carry mortgages of more than 60 per cent of the value of their homes. Consequently, they were unable to borrow more, as their properties were in some instances mortgaged for more than their current market value. Mackintosh suggested creating a general acceptance corporation, similar to those established by American subsidiaries of building-supply firms, in order to finance installations of equipment and building improvements. He also proposed a "special project" to enable people in rural districts to undertake rehabilitation.[8]

Mackintosh's outlines for proposing various programs for housing rehabilitation were considerably narrowed by the intervention of Windsor Liberal MP Norman McLarty. On 23 July 1936 McLarty made a detailed submission to the NEC proposing his remedy for the unemployment that had devastated Windsor. The collapse of the automotive industry, as McLarty pointed out, had crippled the city in numerous ways. Welfare rolls had soared. American employment laws that restricted work in Detroit and an influx of migrants who mistakenly sought work in the closed auto plants had increased unemployment. The resulting higher tax burdens had created so many arrears that Windsor had defaulted on its bond payment. Homes were mortgaged for more than they were worth due to previously high land prices caused by the automobile industry's boom in the 1920s. These factors had caused loan and insurance companies to refuse to make any loans in Windsor.[9]

The "red-lining" of Windsor had caused the construction trades to stagnate. Not a single DHA loan had been made. As a solution to this predicament, McLarty drew the NEC's attention to Title One of the United States Housing Act of 1934. Under this plan the Federal Housing Administration guaranteed up to 10 per cent (originally 20) of the aggregate loans made by a participating lending institution. Loans ran from one to five years and were one hundred to two thousand dollars. Interest charges amounted to 9.72 per cent. This was considered low for instalment credits and small personal loans. From correspondence with FHA McLarty had learned that the federal guarantee reduced lenders' loan-servicing costs by eliminating the need to set aside part of their returns from loans as reserves against losses, which averaged only .5 per cent of the loans made.[10]

McLarty also enclosed an FHA modernization promotional pamphlet. It showed little zeal for basic home repairs like mending leaky roofs or remedying structural damage, but instead focused on items like removing old-fashioned verandas, "built-in" kitchen conveniences, and up-to-date bathrooms, basements, and recreational rooms. As an alternative McLarty outlined a proposal for implementing the FHA's modernization program in Canada. He stressed the need for local reconstruction-finance corporations to manage recovery campaigns. Their members would include representatives of the chambers of commerce, the Trades and Labour Council, the builders' and architects' associations, and a government representative named by the NEC.[11]

McLarty's proposals were tailor-made for King's government. Private lenders would not be disrupted but subsidized through a federal guarantee. This proposal for involving local associations promised enormous potential for the voluntary-corporatist action so favoured by King, Purvis, and Moore. The advertising campaign Purvis envisaged for the DHA could be applied to HIP. Consequently, the NEC enthusiastically and quickly embraced McLarty's proposal. It approved the idea of a privately funded HIP advertising campaign. A target of $50 million in HIP loans was set, of which 15 per cent was to be insured against loss. This was highly generous considering the American experience of .5 per cent loss. However, the commission concluded that this "excessive guarantee" would provide "cheap stimulus."[12] The commissioners were aware that many homeowners, especially farmers, could not afford to pay 9.72 per cent interest, but this concern was passed over in the hope that the provinces would subsidize homeowners in impoverished regions.

McLarty was selective in his borrowings from the New Deal's provision of housing-rehabilitation loans. Consequently, a broader program, capable of meeting social needs for improvements to basic housing structures and facilities, was not adopted. Both the NEC and McLarty rejected a related FHA program for insured loans for more expensive structural repairs, which were paid off over a long term. Only the FHA's "modernization" loans were copied. These were limited to amounts up to one thousand dollars and had to be paid back within five years.[13] Despite his attacks on Bennett for similar procedures, King allowed the HIP plan to proceed without parliamentary approval. Support from the Canadian Bankers' Association was equally swift. Its president cabled Dunning to pledge that "all banks will co-operate with Dominion Government in housing modernization loans along lines indicated by you through Deputy Minister of Finance yesterday afternoon." The scheme also received favourable press reaction, even from the Conservative *Mail and Empire*.[14]

Windsor was selected by the NEC as the first testing-ground for the HIP scheme. It unfolded according to the corporatist visions shared by the prime minister and the commission. Purvis appointed Wallace Campbell, president of the Ford Motor Company of Canada, to be the chairman of a Windsor advisory committee. This committee also included the presidents of the chamber of commerce and Windsor Trades and Labour Council, an area newspaper publisher, and McLarty. This body had five subcommittees. One dealt with the HIP scheme; three others handled publicity, finance, and trouble, and a fifth was the the women's committee. The HIP committee chaired by CCA leader Harry Mero included the secretaries of the Windsor trades and labour council and the chamber of commerce. The publicity committee primarily involved the chamber of commerce, which arranged radio talks, mass meetings, the selection of speakers for service clubs, new items in the local press, streetcar signs, posters, billboards, and window displays. The financial committee investigated the number of homes where improvements could be made and advised suppliers how to approach householders. The women's committee was thought to be critical since it was women who "exercise the greatest amount of influence in persuading the family to spend the family dollar for improvement of the home." It was composed of "local representatives of social and welfare work." Most important in the NEC's view was the trouble committee, headed by "one of the most solid businessmen in the community, who enjoys the confidence of the banks ... has broad human sympathies," and was sympathetic to the HIP. The trouble committee served to investigate cases where banks turned down HIP loan applications, in order to "adjust these technicalities and overcome difficulties that are not major, in order that the loan may be made." After this framework of committees was established, the first HIP loan in Canada was made in Windsor on 16 October 1936.[15]

Although the NEC used Windsor as a model for a national program, it proposed possible exceptions such as having a "local Rotarian, Kiwanian or man prominent in commercial or financial fields" replace "the president of the Chamber of Commerce or ... Member of Parliament." What was needed were persons "able to enlist the active support of the whole community."[16]

The corporatist principle employed in the Windsor campaign was extended throughout the country. The commission took the view that "the *idea* of home improvement must be advertised and merchandised in exactly the same manner and through very similar channels to those used in the Victory Loan campaign." As with that drive, the HIP organization would become, in every locality across Canada, "a smooth-working unit of machinery in which every gear will work so as to direct its maximum power at the very point of sale."[17]

The NEC placed great stress on its national advertising campaign. This spread a gospel of privatism. Homeowners would be taught that "in helping themselves they are helping others." Manufacturers likewise would be encouraged to "further their own ends through helping an unselfish project." In every town, it was felt, the efforts of local committees in raising publicity funds would serve to "educate and inspire local and interested dealers not only to increase their own efforts, but also to seriously seek new loan prospects, for the banks." The advertising campaign would "assume the role of the weapon with which each member of our organization is armed."[18]

In keeping with its corporatist philosophy the NEC looked to national business and "social and service organizations" to carry out much of its advertising campaign. The "two railways, the banks, the insurance companies, oil companies, transportation companies" would all be employed, thanks to their "intricate networks which run directly into the very territory" the NEC sought to cultivate. The commission felt the self-interest of these companies would result in their providing "complete and ideal facilities for the dissemination of every type of propaganda, whether it be in the form of literature or whether it requires actual personal contact by any or all of their employees." Similar results were expected from social organizations. The NEC sent a booklet "to all Community Clubs, Churches, etc., giving them the message we wish delivered to their members, and telling them how they can co-operate and assist in the general scheme of things."[19]

In all its plans the NEC placed great emphasis on the public-relations efforts of its advertising firm, Cockfield, Brown & Company Limited, and its director of publicity, Ray Brown. Together they would be responsible for supplying "a full and interesting stream of news items ... to all the papers of Canada continually." Consequently, "stories about the success of the plan in one town should be spread all around the other towns." An assistant publicity director served "as an advance agent wherever Mr Purvis or the Minister of Labour should travel, in order to prepare the Public for their reception, and in order to add prestige and interest to the whole movement." Cockfield, Brown & Company set about to organize radio activity, prepare advertisements, organize material for speeches, contact manufacturers, and provide Purvis with publicity for his trip to western Canada. Purvis, meanwhile, appointed chairmen for provincial HIP committees and contacted business groups for fund-raising for the publicity campaign.[20]

Cockfield and Brown devised a highly ambitious advertising program totalling $446,000. Some $138,500 was to be raised from the trade associations in various industries, such as iron and steel, cement and concrete, air conditioning, and copper and brass. Some $327,500 was to be raised from a "special list" of Canadian companies. These included the Asbestos Corporation, International Nickel, Imperial Oil, British American Oil, the CPR and CNR, Bell Telephone, Molson's Brewery, the four large chartered banks, and six smaller institutions, Eaton's and Simpson's and the Noranda and Hollinger mining corporations.[21]

A sum of $188,200 was budgeted for advertising in publications. The largest share was $95,000 for thousand-line advertisements in dailies "presenting this amazing opportunity to the men and women of Canada, with the corollary that they will be putting Canada in the list of nations that have emerged from the depression." Advertisements in "foreign-language papers" served an ulterior political purpose. The NEC feared "riots which may arise through non-English people," and the HIP advertisements would have an "ameliorating effect ... on the minds of their publishers." Budget allocations were made for promotional display "(i.e. window trim, mats, store display cards), contingencies (i.e. special 'spectacular stunt appeals')," posters, and radio. Another $10,000 was budgeted for motion pictures.[22]

Purvis's fund-raising task among his business colleagues proved demanding. He told the annual dinner of the Canadian Institute of Plumbing and Heating that it was "the kind of job I hope never to have to tackle again." In this speech he thanked the institute for their $50,000 contribution. This was particularly gratifying since, in addition to the "large proportion" spent to increase sales in the plumbing and heating industry, it would aid in "backing up the nation-wide community organization." Often, industries proved more difficult in negotiating with Purvis. On 12 November 1936 he informed Dunning that "my effort to get half a million from private industry is requiring quite an effort." To obtain $25,000 from steel magnate Ross McMaster, Purvis asked for the inclusion of farm fencing in the HIP scheme. Dunning's response, Purvis noted, was enthusiastic, for "very quickly" the answer came back, "If it will help to put the Plan across, we will include them."[23]

Industries more immediately tied to home improvement proved more malleable than the steel industry. The Canadian Construction Association told its members: "A million dollars worth of work a week for fifty weeks is the objective. Are we awake to the chance?" J. Clark Reilly, manager of the association, noted how one contractor urged "his whole sales force to get behind HIP." Another "worked out explanation to plan and methods of selling it to his customers by his sales force." CCA members assisted HIP throughout the country. On 30 November 1936 Reilly observed that "President Hunt and Secretary Paterson pledged the resources of the Montreal exchange. President Culley and Secretary Nicol (Hamilton) have had to convince civic authorities of the benefits and have done so. President Wilde and Secretary Perkins (Toronto) have largely attended meetings, at which merits of plan were discussed. Godsmark writes from Winnipeg and Lecky from Vancouver that the men out there are studying the situation and getting ready to do business."[24]

The Plumbing and Heating Institute undertook an extensive campaign. A French-speaking committee of the institute was formed to deal with translation matters and the French media. The four trade journals of the industry were organized to facilitate the plan's news features and advertising. On 17 December 1936 a booklet was mailed to each member of the industry, entitled "Santa Claus Comes to the Plumbing and Heating Trades." The institute's president, Professor W.W. Goforth, expressed his satisfaction that his industry was "seeing our own particular part of the Purvis bridge being firmly built, piece by piece, and span by span."[25] Indeed, many businessmen responded with marked enthusiasm. At an 8 December 1936 meeting to discuss HIP at Cockfield & Brown's offices F.W. Nicolls noted that "some of the manufacturers were far ahead of the Home Improvement Plan organization; they have men driving around the city, making notes of homes that need improvements, etc. and approach these prospects with suggestions re: the Home Improvement Plan."[26]

Purvis appointed provincial chairmen with a broad mandate to organize the HIP campaigns in their localities. They would appoint provincial boards and initiate local committees. Purvis was quite pleased with the "non-political" background" given HIP in Nova Scotia, where provincial chairman Ralph Bell succeeded in having the Liberal premier, Angus Macdonald, and the former Conservative premier, W.L. Harrington, serve as honorary chairmen. The committee was composed of the leading industrialists of the province. Purvis noted that in the first HIP broadcast, "the Government representatives who spoke were all Liberals." Consequently, he endeavoured to obtain "some public utterances from prominent Conservatives."[27]

A suitable kick-off for the NEC's HIP publicity campaign was occasioned by a national radio broadcast on 30 November 1936 from the Chateau Laurier banquet that launched the Ottawa HIP committee. King himself was the key speaker, and his address was written by NEC director of publicity Ray Brown. The address was a consummate expression of both King's and the commission's philosophy. King stressed "the word *co-operation*" (emphasis Brown's). He urged that "if the people of Canada put behind the Home the irresistible force of their co-operative sympathy and effort," success was ensured. King observed that "this plan, which has been created with the twofold purpose of giving employment and improving Canadian homes, depends for its success on the spirit with which it is received and carried out. Co-operation between the Government of Canada and Canadian financial institutions brought it into being. Co-operation of municipalities, Chambers of Commerce, Service Clubs, building associations, Women's Organizations and other social agencies is carrying it along."[28]

The appeal that the NEC made to the prospective members of its national HIP campaign is summed up well in its booklet *Profit for You*, which was "dedicated to Canadian businessmen." Its campaign was launched through the Ottawa HIP committee's full-page newspaper advertisement. It featured a NEC statement surrounded by advertisements for local insulation, refrigerator, air conditioner, electrical heating, and plumbing suppliers. A great variety of suggestions for home improvements were included. Businessmen were urged to set a good example "by doing the needed maintenance work in your own home." Every medium of advertising was encouraged:"newspapers, radio, movies, direct mail, magazines, billboards, window displays, car cards, handbills, etc." All of these could "bear the Home Improvement Plan emblem and reflect the 'better living' theme." Again, co-operation with banks and local newspaper publishers was stressed. Such publishers were themselves encouraged by the expectation that HIP would open "an enormous field for the increasing of advertising lineage." Women's pages, it was pointed out, could "be stimulated by home decoration advertising and by the advertising of retail stores." "Furniture and home furnishings" were not part of the HIP loan scheme; however, the commission predicted that such buying would be stimulated, since "when homes are newly painted and decorated the need for such new furnishings becomes obvious." Somewhat reminiscent of the style of the American New Deal's National Reconstruction Act was the NEC's suggestion for ads with headings such as: "The following firms are co-operating in the Home Improvement Plan – patronize them."[29]

By 12 March 1937 the provincial and local HIP advisory committees were sufficiently well established to begin an intense three-month national advertising campaign. In big cities advertisements were quarter-page in size, while in smaller town weeklies they would be of three hundred lines and "have a rural flavour." Close co-ordination was needed, as "proof of advertisements, and a schedule of insertion dates" had to be sent "to Provincial Chairmen in sufficient time to allow for local tie-ups and promotion." Full-page advertisements were placed in such national magazines as *Chatelaine, Maclean's*, and *National Home Monthly*. A series of thirteen radio spot announcements was prepared, and provincial chairmen were wired of the exact day of release. A two-minute animated film cartoon told "very clearly the story of how a Home Improvement Loan can transform an old house into a new one," and motion-picture theatre owners agreed to run this film free of charge. Billboards were to be used in all towns and cities with populations over fifty thousand, and a series of six street-car cards was produced, which would "appear in every street car and bus throughout Canada."[30]

A group of poverty researchers employed by a Senate committee in the 1960s characterized the HIP scheme – which had been restored in 1954 after its termination in 1940 – as having provided "rumpus rooms for the bourgeoisie."[31] This critical appraisal was based on the researchers' findings that the rates of interest under the plan were such that only a wealthy minority of Canadians could take advantage of them. Such an appeal to the rich, and those middle-class groups who were fortunate enough to be able to borrow money for luxuries, is evident in ads that show fireplaces of ivory and black marble, air conditioning, panelled walls, Georgian furniture, and bathroom mirrors of "flawless crystal." Typical suggested renovations were "removing overhanging eaves, eliminating verandas or porches and other 'meaningless doodads.'"[31]

Arthur Purvis himself admitted that HIP might "start a little too high" for Canadian farmers.[32] Farmers flooded the NEC and the federal government with complaints of being unable to afford repairs that were needed to prevent the collapse of their homes. Whole farm districts were not serviced by banks since, as the manager of the Imperial Bank of Commerce in Portage la Prairie put it, "farming operations" had not been "too remunerative...in the last few years."[33] The 197 loans made on non-farm dwellings in Prince Edward Island meant that Canada's smallest province had more success in serving rural areas through HIP than did the whole of the prairies.[34] Alberta HIP chairman H.M. Evans noted that the banks in the province were "not desirous of making loans under the Plan." Many borrowers preferred to deal with local merchants, being able, they believed, "to obtain more if they fall into debt."[35] In Nova Scotia the HIP committee told Purvis that "many larger contractors" were "suspicious of the Plan fearing that large scale modernization will tend to decrease new construction." Landlords in Halifax tended to avoid HIP, since they feared it would "compel them to either improve their properties or lower their rentals."[36]

Working-class urban communities were as neglected by HIP as were the Depression-ravaged prairies. Conservative MP T.C. Church urged that homeowners receive the same debt reamortization extended to farmers and fishermen. He pointed out that most Canadian workers could not afford to pay back $750 dollars in five years under HIP's interest-rate schedule. Likewise, Saskatoon's municipal council complained that, under the plan, it was "virtually impossible ... to make any extensive renovations."[37]

After the NEC was terminated in 1939, the local HIP committees it sponsored rapidly collapsed. Without Purvis's flair for combining advertising hype, patriotic fervour, and corporatist consensus, business contributions to provincial and national HIP campaigns quickly dried up. The last committee folded in September 1939, shortly before the outbreak of the Second World War.[38]

HIP was officially terminated, with little debate, following the declaration of war by Canada. The interdepartmental Economic Advisory Committee (EAC) viewed it as a drain on the war effort, serving only to "stimulate much inessential repair and improvement work."[39] Its abolition as soon as economic stimulus ceased to be a prime concern of the federal government reveals clearly that the program was never intended to assist low-income groups. Of the $59,051,021 in HIP loans made during the Depression, only $174,716 was lost through default. Of this, $146,800 was eventually recovered from borrowers. The federal government consequently paid for all defaults, which came to less than one-quarter of 1 per cent of loans. This was substantially below losses experienced in the United States under a similar modernization scheme. No change had been made in the pattern of bank loans, despite the federal guarantee – the banks were not about to be bribed to service what they felt were high-risk groups.[40]

Despite his flair for HIP promotion, Arthur Purvis had long realized that the scheme was no Aladdin's lamp for the ills of the Depression. Attracted to government service by the desire to enact major reforms, he was a strong supporter of carefully planned public works to combat the Depression and of the federalization of relief. Sharing the outlook of the enlightened construction industry leaders of the CCA, he was also a strong supporter of public housing. In 1936 rumours of his threats to resign over the housing issue had posed a danger to King's government.

Purvis met with the committee that had produced the influential Bruce Report on public housing on 8 July 1936. The entire NEC followed up with a meeting with the Toronto board of control. The NEC received a brief from the Toronto city council that strongly recommended public housing. Purvis hired a member of the Bruce committee, David Shephard, to serve as a low-income-housing consultant. He engaged in extensive studies, which included a tour of housing projects and meetings with experts in Great Britain such as the distinguished planner Raymond Unwin.[41]

On the basis of Shephard's research, NEC commissioner W.A. MacKintosh presented a draft low-rental housing scheme on 6 January 1937. This was adopted by the NEC and presented to King's government on 9 February 1937. Its proposals called for the construction of ten thousand units of public housing, to be financed by capital loans from the federal government. These would be in a dozen or so experimental projects. To ensure affordability for low-income families, the federal government would pay half of the estimated $120 per unit subsidy, the remainder to be paid by provinces and municipalities.[42]

The proposal was shelved by the federal government after Clark wrote a long memorandum attacking it on 7 March 1937. Clark stressed the need to assist low-income tenants "on a less objectionable basis" than "direct subsidies." He challenged the NEC's assumption that subsidized shelter was needed because "wages or other incomes of certain groups of people in our urban centres are below a level which would enable them to pay the ordinary economic cost of family housing." He maintained that to accept this outlook could amount to a dangerous "radical innovation in government programs in Canada." He even downplayed the ways in which public housing stimulated employment by telling finance minister Charles Dunning that "many on relief do not want jobs now." He warned King's government not to be misled by the initial small scale of the NEC's proposals. After the first projects had been completed, the political pressures to expand the program would so become "irresistible" as to "cover more than the favoured few."[43]

The shelving of its social-housing recommendations did not cause the NEC to restrain its concern for the issue. Indeed, one of the reasons the commission had released its interim report in August 1937 was to encourage public pressure on King's government for a more interventionist approach to the problems of the Depression.[44] The NEC's interest in low-income housing was further increased by its study of the shelter of relief recipients. Commissioner A.N. McLean examined the homes of one hundred such families in Ottawa, Toronto, and Montreal. He was shocked that public relief funds paid for homes that were "often damp and leaky, overrun with vermin, cockroaches, bed-bugs and the breeding grounds of disease." Tenants in such homes were so poor that children often lacked "clothes enough to go to school in the winter time."[45]

The impact of the NEC's description of the poor housing conditions of families on relief was, however, reduced by the explanation for its causes put forth by its social-welfare expert, Charlotte Whitton. Whitton refused to concede that the problem was related to the inadequate income of either the tenant or the landlord. Instead, she claimed, relief expenditures were gobbled up by "housing harpies" or "vultures" who preyed upon the "apathy and despair" of tenants. Such ruthless entrepreneurs bought up "abandoned or deteriorated living quarters" and crowded "them with humanity whose desperation seeks any accommodation at all, so long as it be cheap."[46]

Whitton's melodramatic description of the landlords who provided shelter for relief families was largely inaccurate. The majority of landlords housing low-income persons in the Depression years were not professional investors ruthlessly seeking to maximize the returns on their investment. The NEC found that the average rental allowance per family ranged from a low of $6.50 in Quebec City to a high of $11.50 in Winnipeg. Any profits made from housing such families must have been small indeed. The most comprehensive study of the relationship of these figures to the costs of the landlords was provided by the 1935 Montreal housing survey, which found that, although relief payments per family could not exceed $12 a month, a return of $30 to $35 a month would be needed to give small proprietors a fair return on the costs of maintenance. Such poor prospects caused landlords to discriminate against relief families when renting accommodation. In Vancouver organizations of the unemployed found that shelter could only be found for such persons by subletting from other working-class families.[47]

That specific monies were actually set aside for shelter allowances reflected the growing pressure exerted by the unemployed, and increased sophistication in the application of social-welfare measures as the Depression deepened. Originally, such payments were made not by municipalities but by private charity agencies. These agencies also served as intermediaries in disputes between tenants and landlords, frequently helping tenants to "find new quarters owned by a more kindhearted landlord." The willingness of landlords to forgo rents in order to have families maintain their dwellings was crucial in preventing more widespread homelessness. In 1932 the Ontario government committee on unemployment and relief concluded that "were not it for the willingness of landlords to permit tenants to go many months without paying rents, there would have been, before this, wholesale evictions of unemployed families."[48] None the less, Whitton did not like the trend to increased public expenditures for shelter expenses of relief families. She labelled the sum of $14 million to house all the families of relief recipients a national disgrace.[49]

To Whitton's overdrawn picture of rapacious landlords was added Clark's characterization of the residential construction industry as inefficient. Clark told Arthur Purvis that housing subsidies were necessary only because of the inefficiency of the small-business-dominated home-construction industry.[50]

In order to discourage new legislation for low-rental housing, Clark promoted the idea of a limited-dividend housing project in Winnipeg. To assuage conservative opposition to this proposal, Clark's close associate, David Mansur, the Sun Life inspector of mortgages, wrote confidentially to his Winnipeg branch office. Mansur explained that "Dr Clark [was] particularly anxious to arrange some such scheme ... rather than face the necessity of bringing in further legislation to provide for real low cost housing and slum clearance." Despite such assurances, opposition from the city's property industry prevented the project from going ahead. Local mortgage lenders and real estate interests took the view that it would "be advisable that no houses" should be "built in Winnipeg until every house in that city for sale had been sold to a satisfactory purchaser."[51]

Although local lobbies against public housing could be formidable, they were not a presence at the national level during the Depression. In Ottawa powerful ministers and civil servants were as strongly opposed to public housing as was the most forceful lobbyist. When, by the spring of 1938, political pressure produced a government commitment to a modest public-works program to counter unemployment, Clark's skilful drafting of the legislation made it totally ineffectual in practice.

Faced in the same year with the NEC's draft legislation of provisions for low-income housing in the National Housing Act, Clark made similarly important modifications. No subsidy to ensure low-income affordability would be provided by the federal government. As well, the bonds of housing corporations would have to be guaranteed by a provincial government. Taxation by municipalities would be limited to 1 per cent of a project's costs of construction. Price ceilings were rigidly set below the level of per unit costs that Clark believed necessary to build new low-rental family shelter.[52]

The result was that between the passage of the NHA in June 1938 and the termination of its low-rental housing provisions on 31 March 1940, no municipality was able to build a single unit of shelter under its provisions. In Vancouver opposition to NHA initiatives was encouraged by the limitations imposed on municipal taxing powers. In Winnipeg the project's completed design exceeded the cost guidelines specified by the federal legislation. In Montreal action was terminated because the province of Quebec refused to pass the legislation necessary to guarantee the bonds of housing corporations. In Halifax the city council could not meet Clark's demands that the site be purchased and architectural plans be completed before the 31 March deadline.[53]

Although such factors as local lobby groups and the relative weakness of the labour movement played an important part in the defeat of social-housing legislation, the fact remains that Clark simply outwitted and outmanipulated his opponents. The leading advocate of social housing in the period, Humphrey Carver, together with similar reform-minded professionals such as Percy Nobbs, the Reverend S.H. Prince of Halifax, and George Mooney, executive director of the Canadian Federation of Mayors and Municipalities, organized a national housing conference to promote social housing. Writing in the conference invitation brochure, Carver stressed that even if "full advantage" had been taken of the NHA, "the problem of adequate housing for the great mass of low wage families" still would not have been solved. Stressing the need for housing subsidies for low-income groups, he noted in a summary of the event, held in February 1939 at Toronto's Royal York Hotel, that "authoritative speakers at the conference all alluded to the fact that no Housing Act can be expected to reach low income families without direct government contribution to a rent-reduction fund or without a capital grant."[54]

Despite his active work in championing social housing, Carver actually wrote to the chief architect of what he opposed in order to seek advice on how to mount a campaign against that very opposition. On 17 August 1939 Carver asked Clark for advice on "any particular move that would make for progress" in securing better federal housing legislation for low-income Canadians. Clark took the opportunity to encourage this tiresome critic to ignore the federal government. He urged Carver not to pressure Ottawa "but [to] concentrate" on the "development of public opinion in his own community." This included "not only the people but their representatives in the City Council and at Queen's Park."[55]

The reformers' confusion over who their opponents actually were made effective agitation on behalf of social housing a very difficult undertaking. No clear criticism could be made against the influence of Whitton's and Clark's convenient chimeras of "housing harpies" and inept small-scale home builders. These formulas served as the legitimation for refusal to move public policy towards some form of shelter subsidy for low-income Canadians. That this issue should be seized upon by both Clark and Carver as critical is indicative of the way in which low-income housing was perceived during the period as a basic class issue of how large a share of national income was to be enjoyed by different social groups.

Also, in presenting their case the reformers neglected to include the most obvious example of housing hardship, the plight of the homeless. This was in part due to Clark's belief that low-income Canadians were not worthy of special government assistance. The municipal housing surveys neglected to record the most obvious group that needed better shelter, those who were housed in "rough and ready" municipal hostels. In one, men lived in "a very small basement, fourteen by twenty, with double bunks," using only newspapers for blankets. Municipalities usually permitted the homeless to spend nights in the jails, and frequently required that they "lie on floors without bed or bedding other than they supply themselves." One municipal shelter was made by fitting "a large room above a garage" with "beds made from two by four lumber with chicken wire netting." Another was an old jail, opened up after fumigation to "kill off the roaches and insects which infested the building."[56]

In many places shelters were only opened during the winter. The homeless were forced to "sleep out of doors during the summer months and obtain their food from begging." In 1932 an Ontario government report warned that there had been a "great deal more vagrancy, mendicancy and transiency of late," with "bands of idle men roaming about the country" posing a "threat to the peace of the community." The greatest danger was, however, experienced by the men themselves. Transients often lived in conditions of peril. In 1938 alone, 120 transients were killed and 202 injured in railway accidents.[57] Rental-allowance figures for single persons on relief in the Depression remained dangerously low. In Winnipeg they amounted to one dollar a month. The city's board of trade also recommended that the city use warehouses to accommodate the city's single unemployed men.[58]

Since no subsidized low-rental housing projects got under way in the Depression, federal housing policy was pursing an impossible dream in seeking to widen significantly the band of income groups served by the joint-mortgage scheme. While the considerable skill of W.C. Clark was able to prod lending institutions to move beyond the exclusive, often racially restrictive suburbs of Montreal and Toronto, his talents could not reverse the basic reality of costs and incomes. Many lending institutions that signed DHA agreements remained only nominal lenders. Nicolls noted that Sterling Trust had not made one DHA loan. The company made the dubious claim that its customers were "not inclined to subject themselves to the restrictions in government loans." To suggestions that the government pay appraisal fees in districts not serviced by lending institutions, the reply was that they did not wish to be bonused "for accepting loans that they would otherwise not consider." Borrowers, they claimed, would "be glad to pay this difference."[59]

To try to promote the DHA, a committee was organized by Arthur Purvis and Mervyn Brown, whose firm, Alcott and Brown, handled advertising and public relations for HIP and the NEC. The Montreal and Toronto businessmen they recruited, who included Robert Laidlaw, J.J. Ashworth, general manager of Canadian General Electric, and C.F. Sise, president of Bell Telephone, had J.P. Johnson, president of Canada Cement, form a Prosperity Housing Association to promote the DHA. This association claimed the DHA was "the most forward piece of social legislation yet placed upon the Federal Statute books, because it provides for government assistance to encourage private initiative to do its job in national development." It praised home ownership as "a bulwark of the state against Bolshevism and Communism." The association also characterized municipalities with low home-ownership rates as having a tendency to waste and insolvency.[60]

Promotional activities alone could not ease the regional disparities in the DHA. While the joint-loan program never exceeded 12 per cent of completed residential dwellings (peaking in 1939) during the Depression years, lending operations under the new federal legislation remained inoperative in many regions. This resulted in severe criticism by prominent Liberals, such as minister of agriculture and former premier of Saskatchewan James Gardner, British Columbian minister of national defence Ian Mackenzie, and the publisher of the Edmonton *Bulletin*, Charles E. Campbell. Not a single DHA or, after 1938, NHA loan was made in the entire province of Alberta until 1944. As late as 31 July 1938 only two DHA loans had been made in Saskatchewan. Only 30 dwellings had been built under the joint-loan program during the same period. By 1937 British Columbia had begun to receive significant DHA loans; 199 units of family housing were built by August of that year.[61] In British Columbia, however, the joint-loan program quickly became associated with shady and exploitative real estate speculation. Such poor materials were used in construction that concrete and plaster crumbled quickly. Press accounts spoke of homes "supposed to last twenty years" falling apart after "twenty months." After a tour of NHA-financed homes, federal housing administrator Nicolls admitted the homes were "far from satisfactory." Real estate agents also formed rackets to receive undisclosed commissions.[62]

In response to complaints from western Liberals, early in 1936 Clark explained that there were only two approved DHA lenders with offices in the west. Of these, one had not yet made any loans, and the other, the Great-West Life Assurance Company of Winnipeg, indicated it would wait until the spring of 1936 to begin operations. Clark pointed to several reasons for the disinclination of western lending institutions to participate. These included the provincial debt moratoria and adjustment legislation, the fact that old houses could be bought at prices cheaper than new, the costs of real estate taxation and appraisal, and the inadequate inspection facilities of firms that lacked established lending organizations. Clark attempted to clear the last barrier by having the federal government bear the appraisal and inspection fee for "the use of loans approved outside the head office city of the particular company involved." This, he admitted, would be an insufficient remedy. There was so little activity in the west "because the rate of interest paid by the borrower is limited to five per cent." This might be satisfactory "to encourage lending in the East, but there is strong resistance to it in the West."[63]

The determination of lending institutions to block widespread use of the DHA in the west was influenced by fear that its 5 per cent loans would not only disrupt their higher-interest lending operations but would result in pressure to reamortize residential mortgages at 5 per cent, as had been done for farmers and fishermen. The "5 per cent" policy caused head offices to decline applications for DHA loans that had been supported by their local agents. One agent of the Western Savings and Loan Association, H.P. McManus, attempted to promote the DHA in the fall of 1936 throughout Saskatchewan. McManus was favourably received by contractors, lumbermen, newspaper editors, and town councils in Yorktown, Wadena, Prince Albert, and Humbolt. Despite his confidence about the soundness of his loan applicants, all were turned down by the Western Savings and loan head office.[64]

The controversy over the DHA and related matters of mortgage moratoria and readjustment was greatest in Alberta, where no federal mortgage loan was made till after the Second World War. The Social Credit government did pass legislation exempting DHA loans from provincial mortgage-moratoria laws. This action, however, did not appease the lending institutions, who were engaged in bitter warfare with the provincial government. DHA exemption came after a campaign successfully waged by the Edmonton Builders' Exchange and the city's chamber of commerce. A.J. Brown, president of the Edmonton Builders' Exchange, confided to J. Clark Reilly of the Canadian Construction Association that, despite their successful campaign, the lending institutions apparently had "no confidence in this Government and so we are at a standstill."[65]

Edmonton Bulletin publisher Charles Campbell, who was well aware of the lending institutions' aversion to begin making mortgage loans in the west at 5 per cent, exerted considerable effort promoting the DHA in Edmonton. He had the *Bulletin* construct "in the most desirable residential section, a model home, fire proof, air conditioned, every modern electrical appliance, etc., extensively landscaped, costing $12,800.00." Some sixty thousand persons from

Edmonton and northern Alberta visited this new home. The *Bulletin* was "deluged with requests for information and as to how loans could be obtained under the Dominion Housing Act." Campbell told King that he was informed by "the head of the Mortgage and Loan Association of Alberta" that "no loan company was going to make loans in Western Canada on a basis of five per cent." Such a policy would require "that millions of dollars of outstanding loans, now bearing interest at high rates, would have to be reduced to a five per cent basis."[66]

On 13 March 1937 Campbell wired Dunning that it was "imperative that funds be quickly and easily available in Western Canada, under the Dominion Housing Act, if the Liberal party is not to be seriously handicapped." This was true since "no loans or construction took place last year and if building season goes by it will be just too bad for Liberal prospects." Campbell told King that nothing short of making the collection of interest above 5 per cent a "criminal offence" would bring about a "reasonable attitude" among western lenders. In response to King's complaint that he failed to offer a solution to the problem he decried, Campbell urged that if lending institutions would not reduce their rates to 5 per cent and utilize the DHA, the federal government should liberalize the lending policies of the Farm Loan Board and lend to provincial governments at 3 per cent. The provinces in turn would make 3 per cent loans to municipalities for home-repair refinancing and the building of new houses. Such a move would force competition among lending institutions, "the same as the primary producer and the working man must sell his goods and services on a competitive basis."[67]

Housing conditions in Alberta cities deteriorated as Campbell's proposals were ignored and the west remained starved of mortgage money. In 1937 only 15 houses were constructed in Calgary. In that year 195 new houses were built in Edmonton, but only through a relaxation of minimum building standards. The average cost of these homes was only $491. Such houses were characterized as not "much more elaborate than dog kennels or rabbit hutches." Dr G.M. Little, Edmonton's medical health officer, estimated that at least 100 houses should be demolished to conform with the city's health regulations. Poor housing conditions became a key target of the Unemployed Married Men's Association, formed in Edmonton in June of 1939. Patrick Burke, chairman of its grievance committee, estimated that "from 400 to 500 houses of the city's relief recipients warranted condemnation."[68]

Campbell attributed the re-election of the Alberta Social Credit government in 1940 to the popularity of its legislation against mortgage foreclosures. He noted too the political backlash against the mortgage companies' practice of flooding "the province with propaganda and money" during the election campaign. The election also forced the provincial Liberals into an unsuccessful alliance with lending institutions.[69]

In Alberta the federal government's efforts to promote DHA lending were made in a manner that sought to avoid any possible antagonism from lending institutions. For instance, Clark persuaded minister of finance J.L. Ralston to reverse previous plans to encourage the Dominion Mortgage and Investment Association to extend the joint-loan program to Alberta. Such an action, Clark feared, would "seriously embarrass the companies in their negotiations with the Provincial government." These negotiations aimed at the "removal of certain legislation from the statute books, or alternatively, ... some new protective legislation." During these talks the mortgage companies feared pressure from Ralston would "weaken their hands."[70]

The strongest action taken by the federal government to revive the western residential mortgage market was through legislation to create the Cental Mortgage Bank in 1939. The CMB was empowered to issue government-guaranteed debentures to a value of $200 million. Mortgage-lending institutions could join on the condition that adjustments would be assumed by the CMB. These would be made on a 5 per cent basis. This amount, equal to the NHA interest rate, was accepted for farm losses by the Dominion Mortgage and Investment Association – but not, however, for home mortgages paid by urban dwellers. The common outlook of business on residential mortgage readjustment led to substantial amendment of the Central Mortgage Bank Act of 1939 by the Canadian Senate to delete its provisions for "non-farm loans." The attack against this aspect of the bill was led by Conservative Senate leader Arthur Meighen. Meighen argued that "nothing at all has occurred which would seem to warrant the government giving special help to a man in a city or country town who is carrying a heavy burden on a house property. I say, let him look after himself."[71] Unlike farmers, homeowners had not been afflicted by a "natural or special economic visitation," such as a plague of grasshoppers, that would justify government aid. Meighen was unmoved by the plight of middle-class homeowners who, as a result of the deflation of real estate values by the Depression, had homes mortgaged for more than their actual worth. The former prime minister viewed this as evidence of the homeowner's "extravagance." Clark hoped that the bill, by providing for debt readjustment, would remove Alberta's restrictive mortgage-lending legislation, which had antagonized lending institutions into boycotting the province. Meighen viewed such goals as a suspicious bailout of the Social Credit government. He argued that "it would be incredibly unjust that after all this is done and the price paid, the province were to be allowed to go through the antics which Alberta has gone through in the last two years."[72]

After the Commons rejected the Senate amendments, the Senate subsequently relented in its opposition. However, Meighen's arguments eventually prevailed. The lending institutions' opposition to residential debt adjustment was so great that they simply declined participation in the Central Mortgage Bank.[73]

Lending institutions did not want to use the DHA in the communities of northern Ontario where prevailing market interest rates promised higher returns than possible under the act's 5 per cent mortgages. Thomas E. Bailey, district manager for Crown Life in the Timmins area, took the unusual step of writing to King to complain that his own company, London Life Insurance, and every other mortgage lender refused to make any DHA loans in Timmins. Bailey was puzzled since it was a community of twenty thousand people and had the largest gold mines in Canada, possessing substantial reserves. The Halliday Company, builders based in Hamilton, Ontario, wrote to Clark that in northern Ontario mining areas the "few cases" in which DHA loans had been granted were all for "larger homes in choice locations."[74]

Such conditions were not confined to mining communities. For example, in Port Arthur Clark found that lending institutions would not use the DHA since "they were able to get eight per cent on ordinary loans there and did not see why they should break the market by loaning under the Housing Act at five per cent." Clark's response to this situation is indicative of how, by his hostility to federal direct lending, he became a virtual prisoner of the lending institutions in setting housing policy. In response to the Port Arthur situation he blandly advised finance minister H.G. Dunning that it was "unfortunate that the companies take this position, but I do not know of anything further that I can do."[75]

Apart from Vancouver and Winnipeg, the only substantial joint-lending operations that continued during the Depression outside the metropolitan areas of southern Ontario and Quebec took place in Nova Scotia. In Prince Edward Island only 16 units of housing were built, totalling $87,434; all were constructed in Charlottetown. The Summerside Sun Life agent would make loans for new houses only if the lumber was purchased from his firm. In New Brunswick only 157 loans, totalling $636,847, were made.[76]

Nova Scotia, especially Halifax, was a model of success for the DHA in the Maritimes. Originally only Central Trust, based in Moncton, participated in the act. In the first half of 1936 the province received no federal joint loans. Clark changed this situation, in Nova Scotia as elsewhere, through co-operation with major lending institutions such as Sun Life. They utilized the lower DHA interest rates to capture a greater share of the mortgage market. More conservative Maritimes financial institutions disliked the DHA because the 5.5 per cent interest rate it gave private lenders was below the prevailing 7.44 per cent rate. On 10 July 1936 Clark met with Premier Macdonald, Nova Scotia Housing Commission chairman Fred Pearson, and various provincial Cabinet ministers in Halifax. On 20 July he returned to meet with local representatives of Sun Life, Canada Permanent, Eastern Trust, and the Nova Scotia Building Association. Clark's efforts proved successful, for Nova Scotia's share of federal joint loans rose to 9 per cent in 1939, greater than the province's share in national population.[77]

On closer examination, however, this "success" was only a larger share of a feeble DHA program. In 1939 Nova Scotia had obtained only 214 joint loans. Few of these aided low- or moderate-income groups. During 1937–39, building permits were issued for 401 dwellings; of these only 30 could be rented to families with incomes as low as $1,000 per year. Sun Life's use of the DHA aroused considerable controversy. It created, as Clark confessed to Dunning, "something of a mess in Halifax and Dartmouth." Sun Life's principal DHA borrower, a construction contractor named Cleveland, aroused the wrath of local builders over the introduction of "mass-production methods" accompanied by allegedly shoddy building techniques.[78]

The restriction of the operations of the DHA to middle-class residential neighbourhoods proved an even more intractable problem than the legislation's limited regional distribution. So restricted to middle-class communities was the first year of lending operations under DHA that of the forty loans registered in Toronto by mid-1936, almost all went to exclusive areas such as the Kingsway, Stewart Manor, Cedarvale, Forest Hill Village, and North Toronto. Only one loan was made in the entire township of Scarborough. The township of East York received none.[79]

The attempt to expand the affordability of the joint-loan scheme by liberalizing lending terms under the National Housing Act of 1938 brought meagre results. The new legislation gave more favourable terms to borrowers by having the government assume more risks and costs. The NHA beefed up the government bonus for loans "in small or remote communities" from $10 to $20. It also assumed costs of the travelling expenses of the lending institutions' agents in such remote districts. Through an extension of the pool-guarantee principle of the HIP scheme to NHA, the government effectively secured the entire risk of the lending companies in their portion of the joint loans. Although rates of default would prove to be less than 1 per cent, the federal government pledged to pay all participating lending companies' losses as long as this did not exceed 20 per cent of their total NHA loans. Due to pressure applied by labour minister Norman Rogers, provision was also made for joint loans to cover 90 per cent of the value of a home in certain circumstances (in contrast to the normal 80 per cent loan). As an inducement for such loans, the federal government extended its "pool guarantee" from 20 to 25 per cent.[80]

The reformulation of DHA's joint loans in section 1 of the NHA of 1938 was abetted by section 3 of the act. Modelled after a Burlington project spearheaded by Liberal MP Hugh Cleaver, the plan arranged for municipalities to provide lots to home builders at nominal prices. Also, the federal government would pay all taxes on the homes so constructed for one year, 50 per cent for the second, and 25 per cent for the third year after construction. Lots were to be sold at fifty dollars unless municipal regulations prevented sales at below-market prices. Purchasers were to construct homes for their own use within a year.[81]

The NHA did not greatly increase the volume of the federal joint-loans program. From 1 August 1938 to 31 December 1938 only 1,304 NHA loans were made, providing 1,837 units of family housing. These were overwhelmingly concentrated in the metropolitan areas of Quebec, Ontario, and British Columbia. All five of Saskatchewan's loans were in Regina, and only one, valued at $2,500, could be considered low-income housing. On 13 April 1939 Conservative MP Denton Massey ridiculed the government's efforts, observing that the NHA legislation had only helped to build 2,273 units of housing. The government's self-congratulatory pronouncements on the 90 per cent loan program disclosed its tepid success; a finance ministry press release on 3 November 1939 announced that 1,000 90 per cent loans had been made.[82]

Like the rest of NHA mortgage lending, 90 per cent borrowing was heavily concentrated in certain regions. North York obtained about 20 per cent; Hamilton 10 per cent; East York nearly 9 per cent; Vancouver 6 per cent. Together these four municipalities accounted for 45 per cent of the $2,500 program.[83]

Some of the conflicts between private interests that stymied housing efforts under the NHA of 1938 emerge from the internal correspondence of Sun Life, sent to W.C. Clark by D.B. Mansur. One especially revealing letter concerns the efforts of one of Sun Life's local agents to stimulate interest in the NHA in Regina. Like many other Canadian cities where the NHA and DHA were virtually inoperative in the 1930s, Regina had a prevailing interest rate on new home construction of 8 per cent. One company, Western Homes, controlled by the Argue brothers of Winnipeg, had a complete monopoly in the area of mortgage loans. Sun Life's local agent, S.E. Briand, told the company's loan manager, J.A. Gray, that there was "a definite lack of good housing accommodation in Regina." City real estate firms all told him that they had "no difficulty at all to rent houses that became vacant" and that new home construction had stagnated due to lack of financing. It was very difficult to meet NHA minimum specifications in Regina because the city's building lots were only twenty-five feet in width. Regina used section 3 of the NHA to dispose of its most undesirable residential lots. Briand told Gray that he found "upon investigation that these lots are in certain parts of the city such as near the CPR tracks and near the Imperial Oil plant, where we would not care to make loans at all." Briand encountered similar problems in Windsor after Sun Life announced it would "not consider any more applications covering the "Riverside National Housing Act Committee." Its organizer sold lots at 800 per cent above their purchase price.[84]

The government assumption of lending institution's risks and many of their administrative costs did little to change their pattern of investment decisions. The provisions in the NHA of 1938 for losses of 20 to 25 per cent of loans conveyed a misleading impression that the government was venturing into a lower-income, higher-risk area. However, by the time the NHA of 1938 was replaced by the NHA of 1944, only seven borrowers had defaulted. This was out of 22,115 loans worth $88,506,066. The default rate was consequently only .001 per cent. David Mansur confided to Arthur Purvis that, despite the joint loan's "reputation of being benevolent legislation," there had been little departure from normal business practice. As he put it, there was no "intention of the lending company nor the Department of Finance to lose any Money" under the scheme.[84]

Federal housing policy in the Depression was devoid of any social objective, seeking simply to chart an easy path to recovery by redecorating and building new homes for an affluent minority. This policy was never clearly expressed by Clark except in personal correspondence with his trusted associate David Mansur, and was never announced to the general public. Instead, a swirl of hype, government guarantees, and promotional gimmickry helped to prevent innovations that would redirect a greater share of national income to lower-income groups to raise their standard of shelter.

Even the relatively affluent middle class was poorly served by the formula of economic stimulation adopted by the federal government. Its voluntary approach, taken at the last minute, of allowing the Central Mortgage Bank to write down residential debt with the co-operation of lending institutions was a classic example of the perils of an approach that attempted too little, late in the day. Unlike farmers and fishermen, homeowners would not have their debts adjusted to reflect sinking real estate values. Consequently, repairs on many overmortgaged homes would never be made, since these dwellings already had a greater debt burden than their resale value.

The success of Clark's Depression-era housing policies lay in the complicated shuffle from the never-formed Economic Council of Canada to the ignored National Employment Commission, then to the unworkable low-income housing section of the National Housing Act of 1938, all of which served effectively to deflect widespread demands for construction of subsidized, low-rental housing. The price for this success would begin to be paid during the Second World War, when an attempt to cling to the ideology of the marketplace led to massive bottlenecks in the provision of necessary housing for a rapidly growing economy.

5 Between Necessity and Ideology: Conflict over the Federal Housing Role, 1939–1944

The conflict over federal housing policy during the Second World War is personified in the confrontation of two powerful men, W.C. Clark and J.M. Pigott. In background these two individuals reversed the usual roles of career civil servant and "dollar-a-year" man. Joseph M. Pigott was an important Hamilton building contractor who had supported public housing during the Depression. This businessman would use his office to attempt to establish an ambitious social-housing program to serve as a stabilizer for the erratic construction industry, a project widely supported by both labour unions and large-scale commercial builders. Clark, the Queen's-trained economist turned real estate investor, then functioning head of the Canadian civil service, would devote much of his busy time during the war to combatting Pigott's views. These seemed to him to border dangerously on socialism, along the lines of the controversial model of New Zealand that drew so much criticism as CCF influence peaked during the course of the war.

The paradox of an academically trained professional civil servant opposing the socialist musings of an enlightened businessman underscores the basic conflict that underlay discussions of housing policy in Canada during the Second World War. While demands for social housing could be easily, if craftily, dismissed by Clark during the Depression, even his talents could not keep the lid on the pressure for dramatic public intervention during the war. In the Depression housing programs were simply geared to relieving unemployment by taking up some of the surplus capacity of the economy. But housing shortages proved to be an actual drain and impediment to the war effort, and out of necessity the ideological constraints imposed on housing programs by Clark were temporarily swept aside. They would, however, remain a source of conflict, as the battles between Clark and Pigott would well symbolize.

Clark initially regarded the war as an opportunity to eliminate all government-assisted housing programs and so to return the field to private enterprise, as had been the case before 1935. All residential construction he considered a wasteful drain on the war effort. Thus the Home Improvement Plan and section 3 of the National Housing Act (providing for federal payment of property taxes on low-priced houses) were quickly eliminated. The basic NHA joint-loan scheme proved more troublesome, however, as a powerful lobby of real estate agents, residential builders, and retail lumbermen had grown up around it.[1] After 31 December 1940 NHA loans were no longer permitted for homes valued at more than four thousand dollars.

The abolition of NHA received full discussion at a 15 October 1940 meeting of the finance department's Economic Advisory Committee, chaired by Clark. The committee accepted Bank of Canada governor Graham Towers' position that "in the first place the Housing Act had the stimulation of employment as its principal purpose and that other things were secondary. This stimulation of employment was now unnecessary and even undesirable or would be within a reasonable period."[2] It agreed that "this measure to stimulate employment should be dropped now that a labour shortage was impending." Clark's plan to create a housing shortage to provide post-war employment was also accepted. A proposal was made that NHA continue through maintenance of its specifications, inspections, and standards, which would permit lending institutions to arrange 80 to 90 per cent mortgages without a government share. This scheme was dropped after Clark wisely pointed out that under such conditions "the lending institutions would presumably make very few loans."[3]

The EAC rejected counter-arguments for the continuation of NHA made by its administrator, F.W. Nicolls, and David Mansur (Mansur had left his post at Sun Life to head the government's ill-fated Central Mortgage Bank). Nicolls and Mansur defended the joint-mortgage scheme, stressing two basic points. First, the termination of NHA would mean an end to 80–90 per cent loans and a return to usurious second mortgages. Secondly, the end of NHA inspections would encourage speculative builders to return to practices that resulted in "shoddy construction."[4]

Towers and Clark and their colleagues on the Economic Advisory Committee cannot be accused of burying their heads in the sand regarding the impact of their policies on Canadian housing conditions. The committee examined and accepted Nicolls' findings on the extent of residential vacancies in Canada. Out of 155 Canadian municipalities, 42 had vacancy rates listed as "0.00." Another 61 had rates of 1 per cent or less, and 3 rates of 2 per cent or less. Only 11 had rates of over 3 per cent. Of these only 3 were municipalities with populations over ten thousand. The normal Canadian vacancy rate seemed at the time to be 5 per cent. Nicolls, in a 19 August 1940 memorandum to Clark, stressed: "If half the number of families who are now 'doubled up' were to be properly housed in individual dwelling units, there would not be a vacant dwelling in Canada. Vacancies have completely disappeared in many of our municipalities and a large proportion of the remaining municipalities report serious shortages and acute increase in rentals."[5]

Clark and other senior civil servants were fully aware of the hardships their proposed contraction in national housing production would bring. In a report of the Economic Advisory Committee to Cabinet completed on 13 November 1940, Clark analysed the consequences of his strategy in the clearest possible terms:

> If this argument is valid, namely that the war demands for labour and materials is a consideration so important that it must override the normal expansion of house building in most communities at least, then it follows that during the war, Canada must accept an increasing amount of 'doubling up' and overcrowding in existing housing units with all the social disadvantages which are thereby involved. These lowering housing standards are part of the reduction of the standard of living which we must accept as a price of the war. The outlook in this connection is not bright and we should not gloss over the evils that will result and the unrest, and public criticism that will follow.[6]

On 28 November 1940 Privy Council secretary A.P. Heeney told Clark of the Cabinet's acceptance of his proposal to terminate NHA loans. Heeney also told Clark that the Cabinet had still not decided the nature of "any general housing programme." However, Munitions and Supply was given the task of making "provision for appropriate temporary accommodation" for employees involved in war-related production.[7]

The Economic Advisory Committee was clear in its calls for the termination of NHA, but it was vague in its descriptions of the ways in which housing problems that disrupted war production would be solved. Already certain armaments industries were making provision for housing some of their own employees. At the critical 15 October 1940 meeting, Nicolls noted that special provisions had been made at Sorel, Quebec, for housing munition workers, although "it was only the higher paid people who were being taken care of." For the unskilled a forty-room staff house had been built. For "the better class of mechanics and skilled workers, many of whom were being brought from England and would be earning $50 to $75 a week," 115 houses were constructed. These were permanent brick houses that were expected to be used after the war. Clark viewed these developments with suspicion. He told Angus Macdonald, acting minister of munitions and supply, that he had learned "of two housing projects already undertaken where I consider that the present national war interest has suffered by too much architectural refinement, too expensive a type of construction, and by paying too much for the housing accommodation secured."

To persuade Canadians to contribute one and two dollars weekly to war savings bonds, Clark felt that it was necessary "to impress the local population that an economical job is done." Consequently, the government must not "build a more fancy or more costly house than local workmen occupy." Clark called firmly for "resistance to the tendency of architects to plan garden villages, introduce special trim, special doors, special roofs, special porches, all of which increases expense."[8]

Clearly, Clark wanted to do everything possible to keep the government's expenditure on wartime housing to the absolute minimum, consistent with munitions-production targets. He extended this position to a logical extreme by advocating industrial barracks. He told Macdonald that the government should examine "the possibility of using bunk houses ... as in mining, paper and other industrial towns. The families might easily remain in their home localities, as do soldiers' families, and thus decrease the dislocation of those towns, and thus facilitate the post-war return of population to its pre-war domicile."[9]

Clark proposed the creation of a wartime Crown corporation to build temporary housing. However, he stressed that many of the additional war-related housing vacancies would be solved by encouraging the taking in of lodgers and by conversion of older houses to yield more dwelling units. Also envisaged was a subsidized transportation system to move war workers to areas of higher residential vacancy rates. Such subsidies, Clark believed, would prove "much more economical than government housing on the one hand, or higher wages to cover fares, on the other."[10]

Private interest groups that benefited from the NHA launched a massive campaign for its retention. The Ontario Retail Lumbermen's Association urged its six hundred members to wire their members of Parliament and to follow up their cables with phone calls. Likewise, mortgage-lending companies who were major users of NHA, such as Sun Life, protested. These private interests also managed to appeal to broader groups with social concerns for housing. The impact of this combined opposition can well be seen in a 23 May 1941 letter from finance minister J.L. Ilsley to C.D. Howe, minister of munitions and supply. Ilsley told Howe that "the pressure for continuing the NHA is very strong indeed (I have resolutions from no fewer than 52 cities and towns, as well as a great many communications from religious or welfare organizations, building interests, real estate men, insurance companies, etc.) and it is going to be extremely difficult to convince them of the soundness of our decision."[11]

After the creation of Wartime Housing Limited, the dispute over NHA began to take on the aspects of a bureaucratic power struggle. Wartime Housing and Construction Control (both part of Munitions and Supply) argued strongly for NHA's termination. Construction controller C.B. Jackson took the view shared by many Canadian architects and planners that the legislation encouraged wasteful suburban sprawl. He viewed housing built under the act as subject "to speculative demand without proper consideration of the essential necessity of the supply of building resources."[12] Victor Goggin, general manager of Wartime Housing Limited, sent Howe a fourteen-point memorandum urging the discontinuation of NHA. Goggin repeated Jackson's concerns about waste and speculation and added several others. These included the low quality of housing built during wartime for home ownership, due to the inferior construction materials available, the waste to the war effort of workers' investing in houses, the war's disruption of the private rental-construction industry, and the lack of desire of the often transient munitions workers for home ownership. Goggin told Howe that the campaign for retention of NHA was led by "speculative builders and material supply companies and municipal authorities who make capital out of what these interested parties tell them." Such parties saw "prospects of bonanza in the present swollen pay envelopes of our workers." Goggin also disputed this lobby's claim that his agency's housing was temporary in nature. Rather, it was "as permanent as any housing that ever was built of wood," being good for thirty to fifty years "if properly maintained."[13]

Wartime Housing's effort to curtail NHA injured the agency because of the counter-attack it inspired from Nicolls, speculative builders, and their allied interests. Nicolls had developed close ties to the lobbyists in his appeal to save NHA; he was quoted in the

Daily Commercial News as having told a public meeting in Port Arthur that "only the force of public opinion" could "induce the government to continue the NHA." He challenged Wartime Housing's encouragement of rental tenure as an erosion of the vital goal of home ownership, which to him was a pre-condition sine qua non of social stability. He warned that "it would be to the advantage of our democratic form of government to foster home ownership at any cost, in place of providing for rental housing, the obvious reason being that home owners make the kind of citizens that we want, whereas tenants are good prospects for the various kinds of 'isms.' At the end of this War we may be in a very serious situation. By encouraging home ownership we will be fortifying our Government against this possible menace."[14]

The various conflicting interests in the NHA-repeal debate were adroitly brought to a compromise through vote 452 of the Appropriation Act of 1942. An additional $1 million was voted by Parliament for the government's share of NHA loans. The ceiling for an NHA loan was lowered from $4,000 to $3,200, and the minister of finance had to be satisfied that the construction served to "relieve the serious housing shortage without creating a post-war surplus." Homes built using NHA ficancing were to be "of non-essential materials and constructed upon lots already serviced by local improvements in order to conserve labour and essential materials."[15]

The compromise over NHA arranged by the government evidently went a long way to meeting the concerns of Munitions and Supply about building in the wrong way and in the wrong place and the lending institutions' desire for the continuation of NHA's innovation in mortgage finance and building standards. The volume of loans granted under these guidelines was not great. In 1944, for example, only 1,393 housing units were constructed under the modified wartime NHA.[16] Regional disparities were more marked than under the Depression-era DHA. These appear to have been related more to builders' and lending institutions' preferences than to wartime needs. No loans were made in the province of Quebec despite overcrowding in its urban centres. Similar conditions prevailed in Nova Scotia, but only six loans were made in the province. As during the Depression, no NHA loans were made in the province of Alberta.[17]

The Second World War intensified housing problems in communities experiencing rapid urban growth. As such pressures grew, housing construction was simultaneously curtailed to encourage munitions-making and post-war employment. From 1941 to 1944 some 286,000 farm residents moved into urban centres. Over 125,000 persons moved from the prairies to industrial centres in

Quebec and British Columbia. Approximately 26,000 persons moved from Prince Edward Island and New Brunswick. Most of these migrants moved to the burgeoning naval centre of Halifax. Employment in Montreal rose from 170,000 at the outset of war to 290,000 by 1 July 1944. Hamilton's iron and steel employment rose from 6,000 in 1939 to 11,000 by 1944. Entirely new industries were created. For example, 3,000 persons were employed in making shells and bombs in Hamilton. Windsor's automotive employment rose from 8,000 to 23,000. Wartime employment in many Canadian cities increased by a third and in some, such as Vancouver, Windsor, Quebec, Victoria, Brantford, Kingston, Fort William, and Halifax, at least doubled.[18]

The impact of the war on shelter demand was expressed eloquently by federal rentals administrator Owen Lobley, who observed:

> Farmers benefiting from war prosperity have bought hundreds of houses in such places as Regina, Saskatoon and North Battleford; they have moved from their houses and have evicted the tenants therefrom. War workers who have migrated from the country to the cities, have saved up war wages and bought houses over the heads of such tenants as milkmen, postmen, servicemen and the like, whose incomes have not benefited from wartime wages. Manufacturers, great and small, who have obtained lucrative war contracts, have purchased homes over the heads of such tenants as middle-class salary earners in banks, insurance companies, department stores, railways and the like, whose salaries have been frozen and who are thus unable to defend their tenure by buying the house.

Lobley found that the war had caused "a greater number and volume of sales of houses than ever before in the history of this country." Prices for individual homes in major urban centres exceeded those at the peak of the 1929 boom.[19]

One measure of the reduced availability of housing, the low vacancy rate, has previously been illustrated. Another was the rise in multiple-family households. These were defined as two or more families living in the same dwelling unit and sharing the same bathroom and cooking facilities. Both families might have additional boarders. There was also a dramatic increase in doubling up, with 17.2 per cent of families in multiple households by 1941, nearly double the Depression-era level.[20]

The declining rental vacancy rate and the resulting overcrowding were originally viewed by the government simply as an indication that necessary wartime sacrifices were being made. Federal rent controls did not mark a fundamental departure from this policy. They simply restrained the economic hardships faced by tenants and provided some security from eviction. The usual objection to rent controls – that they discourage the production of new housing – made them an asset in wartime. Such controls were introduced into 150 Canadian communities during September 1940.[21]

Concern for the need for rent controls began to emerge within the Wartime Prices and Trade Board (WPTB) by May 1940. Information came to the board from some two hundred real estate agents and "other informants" reporting to the Dominion Bureau of Statistics in May of each year. Their figures determined the basic rental component of the cost-of-living index. Homes were divided into "a typical workman's dwelling" and a "medium grade dwelling." The desire of real estate managers to maximize their rental income in the setting of rents was noted in the WPTB's reports. Differences in levels of rents in a city were frankly ascribed to "the fact that some agents are much more efficient than others in gauging changing conditions and collecting from tenants the full economic rent."[22]

By May 1940 the impact of war on rents in some Canadian cities had become quite significant. In Halifax one agent reported that rents were "already 8 to 20 per cent above pre-war." In New Glasgow, Nova Scotia, rents had risen on the average by "20 to 33 per cent." All informants in Kingston agreed that "everything is getting scarce," with "rents already 6 to 35 per cent higher."[23]

In medium-sized cities such as Kingston, New Glasgow, and Halifax the impact of war on the rental market was rapid and dramatic. In larger communities change was slower. By May 1940 the main difference in major metropolitan centres was that the favourable terms that the saturated Depression-era rental market had afforded tenants were no longer available. Typically, in Toronto, the WPTB noted, while "most informants report no increase as yet, several report that workmen's dwellings are getting scarce, and some report no vacancies and no arrears – a forerunner of advanced rentals."[24]

In communities immediately affected by war-induced housing shortages, demands for rent control quickly emerged. Halifax was a prime example. As early as 20 October 1939 the Halifax Trades and Labour Council newspaper, the *Citizen*, complained that an influx of soldiers had "gobbled up all rooming houses and residential premises." It estimated that fifteen hundred persons in the city were displaced. The *Citizen* charged that "wealthy corporations and foreigners have scooped up all available real estate." By 11 April 1940 the labour council had sent telegrams urging rent controls to prominent politicians.[25]

By July 1940 rental increases came to the attention of the minister of labour, Norman McLarty, who asked W.C. Clark's Economic Advisory Committee for an opinion on the advisability of federal rent controls. McLarty noted that at first both he and the Wartime Prices and Trade Board had taken the view that because of "the localized nature of the problem ... it might be well to invite either the provinces or municipalities affected, to deal with the matter."[26]

On 26 July 1940 the committee responded that, considering the reasons for the rent increases, federal responsibility should be accepted. However, the committee rejected a program of overall national rent controls. Instead, federal rent controls would only be "extended to localities designated after investigation had shown an acute situation to exist." Such a locally based program was justified by the low average 2 per cent rise in rentals across Canada. The EAC recognized the need to prevent increases of a more substantial magnitude. It noted that nothing could "be used so quickly as the basis for increases in wages as the rise in rents, for house rents constitute the biggest single expenditure of the wage earner."[27]

The EAC split over assigning responsibility for federal rent controls to the labour ministry or the Wartime Prices and Trade Board. The EAC believed it "highly probable" that municipal and provincial authorities would be "biased in favour of real estate." It also warned that "no system of rent control should be dominated by a Property Owners' Association." To prevent this, appeal boards were to have a balance of interests. Such a board would be composed of a county judge as chairman and a "local assessor or real estate broker" and an "appointee of the district labour council."[28]

In determining the optimal level of rents, Wartime Prices and Trade Board officials noted in the summer of 1940 that market rents were usually based on "the old rough rule to obtain 10% of the property value." For an alternate ceiling the WPTB felt maximum rent should cover carrying charges, a hundred-dollar repair and decorating allowance, and a profit of "6% of assessed value, less encumbrances." Also included would be the "charges of a collecting agent."[29]

In September 1940 a series of orders-in-council established the first federal rent controls. Privy Council order 4616 specifically included housing among the "necessaries of life" to be regulated by the WPTB. On 24 September 1940 order 7 specified the board's powers to fix rentals and define conditions of leases to prevent evictions. Local committees were set up in the thirty controlled areas, which were mainly in Nova Scotia and Ontario. The maximum rental was fixed at the level in effect as of 1 January 1940.[30]

The WPTB had little grasp of the magnitude of the housing problem it was facing; an early board memo predicted that "the increased demand for housing was not likely to outlast the war."[31] But the extension of rental controls soon became a mounting inventory of communities suffering war-induced housing shortages. On 11 February 1940 rental controls were extended to eight additional communities. By June 1941 the list was extended to another 151 municipalities. For these localities, rents were fixed at the level prevailing on 2 January 1941. By October 1941 a policy of general wage and price ceilings had taken effect across Canada. As part of this policy, new maximum rent ceilings were set on all real property (including commercial premises) on 21 November 1941. These were fixed at the level charged on 11 October 1941.[32]

By the time of the adoption of general wage and price controls, rent restrictions were already an essential part of federal anti-inflation policy. A WPTB memo written during this period predicted that inflation in the "later stage" of war, "when production is under way," would "break down any system of priorities or order of allocation of materials." This would "slow down the whole war effort ... as well as increasing the cost of the war to the government and the fixed costs payable after the war." Inflation would "promote social disorganization." Not only could it cause a "continuous succession of industrial disputes"; it would "bring forth demands for radical measures either costly or destructive."[33]

Rent controls had already served as one of the most effective areas of the government's anti-inflation policies. Uncontrolled in most of Canada from August 1940 to September 1941, rents had risen in the period by 5.7 per cent. After controls were imposed on most Canadian communities, the increase from May 1941 to September 1940 was 0.0 per cent. Most of the average 4.8 per cent increase in inflation in this period was caused by higher food prices.[34]

The pressures that led to rent control also eventually encouraged the creation of Wartime Housing Limited as the housing crisis became severe enough to be considered a detriment to the war effort. At first industries built homes for their own skilled workers, such construction totalling $500,000 by 8 February 1941.[35] In January and February of that year F.W. Nicolls, director of NHA, was pressing vigorously for the approval of an emergency wartime housing program. Nicolls pointed out that while the United States had spent $520 million on munition plants, it had still managed to spend $300 million for housing. He told Clark that, "whether we recognize it or not Canada has a housing problem on her hands which, if not attended to promptly, will result in the curtailment of our war industries." Nicolls gave concrete examples of how the labour shortages encouraged by a lack of housing had handicapped the war effort. Such conditions, he argued, had led to delays in unloading British warships engaged in convoy work. Nicolls told Clark that, even though "many thousands of men" were "required within the year there are practically no houses available."[36]

Nicolls reported a number of disturbing statistics. Hamilton had no vacant dwellings and would require 20,831 men on war work in nine months; Valcartier, Quebec, expected 11,000 workers and similarly lacked any vacant dwellings. Similar situations existed in communities across the nation, such as Brantford, Fort Erie, Malton, Nobel, Ottawa, Welland, Winnipeg, Calgary, Montreal, Sorel, Trenton, Arvida, and St Catharines.[37]

Nicolls also told Clark that, in the absence of a building boom similar to that of the early First World War, Canada was "slipping behind in our housing building by 20,000 to 25,000 houses per year by the past two years." Each government department that had in "any way encountered a housing shortage" had been forced to take to itself "the duty of preparing a housing program." Munitions manufacturers, he found, were worried "not about tools" or plant facilities "but about housing." Nicolls urged that ten thousand units of housing be constructed in an emergency wartime program.[38]

Nicolls' memorandum did sway Clark, for by 18 February 1941 he had drawn up a charter for a "National Housing Company." This was similar to the charter of the soon-to-be-created Wartime Housing Limited. Order-in-council 1286 created the latter on 24 February 1941.[39]

Many features suggested by Nicolls were incorporated into the new Crown corporation. These included conducting surveys, exemption from rent controls, and, of course, the power to "construct housing units and staff houses." Many features, however, were modified. Most importantly, the responsibility for Wartime Housing was placed under the Department of Munitions and Supply. This was a major departure from the previous control of federal housing programs by the Department of Finance's National Housing Administration,[40] and freed it from the rigid ideological supervision of W.C. Clark. Munitions and Supply's approach would allow Wartime Housing to move in directions unwelcomed by Clark and the traditional allies of his housing policies, the mortgage lenders, small-scale residential builders, and retail lumber dealers, and it would set the stage for future conflict over housing policy.

Wartime Housing's board contained many advocates of social housing during the Depression. In addition to its president, Joseph Pigott, these included his colleague in the CCA architect William Sommerville, and Halifax Housing Commission chairman Major W.E. Tibbs. One worried conservative-minded civil servant, Cyril

DeMara, rentals administrator for the Wartime Prices and Trade Board, perceptively warned that Pigott's views were "far reaching and rather startling." Pigott indeed acted boldly, quickly focusing on Halifax as the community suffering from the "most critical" war-induced housing problems. With Major Tibbs, Pigott toured Halifax and personally selected several sites for building. The Halifax project had an estimated cost of $600,000 and involved 225 family dwellings and 4 staff houses.[41]

Other areas quickly targeted included the industrial centres of southwestern Ontario. These regions had been given priority in the previous surveys of the finance department's National Housing Administration. Typically, such surveys found that industries in St Catharines had "lost skilled workers because of the lack of homes." In these circumstances "scores of labourers from outside centres ... flatly quit their jobs after a few weeks work," being unable to find accommodation for their families. This situation also prevailed in cities like Kingston, Belleville, Hamilton, and Brantford."Here, housing was so scarce that people were forced in winter"to live in unheated tourist cabins."[42]

Although the conservative Department of Finance could see the need for major wartime housing projects in Halifax and southwestern Ontario, proposed projects elsewhere encountered greater resistance. On 24 June 1941 the Wartime Housing board reviewed the situation in Quebec. It concluded that the only major projects in the province that needed to be undertaken were 150 bungalows in Valleyfield and another 200 in Quebec City. Despite the support of two surveys, local municipalities, and officials of munitions plants, action was blocked by a Colonel Theriault of Munitions and Supply, "who stated that no houses were necessary." Pigott told the Wartime Housing board that he felt the company could not undertake a housing project in Quebec until "much stronger pressure was felt from the industries concerned." C.D. Howe advised Wartime Housing that the situation in Sorel had "already been taken care of."In Montreal and Longueuil factory managers took the view that" transportation will solve the problem."Pigott noted that the need for more housing would be more apparent when the war plants were operating at full capacity. He felt, however, that Wartime Housing"could not undertake housing projects where the industries themselves were apathetic to the housing program."[43]

The extreme inadequacy of WHL's building program in Quebec could be seen in Cartierville, where three plants employing 16,000 workers were built on pasture land in an area lacking immediate housing facilities. Charles David, an architect and member of the

WHL board, recommended that an additional thousand wartime houses be built in Montreal, but no action was taken. Of the 15,802 wartime houses built before 11 June 1943, only 2,232 were located in Quebec.[44]

In its first sixteen months of operation WHL surveyed housing conditions in eighty-eight Canadian communities. The surveys covered every province but Prince Edward Island. Some cities were repeatedly surveyed to determine changes in housing conditions.[45] Surveys began with courtesy calls to the mayor. The presidents and general managers of all firms having war contracts were supplied with a questionnaire that provided information on such topics as the number and "average hourly rate and average salary" for groups of employees, such as single men, single women, and married men; whether an increase in employees was expected; and "opinions of management as to the relationship between war production and the lack of adequate housing facilities." These estimates were checked against Munitions and Supply's inspections regarding the quantity and time of peak production. Contacts were made with trust companies that managed real estate, local hotels, and the YMCA and YWCA in order to appraise vacancy rates in houses and single rooms. Municipal building inspectors were contracted to provide information about trends in residential construction and conversion. Studies were also made of the possibilities offered by housing rehabilitation. Inquiries were made by having researchers walk likely routes to determine whether housing was located reasonably close to the expanding industries. Transportation was also examined. Data was placed on city maps and studies made of proposed development.

Before completing its surveys, WHL staff took care to discuss their proposals "with representatives of the local Board of Trade or Chamber of Commerce." Such business bodies were told that in "no instance would WHL houses be on the open rental market." Its housing assistance would only be given "to those engaged in war industries."[46]

To meet the housing needs found by its surveys WHL devised standard sets of design for family and single housing units. Through its own architectural and engineering staff the company developed preliminary plans and specifications. These were later examined by independent authorities such as the Royal Architectural Institute of Canada and the Heating and Sanitary Engineers.[47] WHL decided that, while its housing plans should be as flexible as possible to meet the differences of the national environment, the number of housing types would be limited. This would reduce per unit housing costs through standardization. Consequently, it became possible to manufacture component parts in large numbers. Also, they could be made well ahead of field construction.[48]

Wartime Housing's family dwellings were divided into various standard categories. Half of its homes were "the small four roomed houses Types 1, H-1 and H-2." Another 15 per cent were large four-room houses, and 35 per cent had six rooms. All were designed to be easily demountable.[49]

Designed to require as little investment of public funds as possible, WHL accommodations were nevertheless considered acceptable shelter by the families of skilled mechanics who had come from high-quality homes. In deference to the demands of the Department of Finance, WHL homes were officially built for temporary accommodation. In practice this meant that its shelters were designed without basements and were consequently heated with stoves. Conversion to "permanent" housing was a relatively simple task that involved digging basements for furnaces.[50]

WHL found innovative ways to reduce waste. Unusable hallway space was kept to a minimum. Care was also taken to vary outside appearance. Varied materials and finishes, such as clapboard, shingles, and plywood, were used for outdoor trim. Homes were well insulated, colour schemes varied, and houses were not placed in a straight line. Generally, forty-foot lots were provided. Special attention was given to landscaping the surroundings. For this purpose landscape architect H.B. Dunnington-Grubb was retained. Community facilities, such as dining halls and playgrounds, were also provided.[51]

WHL's pragmatism had a darker hue. Concern for the needs of industry prompted its construction of moderate-cost, quality rental housing with a full range of supportive community facilities; however, its innovations stopped with the perceived needs of industry. These did not extend to racial integration. Wartime Housing built segregated "quarters" for "coloured" and Chinese single male workers. Its construction of staff houses for single female war workers aided the entrance of women into industry, but the housing needs of this group were always awarded the lowest priority in the company's construction program.[52]

One area where Pigott's broad social outlook combined well with pragmatic concerns for a contented and stable work-force was tenant relations. Here Wartime Housing sponsored active community-organizing programs. These evolved in stages. One early experiment was a Hamilton men's centre. "Almost vacant at first," the project soon had a waiting list for admission. A Lions' Club provided meals and recreation. Another early venture was WHL's construction of a community hall in Windsor in response to requests from tenants.[53]

Wartime Housing's Department of Tenant Relations viewed its mission as part of the company's function as a "plant staffing job." The mandate of both was to "keep men and women in the production lines." The department was aware of how often uprooted war workers were viewed as alien "newcomers" by established residents. By integrating these workers into local communities, Tenant Relations sought to reduce "unrest, high labour turnover and uncooperative tenant attitude."[54]

By the end of 1943 Tenant Relations employed thirty-nine social workers and counsellors. It published *Homelife* magazine, a thirty-two-page booklet with a circulation of fourteen thousand. The department encouraged 162 active tenant groups. These included "prenatal clinics, well baby centres and libraries, supervised play and young people's organizations, garden clubs, home improvement associations and Community councils."[55] Tenant groups made some remarkable achievements. One built a community hall from airplane crates. Others achieved success in reducing juvenile delinquency, developed post-war planning committees, cooking and sewing classes, first aid and public-speaking courses, credit unions, kindergartens, nursery schools, and group health-insurance schemes.[56]

Pigott protected Tenant Relations from the suspicions of more conservative members of Wartime Housing's board of directors. Some viewed its social workers as "very dangerous" and likely to "stir up trouble." Director Hedley L. Wilson, manager of the Maritime Trust Company, wanted the service to be terminated.[57]

While conducting its surveys to solve the housing problems faced by Wartime Housing, officials invariably encountered a "serious degree of overcrowding and [an] ominous housing shortage" of a general nature. This general shortage was outside the housing corporation's mandate, which was strictly limited to facilitating the staffing of war industries. However, the company wished to address the national housing shortage. This outlook reflected the enlightened attitude of most of the Crown corporation's directors, especially those of its president.[58]

Pigott's views, although forged by his long experience with the Canadian Construction Association, had been sharpened more recently by his critical role in a national joint conference of construction industry employers and employees. This gathering, sponsored by the federal Department of Labour, took place shortly before Pigott's appointment to head the newly formed WHL. Pigott represented employers at this conference, which was remarkably similar to another such gathering held in 1921. Both meetings created a national joint conference board to represent unions and employers in the construction industry.[59] Pigott served on the later board along with like-minded corporatist leaders of the labour movement, such as former National Employment Commission chairman Tom Moore. The conference committee drew up a comprehensive program for the construction industry, including reforestation, prairie farm rehabilitation, conservation, the construction of sewage-disposal plants, farm electrification, school modernization, construction of public baths and swimming pools, the extension of unemployment insurance to the construction industry, and the development of a national apprenticeship training program.[60] Far-reaching as these plans were, housing was placed at the top of the list of the fourteen programs that the conference envisaged for the future. The package of housing reforms was quite sweeping. It called for the liberalization of and return to HIP mortgage lending, slum clearance, and the "development of modern housing and town planning schemes, landscaping and garden home plans, with play-ground and park improvements."[61]

The CCA's proposals for public rental housing with extensive community services would only be accepted by the federal government for munitions workers – and, after considerable pressure, the families of servicemen and returning veterans. For other groups such innovations were seen as dangerous precursors to the socialization of the residential construction industry and real estate markets. The housing problems caused by the war for social groups not directly tied to the war effort would be set aside unless tenant protests, in reaction to evictions, for instance, forced government action.

J.M. Pigott and W.C. Clark soon became the lightning rods of conflict over housing policy in the federal government. Pigott told Wartime Housing's board of directors that his efforts to encourage the construction of permanent dwellings for general housing needs had been rejected "because of the attitude of Dr Clark, the Deputy Minister of Finance." In contrast to "the views of the Department of Finance" Pigott wished it "to be known" within government circles that "there should be permanent houses, whether or not we build them ourselves, where there is now and will after the war still be a need for them."[62]

Clark and Pigott first collided over a Vancouver Wartime Housing project proposed in June 1942. After Clark was informed of this project by the deputy minister of munitions and supply, A.K. Shiels, he complained "that men, material and credit were needed for other things." In early February 1942 Pigott supported a request of Hamilton builders for priorities in materials for two hundred homes to be financed under NHA. His request was answered on 14 February 1942 by munitions and supply minister, C.D. Howe. Howe told Pigott that his request was denied since there "was nothing we could do in the face of the attitude of the Department of Finance." The only exceptions to the rule that housing should be for munitions workers only were in Halifax, Windsor, Hamilton, and St Catharines. Here families of overseas soldiers facing eviction were permitted to rent homes owned by Wartime Housing, through arrangements whereby municipalities leased the dwellings or guaranteed the rent.[63]

The conflict between Finance and Wartime Housing actually preceded Pigott's and Clark's disputes. An opening salvo was fired by F.W. Nicolls on 23 December 1941 in a memorandum against Wartime Housing. Nicolls told Clark that in his own opinion and that "of the lending institutions," Wartime Housing was "endangering our security and the investment of our borrower-owners," by which he meant NHA-financed residents. Likewise, one of Nicolls' assistants, J.D. Forbes, called for the exclusion of Wartime Housing from the metropolitan areas of Hamilton, Winnipeg, and Toronto. Wartime Housing, he felt, should be "confined to remote and small districts where munitions factories have been established."[64]

Clark was naturally sympathetic to Nicolls' critique of Wartime Housing, especially as the agency had been born partly out of Nicolls' own urgings in January 1941. Clark was not inclined to view as serious the mounting housing problems induced by war. These shortages simply followed his earlier predictions to the finance department's Economic Advisory Committee. Clark told deputy minister of munitions and supply A.K. Shiels in August 1941 that "doubling up in wartime is one method of making the necessary savings which the civilian must make if he is not to sabotage the war effort. Furthermore, the deferment of this construction of permanent housing until after the war will make a fine contribution to the support of the business structure and of improvement in the postwar years."[65]

By the summer of 1942 Clark and Pigott were set on a collision course. Pigott, pushed by municipalities suffering from severe emergency housing problems, began to formulate an ambitious program of subsidized rental shelter. His first model for such a program was drawn up between Wartime Housing and the city of Halifax. Finalized by February 1942, the agreement called for WHL to lease fifty homes to the city. Twenty of these houses were reserved for servicemen's families facing eviction. The rest would be reserved for hardship cases. But the city was faced with 422 applicants (involving 1,757 persons) for the remaining thirty homes. Seeking to alleviate such shortages, the Halifax council sent a delegation to meet with Howe and Ilsley in Ottawa. The outcome of this exchange was agreement upon a 199-unit wartime housing development. The Admore Park project, like the earlier fifty homes, was to be administered by the city for rental to families suffering from conditions akin to homelessness. The project still, however, covered only a minority of such cases; Halifax had to select the 199 tenant families from a list of over 1,000 applicants.[66]

Pigott decided that the terms of the Admore Park project would cause "a terrific loss" to Wartime Housing. This caused Wartime Housing to devise a model agreement that made the new form of federal subsidy less injurious to the Crown corporation. In place of the heavy cash deficit it incurred in the Halifax agreement, shelter subsidies would be provided at low interest rates over a thirty-year period.[67]

Hamilton became the city to pioneer the model agreement. Like Halifax, it suffered from an extreme shortage of emergency shelter. Some 130 children lived in an old shirt factory. As many as nine children and their parents lived in a single room. The agreement negotiated between Wartime Housing and the city of Hamilton had the corporation lend $1,380,500 to the city for the construction of 300 permanent houses (with basements and chimneys), to be rented for twenty-four dollars a month to low-income families suffering from housing hardship. The federal loan would be amortized over thirty years at 3 per cent interest.[68]

Pigott was shrewd enough to lend at 3 per cent, the rate at which the federal government lent money in the joint-loan NHA scheme. However, details of the Hamilton agreement caused an immediate furore among mortgage companies, retail lumber dealers, real estate agents, small residential builders, and their allies in government. The WPTB Rental Administration (composed of former real estate managers) was particularly hostile. Rental administrator Cyril De Mara sounded an alarm on 2 November 1942. He warned Clark and other senior officials that Pigott was "of the opinion that now and in the future, permanent housing for workers will have to be subsidized by the government, in order to provide homes for workers, at rentals within their earning capacity." "In a nutshell" this amounted to "the New Zealand plan of wide scale state-owned housing for low income groups with capital provided by Government at low interest rates."[69]

In posing his objections, De Mara used the same analysis employed by T. D'Arcy Leonard against public housing in 1935. Unlike D'Arcy Leonard's hasty public testimony to a parliamentary committee, De Mara's was conveyed in a confidential memorandum. Consequently, the arguments were phrased with greater clarity and force. De Mara predicted that should Pigott's plans be "permitted to become nation-wide in scale," the "capital values of all existing housing offered for rent" would be "forced down radically." This would mean "the socialization of all our housing." To the rental administrator this amounted to "the most dangerous and far-reaching programme that has ever been suggested in any of our present wartime endeavours to meet emergencies." He felt it would "inevitably carry through into peace-time conditions with probable disastrous results to our present economic policy of private home ownerships."[70]

Clark hastily informed WPTB chairman Donald Gordon that he was "afraid" De Mara's alarming memo "reflects pretty accurately the views held by Mr Pigott." Finance minister Ilsley warned munitions and supply minister Howe that if the Hamilton project proceeded, it would be "impossible for private industry, either through the National Housing Act loans or otherwise, to compete." This would cause "the private construction and financing of houses" to cease utterly. Like falling dominoes, the collapse would take place "first in the cities with which agreements are entered into, then in other areas which would expect to receive the same treatment from the Government in so far as they have any housing shortage at all."[71]

The business interests favouring the same position as Clark, De Mara, and Ilsley quickly mobilized their opposition to the Hamilton agreement. Pigott told his fellow Wartime Housing directors that "loan companies, builders, lumber companies and others [had] started a campaign across the country, sending representatives to Ottawa and Toronto." This had caused the Ontario government to step in and refuse "to pass the necessary legislation on grounds composed of fictitious figures and misstatements."[72]

Pigott's complaints contained little hyperbole, for a formidable lobby against the Hamilton project had been created with startling speed. A "Greater Toronto Permanent Housing Committee" was formed for the sole purpose of forcing the repeal of the Hamilton agreement. The committee included the Toronto Real Estate Board, the Ontario Retail Lumber Dealers Association, the Toronto Property Owners' Association, and the lumbermen's section of the Toronto Board of Trade. These Toronto-based groups were joined by the Ontario Association of Real Estate Boards, the Windsor Real Estate Board, the Property Owners' Association of Ontario, the Ottawa Home Builders, the Brick Manufacturers' Association, and the Canadian Lumbermen's Association. Together they wrote an eight-page brief to Prime Minister King. It was endorsed by the suburban municipalities of Leaside, Scarborough, and Etobicoke.[73]

In their letter the anti-wartime-housing lobby repeated the concerns of conservative civil servants about an impending "public monopoly on the construction of new dwellings." To diminish the need for special wartime projects, the adoption of "billeting arrangements" was urged. Private rental construction could be encouraged through 90 per cent NHA loans with a "special guarantee" to secure the participation of lending institutions. The Dominion Mortgage and Investment Association issued a similar protest, maintaining that "construction not essential to the war effort should be postponed." Only "the bare minimum of housing essential to the efficient production of the war" should be built. "Doubling-up" was, therefore, "desirable in order to conserve materials and manpower." Further, the "construction of permanent houses" was to be "left to private industry." The DMIA also complained that the Hamilton project's 3 per cent interest rate was unfair competition against the higher combined NHA rate, especially since firms belonging to the association were "conserving their financial resources to purchase Victory Bonds at comparatively unremunerative rates."[74]

The hurricane of controversy unleashed by the Hamilton agreement precipitated a high-level meeting on 5 November 1942. Howe, Ilsley, Gordon, Clark, Pigott, and Wartime Housing's vice-president and general manager, Victor Goggin, attended. They decided that Wartime Housing was "to leave the field of Permanent Housing to others unless they are required by the Government to take over construction and management of projects that cannot be handled by other means." A new office under the Wartime Prices and Trade Board, the real property controller, would ensure that the most efficient use was made of the existing housing stock through surveys and "a campaign to bring out voluntary offerings of rooms, houses, etc."[75]

An elaborate system of safeguards and red tape was set up to monitor and control the activities of Wartime Housing. Any proposed housing projects not related "directly to the manning of war industries," Howe reassured Ilsley, would "be turned over the Real Property Administrator, Wartime Housing to act only if requested to do so by the administrator." The deputy real property controller, Norman Long (past president of the Toronto Property Owners' Association), would meet with Wartime Housing to remove "certain centres" from Wartime Housing's ongoing housing surveys. If, through their own surveys, the office of the controller encountered conditions where a municipality "required additional housing in some quantity," it would request a further survey and a concrete proposal from Wartime Housing. In order to prevent a repeat of the Hamilton agreement, this proposal would include "a budget for the financing of the project." The finalized plan would then have to be approved by a newly created Housing Co-ordinating Committee (HCC). This watchdog body was composed of real property controller Russel Smart; munitions and supply co-ordinator of controls Henry Borden, who served as chairman; Nicolls, representing the Department of Finance's National Housing Administration; construction controller G. Blake Jackson; deputy minister of labour A. Macnamara, and Pigott, representing Wartime Housing.[76]

By establishing the office of real property controller and the deceptively named Housing Co-ordinating Committee (which was a regulatory body rather than an agency to co-ordinate federal activity related to housing), Clark had clearly triumphed over Pigott and Wartime Housing. New wartime housing projects were stopped even in Halifax, the most congested centre in the nation. On 7 December 1942 Howe told Ilsley that he had rejected WHL's proposals for two hundred houses to be built for dockyard workers and another three hundred for shipyard employees in Halifax. Also, no further requests would be considered unless they had been "advised by the Administrator of Real Property." Howe felt that since "expansion of industry has been brought to conclusion, and since, with the exception of the aircraft industry, our plants are in general fully staffed, it is in my opinion that exceptional reason must be given to justify further wartime housing."[77]

Clark's desire to delay the construction of "permanent" housing until after the war was given further expression by new construction-control regulations that required permits for all residential construction in excess of five hundred dollars. Pigott informed Wartime Housing's board that "the main effort" of the Housing Co-ordinating Committee, the controller of construction, the National Housing Administration, and the real property controller was in "the direction of stopping all further building." In such circumstances the 36,800 dwellings constructed in Canada during 1943 would be the lowest number of new dwellings completed since 1936.[78]

The function of the Housing Co-ordinating Committee, ambiguous in the Cabinet order-in-council establishing the body, became evident at its first meeting on 16 December 1942. Committee chairman Henry Borden and deputy minister of labour R. Macnamara threatened to resign rather than allow the committee to be "concerned with any planning of general policy of this kind" – that is, any planning aimed at setting housing production targets. They also refused real property controller Russel Smart's request for the release of information dealing with construction-materials priorities. After the meeting Smart told Gordon that its members "were not concerned with any general planning to deal with the housing problem." Instead, they viewed their function as simply "considering any plan for Government building which may be put before them."

After Wartime Housing's initiatives were handcuffed with red tape, Russel Smart was given the unenviable task of attempting to squeeze more accommodation out of the existing housing stock. Although a man of socialist sympathies, as witnessed by his past support for the League for Social Reconstruction, Smart attempted to carry out his policy mandate faithfully until bitter experience convinced him of its futility. Shortly after his appointment he launched an "intensive campaign" to "encourage householders to re-let their back-room flats" and divide "up existing houses and other buildings that lend themselves to conversion."[79]

Smart's views at this time were shared by construction controller C. Blake Jackson. Jackson maintained that "before there is any thought of new housing, we should be certain that present dwelling places are used to reasonable capacity." Jackson denied the reality of a "housing shortage of great magnitude." There was simply a demand for better housing than that provided by the "doubling up" conditions of the Depression. This demand, he felt, was caused "by the increased spending power of the working classes."[80]

Smart considered that home conversions would meet the pressing need for facilities for light housework (as opposed to boarding rooms, of which no great shortage existed). Conversion also had been the aim of the Department of Finance's Home Extension Plan (HEP). This scheme operated on the same basis as the Home Improvement plans. It guaranteed and assisted loans made by chartered banks and was intended to be used to create more rental units by building extensions to existing dwellings. This scheme, however, produced meagre results. Only 21 loans were made in 1942. This figure was raised to only to 28 in 1943 and dropped to 8 during 1944.[81]

H.L. Robson, assistant secretary of the Canadian Bankers' Association, outlined to Nicolls why financial institutions felt that HEP had failed. Robson believed that the $2-million ceiling on loans had caused "the impression that the authorities in Ottawa are not seriously interested in the plan." All applications for loans over five hundred dollars had to be routed to Ottawa for the approval of the construction controller. Also, prospective borrowers were plagued by a lack of access to building materials. Contractors consequently could make no cash estimates. In addition, the "high price and inferior quality of certain building materials, particularly plumbing and hardware," served as deterrents. Indeed, the inadequate funding of the program and the failure to assign material priorities to it show that the minimal Home Extension scheme fit closely into Clark's plan to curtail residential construction during wartime.[82]

While the new construction required in home extensions still posed problems of supply allocations, Smart's home-conversion scheme provided an opportunity to increase the number of housing units with minimal demands on scarce resources. Under the scheme, the minister of finance leased for five years large dwellings for conversion into rental units. During 1943, its first year of operation, Smart's scheme involved 36 conversions. These increased to 1,209 in 1944. They dropped to 778 in the last year of the war.[83]

Smart had proposed a far more extensive home-conversion plan for 1943 than was adopted by the government. Unlike previous government housing programs, home conversion was never passed by parliamentary legislation. Orders-in-council had to be drawn up for various projects in different cities. Consequently, a cumbersome approval process developed, one that required endorsement of the Housing Co-ordination Committee, Treasury Board, and Cabinet. The 3,000 units Smart envisaged for 1943 were reduced to a mere 36. Like the ill-fated Hamilton agreement, home conversion posed the spectre of a dangerous "socialization" of the housing industry. Fear again was expressed of a dangerous "precedent which would quickly spread to other parts of Canada."[84] When a scheme for fifty house conversions in Toronto was narrowly approved by the committee, its chairman, Henry Borden, wrote a dissenting letter to finance minister Ilsley. Borden successfully persuaded the government that the project's financing "should be done by private individuals."[85]

Frustrated in his efforts to increase the nation's rental housing supply, Smart met no opposition from government in his campaign to encourage people to take in boarders and rent vacant rooms, a project that posed no threat to the property industry. The campaign was directed by deputy real property controller Norman Long, who established "local housing committees" in districts suffering from war-induced housing shortages. The committees appealed "to the patriotism and enlightened self-interest of householders who are 'hoarding' living space – to make a more direct appeal to them to offer their place for rent." Long saw the campaign as an extension of the earlier advertising efforts of the Women's Advisory Committee of the Wartime Prices and Trade Board to increase the number of rooms for rent. An additional feature would be the use of compulsory surveys. In designated areas of greatest housing shortage, householders would be required to declare whether they were willing to rent out part of their homes. By 18 February 1943 Gordon reported to Ilsley that such compulsory surveys in Windsor, Kingston, St Catharines, and Sarnia had made available between 550 to 1,100 additional rooms in each city. Housing registries had opened in sixteen centres and active committees were operating in sixty others.[86]

Smart's optimism over the possibilities provided by existing shelter soon vanished after the newly opened registries began to reveal chronic housing shortages. Pigott had predicted earlier that Smart's proposed house-to-house surveys would, when applied to a country facing a shortage of 300,000 homes, serve "to bring demands" rather than cause the situation to "quiet up." When the compulsory survey was taken in Vancouver, many householders replied that they were already suffering from being "doubled-up" with other families.[87]

In cities across Canada the registries demonstrated the inadequacy of the efficiency expert's approach to housing problems. In Montreal the housing registry's opening was delayed, since the city's housing shortage made it impossible to obtain sixty stenographers at once. In Victoria a "shortage of room and board" was found; the city's landladies had begun to discontinue meals, taking advantage of the tight market. In Ottawa many families had to "be divided, some staying with relatives and the small children placed in the care of the Children's Aid." Lethbridge, Alberta, reported protests against "landlords who refuse to give accommodation to families with children." By 4 May 1943 the Halifax registry decided to terminate its publicity efforts. These were now futile since there was no longer any space "to be uncovered in Halifax." In Edmonton efforts to stretch the city's housing stock had also come to a dead end. Families had quickly found themselves to be "stretching their hospitality to the limit" by setting up cots in their living rooms to accommodate servicemen over weekends. Similar conditions were reported in Toronto and Quebec City. In Toronto, after the registry had built up a waiting-list of over nine hundred names, it began to press for more housing construction. The Quebec City office stopped its publicity efforts after it failed to cope with a flood of evictions on 1 May 1943. It could find only 43 spaces for 363 applicants. Likewise, in Moncton – despite appeals from the pulpits, radio stations, and the local press – it was "impossible to find accommodation for girls wanting room and board, and for families deserving 2 or 3 rooms for light housekeeping."[88]

The rejection of his proposals for home conversion and the lessons of the housing registries taught Smart to alter his position. He became convinced that housing committee chairman Henry Borden was inclined to oppose all "positive" housing measures. Smart also felt that Munitions and Supply should not continue to control construction priorities. He told Gordon that similar agencies in Great Britain and the United Stated lacked the same authority. These countries, he argued, had more "sufficiently in mind the needs of the civil population, which, if not satisfied, are likely to impair the over-all war effort." Smart now believed that (contrary to the limited mandate given Wartime Housing) there was "little distinction between those who are actually making munitions and those who are carrying out other necessary work in the economy." This was because "a munitions worker could not get along without bread." By 24 September 1943 Smart concluded that the housing shortage was developing "to a point where it will seriously interfere with the war effort and be likely to produce a social or political disturbance."[89]

Smart embraced Pigott's policies almost a year after he had been hired to develop an alternative to them. He plainly told Gordon that "the majority of people who need houses in Canada want to rent, not to buy them, and it is these persons who are suffering extreme hardship at the present time, due to the fact that there has been a cessation of building for rental purposes over a long period of time." Smart also found that "very few people for six or seven years before the war were investing money in accommodation to be rented." He recommended the restriction of the sale of scarce bathtubs and heating units to the construction of rental housing. Failing this, he urged, the government must release "more of these critical materials, even if it involves the transfer of some labour from direct munition work." Like Pigott, Smart also identified the focus of housing concern as "the low income group who are, and will be, unable to find places at the rent which they can afford to pay."[90]

Smart's shift in analysis was accompanied by a taste of the extreme means employed by the Department of Finance to block Wartime Housing projects. C.D. Howe fully shared Smart's exasperation over Clark's delaying tactics. Wartime Housing's General Manager, Victor Goggin, had told Howe he felt the committee would "serve no useful purpose as long as Mr Nicolls' actions can nullify the decisions of the committee or be a motivating factor for holdups or vetoes in the Treasury Board." For Howe the last straw came when Nicolls tried to kill a Kingston wartime housing project. Howe was outraged that Nicolls had told the mayor of Kingston that a proposed Wartime Housing project the city had requested "would bring about some kind of blight on the City generally in so far as mortgage loans are concerned." On 13 April 1943 Howe suggested that Nicolls be removed from the Housing Co-ordinating Committee. Nicolls told Smart he had repeated what Kingston's lending institutions had said. Smart suspected that the lending institutions had developed this idea after "a hint from Nicolls."[91]

Municipalities appealed to Smart to use his authority as real property controller to exempt their cities from the ban on wartime housing projects for non-munitions workers. On 13 May 1943 Toronto's city council made a detailed submission to Smart in order to receive three hundred wartime houses. The council pointed out that many war workers had obtained housing by purchasing homes as "the only means of securing accommodation." "An endless chain or wave of eviction notices and orders" had ensued. This meant that increasing numbers were "on the verge of being left homeless." The city welfare department calculated that 200 families faced eviction and that 15 court orders to vacate were still pending. In all, 603 cases of extreme housing hardship were documented.[92]

Despite Smart's sympathy for Toronto's plight, the ban on Wartime Housing's new construction continued. It was finally lifted in Edmonton, where the city council made a request on 18 November 1942 for wartime housing. By April 1943 evicted families with their children had moved into tents. Noting the refusal of landlords to rent to families with children, the Edmonton Labour Council predicted the creation of a "tent city." By 15 June 1943 the local housing registry had a waiting-list of 1,350 persons. Smart approached Wartime Housing with a proposal for 250 homes. He argued his preference for permanent housing for Edmonton, as it was destined to have "a larger population after the war." He noted that there was no danger of competing with the NHA and private home builders in Edmonton; these had long since withdrawn from residential construction, unwilling to commit their "own money in Alberta because of Provincial debt legislation." On 17 May 1943 munitions and supply minister Howe approved the project in a letter to Ilsley, overturning the usual objections of the Department of Finance that approval would make it difficult "for the Dominion to resist applications from other municipalities similarly situated."[93]

In order to reduce the influence of the Department of Finance over housing, especially on Howe and his department, Pigott had a study of Canadian housing problems and programs commissioned. It was written by Leslie R. Thomson, associate economic adviser to Munitions and Supply. Thomson's three-hundred-page report, completed on 22 October 1942, provides a striking portrait of the divisions over Canadian housing policy both within the federal government and throughout Canadian society.[94]

Thomson's report was an impressive call for a major shift towards a socially responsive federal housing policy. It concluded that "one million people today are undergoing definite and unnecessary hardship because of the lack, in varying degrees, of decent, sanitary and adequate shelter." Using 1941 census data the report showed that 74.6 per cent of Montreal's lower third in income, 59.7 per cent of Winnipeg's, and almost half of Toronto's and Vancouver's low-income residents lived in overcrowded conditions of more than one person per room. A housing shortage of 230,000 was estimated. This amounted to a construction backlog worth $750 million.[95]

Thomson extended Wartime Housing's critique of the Department of Finance's joint-mortgage scheme. His report demonstrated that NHA housing was too expensive for 80 per cent of Canadian householders, and repeated Wartime Housing's call for the termination of NHA during the war. The report noted that, despite its priorities, Wartime Housing often encountered difficulties in obtaining hot-water boilers, nails, bolts and leg screws, lighting fixtures, and flooring. Scarce supplies, Thomson argued, should "not be drawn upon by private interests who are building for resale or for individual ownership."[96]

Thomson's proposals bore the imprint of Pigott's and the Canadian Construction Association's thinking on a number of critical points regarding the future role of government in the residential construction industry. A long-term assumption of public responsibility was seen as the remedy for "violent swings between super activity and stagnation." These were judged "almost inevitable if all construction and private home building are left to private initiative with private profit as the motive and private capital taking the whole risk." With long-range planning, economies would be larger and organized labour would moderate its wage demands, consequently reducing housing costs.[97]

About of a third of Thomson's massive report was spent reviewing the housing experiences of other nations, with particularly favourable attention drawn to the social-democratic states of northern Europe and the socialist municipality of Vienna. The report concluded that it was "the universal experience of western societies that private capital has been wholly unable to provide adequate housing for the low-income group of the population." The housing programs of Vienna were given special praise, an advanced stance for the report to take, for those programs' writing off of capital costs was considered too radical even by some progressive Canadian advocates of social housing, such as Noulan Cauchon and Percy Nobbs. The report noted sympathetically of the Vienna experiment that "the municipality used its own granite works, paving-stone works, tile works, lime shops and repair shops." As the supply and purchasing of materials were centralized, "the middleman, whose efforts are sometimes parasitic, was almost completely eliminated." Indeed, the Vienna housing program closely resembled that of the National Construction Council. The Viennese had realized many of the NCC's most cherished aims, such as the standardization of building parts and the restriction "to skilled contractors" of the execution of building plans. Vienna's program had achieved a key goal of Canadian housing reformers, for here the "speculative elements" were "entirely eliminated."[98]

Other evidence from abroad in Thomson's report assailed the world-view of the Department of Finance, mortgage lenders, and small builders. One "outstanding example" of a successful housing program was the Swedish government's policy of lending at 0.75 per cent interest rates to co-operatives and building societies. Such a practice geared federal housing policy towards financing non-profit housing associations rather than private lenders. Recommending a dramatic turnaround in federal policies, the report proclaimed, "success is to be attained in creating a large number of houses for the use of the low income bracket[;], it will be essential to develop adequate local authorities, rather than to limit the local contacts to a series of trust companies, or the like, whose principal function is, having lent a dollar, to get it back."[99]

Thomson's report was of considerable breadth and philosophical depth. It noted that while "capitalism has made the most astonishing advances in western society," it had failed to solve "the problem of how to distribute the wealth that it can create."[100] The report's central recommendation was the formation of a special research group and a national housing commission to study housing further. These recommendations appeared to be mild, but the proposed composition of these bodies demonstrates that their intention was to set the stage for sweeping reforms.

The special research group was to be placed under the "direct guidance and control" of Wartime Housing Limited but would also be "in liaison" with the Advisory Committee on Reconstruction, chaired by Dr Cyril James, principal of McGill. It would be composed of economists, engineers, architects, sociologists, specialists in tenant relations, and trained administrators. The group's study of "the housing situation in Canada as is now and is likely to be at the close of hostilities" would be given to a national royal commission on housing, which would have representation from "organized labour" and the other groups that had been active over the previous decade in urging a social-housing program. These were the Royal Architectural Institute of Canada, the Engineering Institute of Canada, and the Canadian Construction Association. Thomson also recommended the appointment of one of three internationally recognized female housing experts (none of whom was Canadian). These were Catherine Bauer, Edith Elmer Wood, and Elizabeth Denby. Also suggested were representatives from the Canadian Medical Association and the public-health field, and an industrialist known to be "gifted with good interests as a citizen." Groups such as the Dominion Mortgage and Investment Association, property owners' leagues, retail lumber dealers, and real estate boards were not suggested as potential committee members. Likewise, the omission of certain civil servants from Thomson's list of acknowledgments is revealing of conflicts. Nowhere mentioned were any officials of the Department of Finance. However, many other key figures were mentioned, such as R.H. Coats and H.F. Greenway of the Dominion Bureau of Statistics; Leonard Marsh, research adviser to the James committee on reconstruction; and civil servants in Wartime Housing and Munitions and Supply. Also working with Thomson was J. Clark Reilly, secretary of the Canadian Construction Association, but not his antithesis as a lobbyist, T. D'Arcy Leonard.[101]

Thomson's call for a royal commission was an effort to place the same intense pressures on King's government as had been prompted by the earlier Purvis and Rowell-Sirois inquiries. His Preliminary Report on the Housing Situation in Canada and Suggestions for Its Improvement was never publicly released. However, the study's influence would be evident in the report of the James committee on reconstruction. Indeed, its recommendations for housing and community planning created considerable political pressure on King's government for action.

During the fall of 1943 federal housing policy remained deadlocked. The Department of Finance had failed in its efforts to curtail completely Wartime Housing's building for non-war workers, but both Pigott and Smart had achieved very limited departures from this norm. Inter-agency struggle had become intense. Wartime Housing still wished for the suspension of the National Housing Act. The Department of Finance lobbied at every level against Wartime Housing's proposed new rental-construction contracts. Russel Smart tried to have Construction Control officials produce information about materials priorities, which they would not give.

Smart tried a number of schemes to increase the supply of moderate- and low-rental housing, but was blocked by Byzantine twists and turns. The Department of Finance did approve an Ottawa home-conversion plan, but the federal Cabinet rejected the project on the advice of housing committee chairman Henry Borden. Borden raised the usual alarms about the creeping socialization of the housing stock. He pointed out that it would be "a dangerous precedent to have the Government enter into the direct purchase of buildings for the purpose of creating housing accommodation." Borden, a dollar-a-year man and president of Brazilian Traction, served as a watchdog against measures that tended towards the socialization of the housing market. A Smart proposal that managed to get through Borden's committee was blocked subsequently when the Cabinet declined to approve the necessary order-in-council. It called for the loan of $200,000 to the city of Saint John to build multiple-family rental dwellings. This was rejected on the usual grounds of the danger inherent in "establishing a precedent of making a loan to a municipality."[102]

One proposal of Smart's that was not extinguished until a most tortuous process had been undertaken was for low-rental housing in Montreal. Smart set the train of events in motion in a memorandum of 21 September 1943 to Donald Gordon. Smart labelled Montreal's housing conditions "indescribable." He warned that the "effect on health and morale is now serious and may become dangerous." He chronicled the "run around" endured by Montreal and pointed out that this city, along with the province of Quebec generally, had "received much less assistance in its housing problem" than English Canadian provinces. The building of war industries on "pasture land far removed from any housing accommodation" had intensified housing problems. Now the city of Montreal was offering fifty thousand fully serviced vacant lots free of charge to Wartime Housing. As a "useful guinea-pig," Smart suggested a project somewhat similar to Pigott's Hamilton scheme. The federal government would lend money at 2 per cent interest for homes to be built on city-owned lots. Smart hoped this would be a model for other cities facing housing shortages. He urged that "a sufficient allocation must be made to take care of civilian needs of this kind even if it means the transfer of some workers from war industries."[103]

Knowing the opposition he faced, Smart presented a detailed report on housing conditions in Montreal, prepared by his executive assistant, MacKay Fripp. Fripp alleged that the conditions "of the poorest negro families in Halifax" were "palatial" when compared to those that prevailed in Montreal. Even "respectable middle class families" had "been forced to live in stores." Also, some 420 families lived in "garages, empty warehouses, sheds and shacks." In addition to the 4,000 "doubled up" families, 300 families lived in homes of three families each. People lived in "cellars, summer cottages, tourist camps, trailers and in boats and yachts tied up at local warehouses." Shortages in plumbing materials, electrical fixtures, and hardware meant that Montreal's residential builders were "slowly disintegrating." Fripp also noted that, "due to high rentals," the vacancies listed by the city's housing registry were "only available to white collar workers." Although Montreal always "has had a housing problem," the present shortage forced people "in a position to pay rent for decent housing to live in slum conditions." This had produced "social dissatisfaction, the breaking up of families, absenteeism, crime, child delinquency." Most seriously, Montreal's level of health was falling, as shown by the increase in tuberculosis. Fripp had observed tenants living in and paying rent for "windowless, unheated quarters, infested with rats and vermin."[104]

Fripp also documented Montreal's long-standing requests for wartime housing, first made over a year earlier on 5 August 1942. On 7 August 1942 Pigott had told Montreal's planning commission that no action could be taken until the Hamilton scheme had been approved. Pigott did co-operate with Montreal's officials. He sent a copy of the draft Hamilton agreement to Montreal. It was used by the city to formulate a proposal. However, after the creation of the misnamed Housing Co-ordinating Committee, Montreal was told that its requests had to go through the real property controller. On 2 September 1942, former real estate lobbyist deputy real property controller Norman Long told Montreal that "full use must be made of all existing accommodation before new construction can begin." For the next eight months the matter would be tossed back and forth among Smart, Pigott, and C.D. Howe.[105]

As Smart's proposal was fully supported by Wartime Prices and Trade Board chairman Donald Gordon, it could not simply be dismissed by Clark in the fashion of Pigott's Hamilton project. Moreover, the proposal had the support of Montreal's civic administration and the area members of Parliament. Rather than provoke another heated battle within the civil service, Smart's proposals for public housing were altered to a scheme for a limited-dividend corporation. This would reduce the threat of government intervention posed to the property industry, especially as Clark ensured that the project's standards would not exceed those of similar moderately priced rental accommodation. Like the stillborn low-rental housing provisions of the NHA of 1938, these efforts would not create additional shelter. They served only to deflect political criticism and intergovernmental disputes.

Clark and Montreal rentals administrator Owen Lobley drew up a model for a limited-dividend scheme. It combined the goals of low rents and investors' return, mainly by approving the use of plans that fell below accepted minimal building standards. Nicolls complained to Clark's assistant, Mitchell Sharp, that he found it "hard to enthuse over a proposal to create housing in a community notorious for its bad housing by duplicating the mistakes which have been made in the past." He believed that, because of its avoidance of contemporary standards, the scheme "might not be approved by the Montreal Building Department." If it were, "the reaction of the public, particularly those familiar with the trend in low cost housing...might prove embarrassing to the government." Nicolls predicted that "the Canadian Government would be the laughing stock of the world in building such poorly designed units." He complained to Sharp that the government was "throwing into the discard all the experience accumulated by the Housing Administration," and on "the advice of a mere real estate agent," a reference to Lobley[106].

Although Clark and Lobley could draw up financial plans and architectural specifications fairly quickly, the search for a Montreal businessman to spearhead the limited-dividend company they envisaged proved more difficult. J.M. McConnell, Liberal publisher of the *Montreal Star*, declined Ilsley's request to participate. McConnell took the view that the scheme was, as Ilsley had "intimated," simply "a gesture towards placating those who are clamouring for immediate housing relief." Since its construction would not be completed for nearly a year, it would "certainly not provide the relief presently sought by many unfortunate individuals." McConnell also did not want "any committee of private citizens" to assume responsibility for "the continuation of the antiquated type of French Canadian dwelling." These involved an "ever present fire hazard" from "the out-dated method of heating by stoves, and extension of stove pipes throughout the dwellings." Taking part in such a project, he felt, would not be "charitable work" but participation in a profit-seeking business venture. Ilsley unsuccessfully attempted to persuade McConnell that the government was not seeking to encourage the participation of private capital to provide opportunities for speculation. Its purpose was to guarantee "a nucleus of sound business judgement and management."[107]

But by December 1944 Clark and Ilsley had found in George W. Spinny, president of the Bank of Montreal, a prominent businessman who would accept the task of organizing the limited-dividend company. Spinny assembled a board of directors and $400,000 in capital from his colleagues in Montreal's business community. While McConnell viewed the project as a shady, disreputable venture, other business magnates regarded it as a dangerous "socialistic" scheme that would cause "considerable disturbance in the minds of leaseholders generally." To encourage business participation, Clark agreed to free the project from corporate income and excess-profit taxes, and to exempt it from the sales tax on the purchase of materials.[108]

Smart and Gordon indicated to Clark their preference for a public project over a limited-dividend housing scheme. Smart attempted to have its size restored to 2,000 rather than 900 units and the minimum rents lowered from $22 to $8 a month. He pointed out that "even an increase by a dollar or two in rent cuts out a number of people who cannot afford to pay increases in rent." At the same time, on 25 February 1944, Gordon pointed out that the last lifting of a freeze on leases, on 1 May 1943, had caused 958 evictions in Montreal. Some 500 householders from this group were still homeless. Gordon stressed that if a public rental scheme had been quickly carried out, this problem would likely have been less acute.[109]

Spinny was placed in the impossible position of trying to reconcile what public opinion in Montreal saw as housing of acceptable quality and the economics of Clark's limited-dividend scheme. Between May and October of 1944 the directors of his company grappled with this contradiction. The directors found the use of stoves for heating in the project to be a "retrograde step." However, after they agreed to the installation of central heating, projected costs were forced up by 10 per cent. This caused the board to reduce the project by 100 units and to raise rents to the $29.50 to $39.50 a month range. This latter step, as Spinny told Clark, forced the company to abandon "what we originally started to do, namely, to provide low-income housing." Indeed, 63 per cent of Montreal's population were paying "rentals today lower than the *minimum* figure" projected for units created by the special government-initiated scheme. Spinny admitted that "our plan has now got completely outside the original orbit of low rental housing and we are now missing the original mark," adding that the federal government had refused to commit scarce "building materials and plumbing equipment" to the project.

Spinny's memorandum of 9 September 1944, wherein he told Clark of his desire to end the low-rental scheme, placed the government in an awkward situation. The protracted experimentation of Spinny's board had simply delayed for another year what Smart's housing registries had delayed the previous year; government commitment to a program of low-cost public-rental housing. Both efforts showed the validity of the basic lesson of social-housing reforms since 1935, namely that only subsidized rental housing could improve the shelter needs of low-income groups. The pressures for such accommodation had greatly increased since the early years of the war. Spinny's failure threatened to prove especially embarrassing since the new NHA of 1944, passed less than a month before his memo calling for his scheme's termination, made no provision for subsidized rental housing, only for the limited-dividend approach that had just failed. Clark waited almost a month to give Ilsley a copy of Spinny's decision. Even then, he took special pains to "avoid giving a formal reply in case production of correspondence is later asked for in the House."[110]

Although Montreal's failed quest for low-rental housing was repeated in many cities, it was a particularly bitter and winding trail in Halifax, whose black residents were seen by the WPTB as having the second-worst housing conditions in Canada. During Wartime Housing's period of building, the city continued to experience severe housing problems. At its peak, on 21 November 1942, Smart told Gordon that these efforts were hindered by the lack of "accommodation for the workmen who would build the homes."[111] Shortly afterwards, in response to the Hamilton controversy, the freeze on new wartime housing projects was extended to projects intended for munitions workers, such as the ship and dockyard workers in Halifax. On 7 December 1942 Howe told Ilsley that he had rejected Wartime Housing's proposals to build 500 houses for these workers, and that any further requests for more wartime housing would require "exceptional reasons."[112]

Future housing production would be held back in Halifax by a convenient chimera. This widespread myth attributed housing shortages to an influx of persons who were failing to contribute to the war effort. Also implicit in this view was an undervaluation of the clerical work of women. On 15 December 1942 Prime Minister King received an urgent appeal from Betty Paice, stenographer to the naval provost marshal. Paice informed King that, for her first three months in Halifax, the best accommodation she could secure was a large bedroom shared by six women. Her salvation in finding a room in Waverley House was now endangered by the government's intention to convert it to a home for air force women. Earlier, she noted, the residents of another shared home had been similarly evicted.[113] Paice wrote of the "great deal of bitterness" among Halifax tenants against the "high-handed" actions of the federal government, but her pleading brought no response. This was in contrast to the steps taken in reaction to a memorandum from P.B. Carswell, controller of ship repairs and salvage. On 8 May 1943 Carswell notified Howe that the "refitting of escort vessels for convoy duty has become a serious problem in the Maritimes." This was because "conditions in the Port of Halifax offer serious resistance to our efforts to fully man facilities." Not only was "sleeping space ... not available to workers to rest," but "good food is difficult to procure ... transportation facilities are inadequate and in the event of an air raid panic conditions would follow." Carswell seized on a blunt solution: the "evacuation of large numbers of new inhabitants who stay in Halifax" but contribute "nothing towards winning the war."[114]

Carswell's solution was adopted by Wartime Prices and Trade Board real property administrator H.D. Fripp. He urged that "additional military dependents and non-essentials" be barred from settling "within 50 miles of the city." Those already in would "be cleared out from the area." Fripp found that of the 12,500 dwellings in Halifax, 90 per cent were of frame construction and half were taking roomers. Only 776 "houses and living units" had been built from 1938 to mid-1943. Some companies built bunkhouses for their employees, a measure taken by the National Harbour Board, which had 130 of its 750 workers "sleeping in freight cars." The National Selective Service refused to allocate workers to employers unless they could guarantee accommodation. The WPTB had requests for "5534 additional males and 933 females" to do work in the city.[115]

After a meeting of the Cabinet War Committee on 8 September 1943, the problem of housing in Halifax was delegated to an interdepartmental committee of senior civil servants. Servicemen were restricted from obtaining accommodation outside of barracks and warned against bringing dependents to the city. E.L. Cousins, administrator of Canadian Atlantic ports, was given greater powers by the War Committee to remove 4,000 persons from the area to create "the equivalent of 1,000 houses." This approach was endorsed by Halifax mayor J.E. Lloyd. He argued that "the solution is not more houses" but removing "thousands of people in Halifax who do not need to be."[116]

Cousins reported to Howe that he found existing housing data inadequate to justify the proposed evictions. He consequently had the Civilian Defence Corps conduct a housing survey at an estimated cost of $30,000 to $40,000. Under Cousins' command the navy provided barracks for an additional 3,000 officers and ratings by mid-April 1944. Rejecting entrance control, Cousins had the federal government undertake a publicity campaign against unwarranted travel to Halifax. It involved "the CBC news, moving pictures in every theatre in Canada, general press publicity," and posters "in the various railway stations in Canada." Another 4,000 service personnel were removed from the Halifax area, largely through the RCAF's turning over its "Y" Depot to the navy. Cousins noted that 180 wartime houses originally planned could not be completed because of the city's strained finances. However, he would not recommend more wartime housing, only that this receive "further consideration."[117]

The view that the underemployed caused Halifax's housing problems was abandoned by Cousins in his report of 17 July 1944, based on a Halifax-Dartmouth population census conducted under a special order-in-council. This census found that 19,195 arrivals had been added to the city's pre-war population of 65,000. Of these, with the exception of 501 women married to service personnel, only a "very few" were unemployed or not "members of families whose heads are in business employed in Halifax." These 501 women did not cause housing shortages, as all but 119 lived in rooming houses, which currently had 349 vacant rooms. Even in this select group of 119 persons "living in houses, flats or apartments," 46 were "employed on necessary war work." The 73 women eligible for eviction consequently amounted to ".890 per thousand of the population," whose deportation would provide only "negligible" relief. The census revealed that while scaring away visitors by publicity had been effective, housing shortages continued. This was witnessed in cases of a family of 8 living in a single room, 25 persons in a four-room house, and 36 in a twelve-room structure. Some 270 houses had been condemned by the health department, but tenants could not be evicted because of the prevailing overcrowding. Also, 43 per cent of Halifax's dwelling units were "not structurally good." About 18 per cent had inadequate sanitary facilities; another 400 dwellings lacked inside toilets, and 2,500 had neither bathtubs nor showers. Approximately 5,800 homes were heated by stoves.[118]

Despite continuing hardships the federal government's housing efforts remained decidedly modest. Wartime Housing moved 25 houses from Liverpool to Halifax for emergency accommodation. Cousins gave material and labour priorities to the St Lawrence Construction Company, which permitted it to build 120 private homes, almost double the 67 constructed from January to July of 1944. Cousins recommended that "the Halifax-Dartmouth area could advantageously use a further 1,000 houses, to rent from $30 to $35 per month."[119]

The continuing freeze on the production of new wartime housing projects in Montreal and Halifax during 1943 and 1944 illustrates the extreme lengths to which the Department of Finance would go in order to prevent the growth of social housing. The stillborn Spinny scheme and the contemplated eviction of servicemen's wives from Halifax were all time-consuming, fruitless ventures that served only to buy time for the department's efforts to curtail wartime housing. Although such ingenious measures would limit social housing, they would not suffice to achieve the department's objective – postponing residential construction to the post-war period. By the spring of 1944 greater flexibility would be demanded on this issue, primarily because of the havoc the housing shortage was causing with the administration of rent controls.

The prime difficulty faced by the administrators of rent controls during the Second World War was that so much of Canada's rental housing stock was in the form of detached homes, easily sold for a high price in a heated real estate market. Largely under the management of trust companies, these properties were owner-occupied until the surge in foreclosures caused by the Depression. Until the Depression ended, these homes could not be advantageously sold on the private market, but the war-induced housing shortage quickly reversed this situation, and regulations concerning the eviction of well-behaved tenants became the focal point of disputes between trust companies and tenants.

In November 1942 the administrator of rentals for Quebec and the Maritime provinces reported that a major problem had emerged from the sale of tenanted accommodation and subsequent evictions. At the same time the Wartime Prices and Trade Board became concerned about how its regulations were being interpreted. Instead of being restricted to cases in which the landlord "urgently needed" to repossess housing for his own use or for family members, as the board intended, evictions began to be awarded solely on the owner's "desire" for change. Consequently, on 1 December 1942 the WPTB issued order 211. This allowed repossession only if a landlord "*needed* the accommodation for *personal occupation* as his residence for a period of at least one year." Evictions could no longer be based on the needs of the landlords' relatives, and tenants could require landlords to prove to a judge that the residence was in fact needed as a personal home. An internal, unpublished history of rent controls prepared by the Wartime Prices and Trade Board noted that this measure proved "extremely popular with the tenant-class and equally unpopular with landlords and real estate agents."[120]

The end of profitable opportunities for real estate trading caused by the WPTB order resulted in a storm of protest. The WPTB internal history observed that this was caused by "the real estate fraternity, by property managers and by landlords." Their energetic lobbying, fuelled by opportunities for sales at high prices, caused the government to reverse order 211 by the summer of 1943. Mrs Taylor, secretary of the WPTB, advised Gordon that "during the past two weeks we have received between 75 and 100 letters from Real Estate Agents, Property Owners' Associations and from persons who administer property on a professional basis, protesting usually in very strong terms against the requirements of one year's notice to present occupants in the case of the sale of residential property." H.E. Manning, KC, a lobbyist for the Toronto Real Estate Board (and writer of anti-public-housing tracts) sent telegrams to real estate agents' and property owners' associations "urging them to protest strongly and violently against this regulation."[121]

As a result of pressure from the property industry, the WPTB issued order 294. This allowed landlords to evict well-behaved tenants in six months if a landlord indicated a "desire" to use the property for a residence for himself, "his father or mother, his son, daughter or daughter-in-law." Being undefined, "desire" could be based on "nothing more tangible than a whim of the landlord." Regulations requiring that the landlord or his relatives actually live in the repossessed home were unenforceable since WPTB rental appraisers, aware of the housing shortage, would not refuse sales permits and so allow the property to become vacant. Consequently, "real estate agents seized upon the easy opportunity of selling houses."[122]

Many loopholes were cooked up to evade rental regulations. In Halifax "many kinds of space subdivisions were created, varying all the way from substantial structure alterations down to imaginary lines separating one family from another." Given "the pressure and confusion of makeshift arrangements for housing of a multitude of people," it became impossible for grievance officers to discover or to prove what the rent had been for a room or section of a house at the time controls came into effect. In such circumstances, controls could be evaded by charging the same rent while diminishing the space given to a tenant. Halifax officials seldom encountered "a written lease"; most new contracts were for "weekly or monthly tenancies," and in the conditions of a "black rent market" tenants would hesitate to report overcharging out of fear of eviction. There were "hundreds" of appeals for legal assistance from tenants who received notices to quit.[123]

Pressure on tenants in shared accommodation increased when that category was removed from eviction control on 1 October 1943. Ironically, this weakening of rental regulations had been part of Smart's original mission to ease the housing shortage by inducing "householders to open up their homes for those who were homeless." The WPTB history indicates, however, that this move resulted "in a tremendous number of tenants occupying this type of space being expeditiously evicted by landlords with the comparative ease which is possible under provincial law." In many cases landlords evicted "soldiers' wives with children so that the space could rented to persons without children or to tenants willing to pay higher rentals." As a result of such trends this loophole was plugged on 29 July 1944, when eviction control was returned to shared accommodation.[124]

Still other landlords evaded rental regulations by making the lease of an apartment contingent "upon the sale of furniture or payment of substantial commission to third parties who were more usually relatives of the landlord." Sometimes the caretaker or agent who managed an apartment building would collect an extra charge. Would-be tenants were told that their payments would "fix it" with the landlord so that they "would be first in line for an apartment" that would be vacant on some "future date." On 30 November 1944 the WPTB passed order 459 to prohibit such charging of commissions to find accommodation. It closed another loophole by fixing, in the same order, the "price for used furniture which a tenant was compelled to purchase in order to obtain a lease."[125]

Coming from the real estate industry, many rental-control administrators had evident sympathy for their former colleagues who were still administering rental property. During a moving crisis in May 1942, rental administrator Owen Lobley told the Department of Finance he regretted that some 55,000 people "with nowhere to go" aroused "false sympathy." Lobley viewed the housing shortage as a simple product of the impact of the war's higher incomes on working-class families. Previously low incomes during the Depression had encouraged families to double up. Now he believed that greater affluence had provoked a tight rental market. He observed that even if Montreal's poor were "a little crowded," they were "much better off than in Singapore and Hong Kong."[126]

The most significant change to rent controls during the war was a reversal on 4 January 1944 of concessions made to landlords by the relaxing of eviction regulations during the summer of 1943. But the new and more flexible provisions for evictions were not scheduled to come fully into force until 1 May 1944. The Wartime Prices and Trade Board's own history notes that, by late November of 1943, it had become evident in Montreal that 5,000 notices to vacate would come due in the city on the first of May. Montreal tenants had become "alarmed" and been inspired to take "concerted action in their own defence." One Montreal city councillor, a member of the CCF, "organized meetings of tenants and urged a sit-down strike"; some five hundred tenants attended his first meeting. Later protests staged during December of 1943 proved to be equally well attended, helped along by media attention: during December of 1943 "scarcely a day passed without some sensational article appearing in the newspapers" exposing Montreal's housing problems. The "most sensational" of these stories "contrasted the living conditions in Montreal with those of the worst Negro families of Halifax."[127]

As a result of tenant protests the WPTB issued order 358 on 4 January 1944. This declared notices to vacate in multiple-family dwellings to be "null and void." The board decided that the "anticipated outcry, from the landlord class, while not politically desirable," appeared "a lesser problem" than an anticipated "sit-down strike" by tenants. Evictions of well-behaved tenants now would only take place if the landlord could prove he was "not already in occupation of housing accommodation in that building or in another multiple family building owned by him in the same municipality." The period for notice was extended from six months to a year. All notices to vacate were henceforth to be made on standardized forms, and a copy would have to be given to a WPTB rentals appraiser before it was sent to the tenant. The relatives said to need housing would have to be specified when the notice was filed and were required to sign statements. Also, notices could not fall due in winter.[128]

Tenant pressure had forced a restoration of the security of tenure that had been removed by the lobbying of the property industry. These protests invariably strengthened the position of rentals administrators who were not from real estate backgrounds, such as former League for Social Reconstruction member J.F. Parkinson. They also awakened the conscience of others and proved to be an effective counterweight to the lobbying of landlords. The conflict between tenants and landlords also sent a message to the Wartime Prices and Trade Board that an increase in the nation's housing supply could not wait until the coming of peace.

The controversy over rental tenure for existing units, by highlighting problems caused by the housing shortage, helped to cause a change in rent-control policy regarding newly built units. In February 1944 the WPTB allowed rents in newly built rental housing to rise. 37 to 40 per cent above ceilings for comparable accommodation. This move was made after consultation with potential rental-housing investors but was not generally publicized. In order to prohibit the construction of luxury rental accommodation, rent levels sufficient to permit entrepreneurs to build it were not awarded. The WPTB had come to the conclusion that "the only way to avoid recurring crises of evictions on moving days was to provide more shelter."[129]

At the same time as the WPTB decided to encourage new rental housing, construction-control restrictions were relaxed to encourage more residential building. Apartments were no longer restricted to a specific number of suites but to the "three stories and basement walk up type" for maximum size. Single-storey garages and open verandas were no longer included in the maximum size calculations for single-family dwellings, and the maximum ground-floor area for two-storey houses and bungalows was increased. The *Financial Post* predicted a 20 per cent increase in new residential building; however, it cautioned that "wherever you go in Canada you will find building men clamoring without much hope for good grades of lumber, for quality millwork and better deliveries of such material." Montreal still faced a shortage of "unskilled and common labour," and in Toronto many carpenters were reported to be "permanently lost to other industries."[130]

Any optimism for a modest increase in residential construction soon faded. By April 1944 it had become "evident" that the "projected increase in the heavy ammunition program together with the continued absorption of industrial workers" would "seriously reduce" the production of construction materials, according to construction controllers. Structural steel, bricks, plumbing fixtures, and hardware were all in short supply. There was the danger that approved construction projects would "only be partially completed." J.G. Godsoe, chairman of the Wartime Industries Control Board, asked what projected housing demands would be. WPTB chairman Donald Gordon, on 12 May 1944, replied that the board's reports showed "a very serious shortage position." He asked Construction Control for "an analysis of construction permits" issued by the controller and the formulation of a system of priorities. He saw that the "real bottleneck" was the "shortage of manpower" in industries producing needed construction supplies. However, he did not deem it "advisable" to contact the National Selective Service to obtain more workers for such industries, as there was no government policy on the matter. Also, Gordon felt that removing existing restrictions on "the manufacture of bathtubs, furnaces and certain other essentials" would have negative "implications" for supply and also run contrary to "views which might be expressed in Washington."[131]

On 21 May 1944 a meeting was held involving Godsoe, Gordon, J.G. Fogo, associate co-ordinator of controls, chairman of the Housing Co-ordination Committee G.W. Withwell, and H. Foreman, co-ordinator of capital equipment and durable goods in the WPTB. Godsoe indicated that although the Wartime Industries Control Board desired "a unified housing policy," this would have to wait "until the formation of the proposed Ministry of Reconstruction." Given projected building trends and the residential construction programs of federal agencies, it was estimated by the HCC that 41,414 housing units could be built in 1944. However, because of the shortage of supplies, it was concluded that "under the most favourable circumstances" "less than 10,000 units could be completed" (this was below the 1943 figure of 36,376 units). The committee decided to restrict the sale of furnaces on a permit system, and to prohibit furnace installations for the improvement of existing homes. No effort was made to have the National Selective Service reallocate workers. Builders were warned in their licences that they lacked "priority rights" and that it was "unlikely that the production of bathtubs, heating boilers and hot air furnaces will be sufficient for all demands." To encourage substitution, builders were told that showers and stoves would be available. Applicants for construction permits were warned to blueprint "in accordance with the type of material available."[132]

A target of 25,000 housing units was set for private builders. This was in addition to approximately 5,000 housing units to be built by public agencies. The 5,000 figure included 900 units for the ill-fated Spinny project, 1,899 units of wartime housing, and 3,000 homes to be built under the Veterans Land Act. The 25,000 allocation for private builders was divided among regions by the construction controller. Estimates for these regional quotas were based on local supplies and the year's past construction rate. Based on these calculations, construction-control permits in Vancouver were reduced by a third by June. No effort was made to have more residential construction permits awarded where housing shortages were particularly severe, because the federal government lacked the detailed information such decisions required.[133]

Fogo, Gordon, and Withwell all became impatient over their inability to plan beyond negative construction controls and supervision of federal agencies. In a 31 May 1944 letter to Gordon, Fogo stated that the "whole housing problem" seemed "to turn on the lack of a central body charged with responsibility for housing generally and with sufficient organization and staff to enable it to make proper recommendations as to policy." Withwell urged Godsoe to have the Housing Co-ordinating Committee reorganized to analyse "the housing needs in each area of Canada" and how these would best be met. Gordon wanted the WPTB removed from the HCC. He considered that the "real difficulty" it faced was "the lack of a Housing Policy." If such a policy were developed, it would serve "as an objective for the Committee to aim at." Without such a goal the committee should disband, as the WPTB by itself could not "formulate a policy." Godsoe told Gordon that he agreed with his evaluation and had only "assumed the Chairmanship of the Committee in order to preside at its abandonment." He had, however, decided to stay on until the new ministry of reconstruction was formed; to resign or disband the committee now, he felt, would "prove embarrassing to the government."[134]

Gordon was disturbed at the lack of any federal housing policy. Following months of indecision after the death of Russel Smart in May 1944, it was agreed that his office of real property controller would not be revived. Consequently, no one was responsible for preparing reports on local housing conditions, required before Wartime Housing would embark on any permanent housing projects. Gordon did not want the WPTB to be held responsible for powers it could not use, which was a major factor in his efforts to disband the HCC.[135]

Housing production in 1944 did reach the level of 44,200 units. This was slightly more than the number of units that would have been started if construction controls had not been imposed; careful restrictions on scarce supplies boosted housing production by diverting materials from less essential purposes. Consequently, the anticipated need to cut back residential construction did not materialize. An increase of 22 per cent over the 1943 level of residential construction was achieved in 1944. This was close to the level predicted by the *Financial Post* in March 1944, an increase that construction controllers had feared could not be sustained. Of course the restrictions on furnace sales and encouragement of substitute products assisted this achievement. Although they were part of the 42,000 projected ceiling, publicly sponsored projects were fewer than expected and hence freed more supplies for the use of private builders. Not only was the Spinny scheme stillborn, but of the anticipated 3,000 veterans' housing units, only 395 were ever built. This was done after May 1944 by changing Wartime Housing's mandate to provide for the housing of families of servicemen as well as munitions workers.[136]

Although the decline in new housing construction was a general trend during the war, the drop in new rental housing built was particularly severe. While figures for rental and home ownership are not available, they do exist for units in multiple-unit structures and those built in single detached structures. The number of single dwellings built declined from 29,300 in 1941 to 26,500 in 1942 and 24,900 in 1943. The relaxation of controls encouraged an increase to 28,700 single units in 1944. The number of units built in multiple structures fell from 27,500 in 1941 to 20,700 in 1942 and 11,200 in 1943, rising to 14,100 in 1944. Consequently, from 1941 to 1944 the drop in construction of single dwellings was only 5.4 per cent, while the same period produced a decline of 48 per cent in multiple-unit buildings. The pressures on rental housing stock caused by landlords' sales to homeowners and the influx of war workers into new areas are apparent in the resulting evictions, overcrowding, and homelessness. The 40 per cent increase in sales of new dwellings produced only a modest rise in rental construction in 1944. Even when combined with its veterans' housing program, only 1,591 units were built by Wartime Housing in 1944. This was far below the 1943 level of 6,326 units. Despite the protests of organized labour, municipalities, Wartime Housing, and the Wartime Prices and Trade Board, the Department of Finance had turned Canadian housing production towards its preference for home ownership.[137]

The desire to maintain Canada as a nation of homeowners would continue to motivate Nicolls and Clark during their drafting of the National Housing Act of 1944. Clark's continued predominant influence was evident in the orchestrated chaos of the poorly named Housing Co-ordinating Committee. Despite the protests of Gordon, Godsoe, and Fogo, a weak and divided HCC floundered and could not provide any alternative to the Department of Finance's policy of minimizing housing production. Once Wartime Housing had been blocked from taking any action to solve the housing problems of other than munitions workers, it became a relatively easy task for Clark to assign a minor role to public housing in the post-war period. Clark's civil-servant opponents were continually checked by him in the immediate area of housing – emergence shelter – that lay within their legal jurisdiction. They would have even less influence in charting the nation's post-war housing course.

6 Upholding the Private Market in Adversity: The National Housing Act 1944 and the Birth of CMHC

The last years of the Second World War and the dawn of peace would be difficult for the champions of a market approach to housing policy. Utopian expectations encouraged by wartime sacrifice in the name of high principle, when combined with escalating housing shortages, led to enormous public pressure for bold government intervention. As usual, Clark would promote the appearance of change while blocking the substance of what reformers were urging, a comprehensive housing policy designed to curb speculation and make a high-quality residential environment affordable to all.

Both the passage of the National Housing Act of 1944 and the creation of the Central Mortgage and Housing Corporation were superb political statements. They allowed King's government to sail the rough seas of mounting social expectations without introducing more radical measures that would have alienated some of the Liberal Party's traditional supporters, the downfall of R.B. Bennett's Conservative government with its New Deal proclamations of 1935. The NHA was passed shortly after the creation of family allowances, a significant step in the evolution of the welfare state. The promise of better housing and security for families proved an unbeatable formula for success at the polls and provided Mackenzie King with his final election victory. Likewise, CMHC was a hastily drafted substitute for rent-control and veterans' rental-housing programs that were being wound down – but it had the appearance of yet another important public intervention to ease the housing crisis.

Ironically, the storm of demands for widespread subsidized housing that the King government had to weather was a storm in part caused by a report it commissioned. This episode reflected the tensions and conflicts within the Liberal Party, the civil service, and Canadian society, and in many ways repeated the turbulent experience unleashed by King's appointment of the National Employment Commission. As the NEC was promoted by an enthusiast for social policy, labour minister Norman Rodgers, so would a later thorn in the side of King and W.C. Clark, the James committee on post-war reconstruction, be nurtured by a similar-minded Liberal, minister of pensions and national health Ian Mackenzie.[1]

As chairman of the Cabinet committee on demobilization and re-establishment, Mackenzie was in a good position to promote his desire for a comprehensive system of social welfare and housing. Completing his work with veterans' benefits and gratuities by the end of 1940, he began the delicate task of securing support for bolder measures by March 1941. He gathered together a group of experts who would serve as an advisory committee on reconstruction. It included Principal Cyril James of McGill University, Principal W. Wallace of Queen's, former National Employment Commission vice-chairman and labour union leader Tom Moore, and Dr Édouard Montpetit, professor of commerce at the University of Montreal. The committee's direction was heavily influenced by its research director Leonard Marsh, an active member of the CCF "brain trust," the League for Social Reconstruction. Marsh had previously worked under the architect of the British system of social insurance, Sir William Beveridge. He had founded McGill's School of Social Work and was the author of numerous works on Canadian social problems.[2]

That a socialist supporter of the CCF should be in a key position on a post-war planning committee established by a federal Liberal government reflected Mackenzie's strategy for dealing with the political threat posed by the rapidly growing radical party. He felt the answer to its rise was for the Liberal Party to lead the way in social insurance and to secure full employment in the post-war years. In 1943 Mackenzie had tried and failed to persuade King's Cabinet to accept national health insurance.[3] With greater success he urged the adoption of family allowances, at the same time warning King of the "national political menace" posed by the growth "of socialism across Canada."[4] The impact of such events as the CCF York South by-election victory, its rise to official opposition in Ontario in 1943, and its electoral triumph in Saskatchewan in 1944 was also not lost on another Liberal social reformer, the first minister of national health and welfare, Brooke Claxton. During the 1943 surge of socialist support Claxton warned King that "post-war problems lead all others in public interest" and that to regain its public confidence and stem losses to the CCF, the Liberal Party had to show its ability to master problems of reconstruction.[5]

Although favoured by public desire for social reform and by Mackenzie's patronage, the basic Fabian-minded outlook of the James advisory committee on reconstruction would bring it inevitably into conflict with the Department of Finance. A head-on collision between the committee and the department took place in the fall of 1942, when the former recommended creation of a ministry of economic planning. The Department of Finance struck back through its Economic Advisory Committee, a body chaired by Clark and composed of business-minded supporters such as David Mansur and Bank of Canada president Graham Towers, who persuaded Prime Minister King not only to reject the concept of a ministry of economic planning but to degrade the James committee to an explicitly advisory role.[6]

The battle between the James and Economic Advisory committees in the fall of 1942 was part of the internal war over housing policy expressed so vividly in the conflicts between J.M. Pigott and Clark. Their debate over post-war policy surfaced quite clearly in the Thomson Report, commissioned by Pigott, which called for the establishment of a high-profile national housing commission to build public support for comprehensive housing policies. Although such a body was not established through the usual Canadian procedure of a royal commission, it was effectively created through the actions of the James committee. It developed a subcommittee on housing and community planning, which would be popularly known as the Curtis committee after its chairman, C.A. Curtis, a Queen's professor of commerce and administration. Leonard Marsh, who had been consulted in the preparation of the Thomson Report, served as the Curtis committee's research adviser. The decision to create this important body was itself influenced by Pigott's sending a copy of his massive report to Cyril James.

Although the Curtis subcommittee was not formally composed of organizations concerned with social housing, it contained the mixture of expert advocacy desired by Thomson to make an impressive case. Among its members were some of the most perceptive champions of social-housing innovations in Canada. These included George Mooney, then secretary of the Canadian Federation of Mayors and Municipalities; S.H. Prince, chairman of the Nova Scotia Housing Commission; architect Eric Arthur, author of the influential 1934 housing survey of Toronto, and McGill professor B.H. Higgins. Higgins had served with American public-housing agencies during the New Deal prior to his appointment at McGill. His American experience would later serve him well in polemical disputes with the federal government.[7]

George Mooney articulated the subcommittee's bold sense of vision, determination, and imagination. He called for a "courageous" federal housing policy that assured "every Canadian family decent accommodation." Goals "such as providing employment, restoring property values, rehabilitating the construction industry, providing outlets for investments, and even clearing the slums" were in his view to be made subordinate to the "central purpose of improved welfare and better living environments." A "first maxim" would be that all housing, "no matter what income group it is being built for," was to be "of solid materials and good craftsmanship." He pledged that the "days of jerry-built housing and speculative 'high profit–low cost' real estate promotion should not be permitted in post-war Canada." Mooney envisaged public ownership of land for new communities, parks and parkways, and a high standard of community amenities in housing. He urged "revolutionary neighbourhood replanning" to provide "inner-park area, playgrounds, day nurseries, laundry facilities, indoor recreation facilities, and the like."[8]

As the member of the subcommittee most familiar with the workings of municipal government, Mooney directed the subcommittee's first task, developing recommendations on town and community planning. He reviewed all the planning legislation of six Canadian provinces and quickly obtained the subcommittee's agreement to the creation of a federal agency "concerned with the promotion and co-ordination of town and community planning." It would also serve as a "clearing house and dissemination centre" of planning information. Although there was "much material" on this topic both inside and outside of federal departments, there was "no satisfactory co-ordination" of the information. The federal agency would co-operate with citizen groups, conduct research, prepare minimum planning standards, model planning legislation, and devise policies against urban sprawl and for the future use of wartime training bases and factories. It would also advise federal authorities of the adequacy of local planning and play a similar advisory role to provincial and municipal governments.[9]

The Curtis subcommittee's report presented an insightful analysis of land-use planning problems that went beyond the analysis of turn-of-the-century public-health reformers and that of the later generation of urban planners inspired by Thomas Adams. Its authors repeated many of their predecessors' concerns, noting that land speculation, urban sprawl, and unregulated shack towns on the periphery of Canadian cities still posed serious problems. The subcommittee also noted a variety of central-city ills. These included the presence of large privately owned tax-free areas, inadequate and overcrowded public facilities, "despoiled waterfronts," and a lack of "lake or riverside parks, walks and bathing facilities." The general "ugliness and unsightliness" of urban Canada was deplored. This included such visual features as the lack of "space for trees and landscaping" and serious "menaces to health" such as "smoke, soot, dust, cinders, and noise." In documenting such conditions the Curtis subcommittee, like Adams and other Canadian planning pioneers, relied heavily on American studies. Especially important to its work were those of the United States National Planning and Research Board.[10]

The recommendations of the report also outstripped previous planning proposals of Canadian experts. It called for planning boards to regulate the future of land use for an entire province. This would be done in order to designate lands to secure their permanent retention for "recreation, forestry, watershed protection, farming or other open development" and "protection against destruction of natural beauty spots." The report also recommended that the federal government provide municipalities with low-interest loans for land assembly.[11]

Like the authors of the Thomson Report, the Curtis subcommittee noted the "dire shortage of adequately trained and experienced persons" in town planning. Not until 1943 had a town-planning course been developed in a Canadian university. This was a special extension course on housing and community planning at McGill. The University of Toronto's School of Architecture also provided a course "on technical training with some field work for town planners." A special Dominion grant for courses "in the techniques of town-planning, and in the co-ordinated social and economic subjects which are required to provide the proper background for such training" was recommended. One of the first tasks of the proposed federal town-planning agency would be "to draw up a model curriculum for the training of planning personnel." The Curtis subcommittee's educational proposals were not limited to the training of professionals. A broader program was envisaged that would encourage the citizens of a community "to consider alternatives and voice their preferences."[12]

Although one of the members of the Curtis subcommittee was national housing administrator F.W. Nicolls, who had developed housing policies with W.C. Clark and the Department of Finance, a major conflict did not emerge. Its key meetings were held in the spring of 1943 on weekends at Toronto's Royal York Hotel. While following George Mooney's lead in planning matters, the subcommittee accepted Nicolls' views on government programs to stimulate building for home ownership. These included a reduction in NHA down payments from 20 to 10 per cent and a reduction in the government rate of interest on such NHA joint loans. Amortization of NHA loans would be extended twelve to thirty years and their ceiling raised to $6,000. A 4 per cent interest rate was recommended.[13]

The Curtis subcommittee also recommended major changes in government programs for housing rehabilitation. It urged dropping the interest rate on government-insured home-improvement loans from 6.32 to 5 per cent. An "even lower rate" was urged for farmers. The report noted that the 1941 census had found that 288,000 farm dwellings were in need of external repairs. Not only did most farm homes lack electricity, bathing facilities, and flush toilets; very few prairie farmhouses had foundations. In addition, the insulation of walls, floors, and roofs was "practically unknown."[14]

The most controversial aspect of the subcommittee's work was its recommendation of a major program of construction of subsidized, low-rental housing. Its commitment to such a program was made at its third meeting, on 14 May 1943, when it received a report from economist Dr O.J. Firestone that examined the serious housing problems of families in the lowest two-thirds by income among renters in Montreal and Toronto. For families to pay more than one-fifth of their annual income in rent, would cause them to fall below "minimum nutrition standards," threatening their well-being. Only 7.7 per cent of the lower third of Montreal families paid 20 per cent or less for rent. Even fewer, 6.4 per cent, could afford the rents they were paying in Toronto. Another 54,000 tenants in Montreal and 10,000 in Toronto lived in overcrowded conditions of more than one person per room. Such figures led Firestone to conclude that the "basic problem" of housing in Canada was that many families could not afford "rents which would make house building a commercial proposition."[15]

To fill the gap between what tenants could afford to pay and their need for shelter, Firestone urged a "publicly financed low-cost housing programme to assure decent living accommodation for low income groups at a rent they can afford to pay." The subcommittee accepted Firestone's conclusions. To prevent a repetition of the confusion and wrangling that had made 1938 NHA provisions for low-rental housing a dead letter, the subcommittee agreed that 100 per cent of the subsidy for public housing should be paid by the federal government.[16]

The Curtis subcommittee recommended the building of 92,000 new public-housing units in the immediate post-war period. These would be administered by municipal housing authorities. The federal government would loan money for the construction of these units on the basis of the current rate of interest on long-term government bonds, then 3 per cent. To encourage municipalities to undertake housing projects, the federal government would provide grants for preliminary development expenses. It would also provide annual subsidies to local authorities to ensure that project rents were affordable for low-income families.[17]

The subcommittee urged the scrapping of NHA section 32 of 1938 as the first step in securing an effective federal low-rental housing program. It barely mentioned the 1938 NHA's favoured solution of limited-dividend housing, and then only in the summary of its report, nowhere in its chapter on low-rental housing.[18]

The report also extended support for co-operative housing. It described the progress made by the Nova Scotia Housing Commission in encouraging co-operative groups to pool their resources to build individual homes. A special section in the new NHA, under which co-ops would be given loans at the same low rates as limited-dividend corporations and public-housing authorities was recommended as a way to encourage both rental and home-ownership co-ops.[19]

The Curtis Report fully endorsed Pigott's view that a long-term public-housing program would stabilize employment, restrain wage increases, and win trade union co-operation in areas such as reducing craft demarcations and restrictive apprenticeship regulations. It also called for measures against monopolistic restrictions on building supplies. These included a review of tariff schedules and a special inquiry under the Combines Investigations Act. The envisaged large-scale public-housing program, it was believed, would achieve important economies of scale. It would encourage higher standards "in building practice, housing design and project administration." Architects were urged to concern themselves with public housing, to combine concern with social objectives, the use of newly developed materials, and the incorporation of "aesthetic values into projects which otherwise may all easily be too drab and ugly."[20]

Despite its frontal challenge to existing federal policies, the Curtis Report was too important a document to be dismissed by the government. It was, as Leonard Marsh, who served as the report's "hard-pressed editor," later recalled, a compilation of "the most extensive surveys in every aspect" of housing "ever made in Canada up to the time." The Housing and Community Planning study was three times the length of any other James committee report, and consequently became a vivid symbol of public expectations for post-war social reform.[21]

The Curtis Report had considerable public impact. The Dominion Mortgage and Investment Association expressed concern that as a result of the publicity that surrounded it, applications were being made for home-ownership loans as if its proposed recommendations had already been legislated. This confusion had resulted in the delay of "a considerable amount of projected building."[22]

Clark responded to the renewed public attention to housing in a manner similar to his actions in 1935 and 1938: by attempting to strike the most favourable deal he could negotiate between the federal government and the Dominion Mortgage and Investment Association. At a critical 4 May 1944 meeting between Clark and DMIA representatives, the future terms of the NHA were ironed out. The companies agreed to a reduction of the NHA interest rate by 1 per cent to 4 per cent. They indicated opposition to the Curtis recommendation of thirty-year mortgages; even with "proper community planning and zoning" it was felt that mortgages should not be extended for more than twenty-five years. Clark struck a compromise: thirty-year mortgages would be provided where, "in the opinion of the Minister," there was adequate "community planning and appropriate zoning restrictions." As a compromise, 90 per cent mortgages were placed on the first $2,000 of lending value, 85 per cent on the next $2,000, and 70 per cent on the rest.[23]

Although Clark accepted the Curtis subcommittee recommendations for a modest reduction on interest rates on home-improvement loans, its recommendations for grants and subsidies to farmers for such home rehabilitation were rejected. The Farm Loans Act of 1944 provided only for loans to farmers at market interest rates. As a Curtis subcommittee member, B.H. Higgins, pointed out, the report's recommendations "for special subsidies to lower interest rates on rural mortgages, for assistance with down payments, for low-cost loans to erect cottages for farm labourers, and for building community centres in rural districts" were "completely ignored."[24] Although the NHA did contain provision for home-improvement loans, this section of the legislation was never given royal assent, and except in certain regional cases such loans were not provided for until after the passage of the new National Housing Act of 1954.[25]

The provision under section 5 of the 1944 NHA for studies into economical housing and empowering the minister to give contracts to conduct economical housing experiments were Clark's only response to the Curtis Report recommendations for lowering housing costs. Proposals to stabilize the building industry, break up cartels in building supplies, and review tariffs were all rejected. Although diluted, the Curtis recommendations were implemented in the area of community planning. Section 5 of the act established a permissive framework for the finance minister to embark on the promotion of planning and housing studies when he saw fit. This would only begin slowly after the war's end and was a pale reflection of the Curtis Report's recommendation of a Dominion town-planning agency that would immediately conduct planning surveys, assist provinces, fund and establish curricula for university planning courses, award scholarships, and develop a field staff.

The new act's most important departure from the Curtis Report lay in its contrasting proposals in the area of low-rental housing. Here Clark made no provision for assisting public housing; instead, limited-dividend housing corporations were to do the job. Clark dusted off the 1938 NHA provisions for limited-dividend housing but eliminated its provisions for housing by public authorities. The legislation was simply unworkable; private entrepreneurs would refuse to invest until its terms were modified by the 1948 amendments to NHA. Despite this limitation, the NHA estimated, incorrectly, that $70 million on limited-dividend housing would be expended by the federal government. The failure of the NHA to make any provision for helping municipalities with their own projects was seized upon by opposition critics in Parliament. Stanley Knowles, then a recently elected CCF member, perceptively viewed it as "an insult to municipal government in this country."[26] The low opinion of municipalities held by the federal government was later revealed by Prime Minister Louis St Laurent in his famous speech against public housing at McGill University. St Laurent told his audience that he had opposed public housing in 1944 since it would mean encouraging "a vast Tammany Hall organization with its ensuing corruption."[27]

St Laurent's McGill speech also described how he and his Cabinet colleagues had come to the conclusion that family allowances were preferable to housing subsidies. In taking this view, King's ministers were echoing the memorandums of W.C. Clark. Seeing that some new reform was needed to allay public concerns over post-war readjustment, Clark had seized upon the James committee's support for family allowances. He used the Curtis subcommittee's call for public housing as a socialist bogeyman to win Cabinet acceptance of his family-allowance proposals. Clark told the Cabinet that with "children's allowances on anything like an adequate scale it should be possible to avoid" having "municipally constructed and municipally managed low-rental housing projects."[28]

Family-allowance payments did not provide, as the government claimed they would in the summer of 1944, the means by which the gap could be filled between what tenants could afford to pay and builders afford to charge. Limited-dividend housing remained a dead letter. Municipalities expressed interest in obtaining NHA loans for low-rental housing while private investors, who were eligible, did not. On 31 January 1945 the mayor of Quebec wrote to finance minister Ilsley informing him that his municipality would establish a housing authority and was willing to relinquish any political control over it so that construction could get under way. He told Ilsley that no private investors could be found willing to "subscribe personally the 10 per cent equity required." Ilsley responded that he found it "most difficult to believe that it would be impossible to find in the city of Quebec a group of persons and corporations who would be willing to interest themselves in such a project and make the small investment required."[29] From Vancouver H.H. Stevens similarly told Clark that it was impossible to interest private investors in limited-dividend housing. Clark responded that, despite the low returns on investment, the investors would get "a great kick out of it."[30]

The failure of NHA's limited-dividend provisions to produce any housing led C.D. Howe to conclude that its "workability" was "doubtful." Howe urged that Wartime Housing be encouraged to move into the low-rental housing area. He believed that, as it had proved to be "competent in the building field ... it would seem logical that any government building and rental operations should be entrusted to this organization."[31]

Howe's pressure for change helped lead to what CMHC president David Mansur would term a "monstrous" arrangement; the fusing of the limited-dividend and life-insurance sections of the NHA into a joint government–insurance-company effort known as Housing Enterprises Limited. Ilsley proposed as a model to Howe Metropolitan Life's Parkchester apartments in New York City, which were able to provide housing for middle-income tenants by savings derived from mass production and neglect of landscaping amenities.[32]

Housing Enterprises Limited was the brainchild of a dollar-a-year-man life-insurance executive, Nicolls' successor as NHA administrator, William Anderson. He told a meeting of life-insurance executives in July 1945 that if "private enterprise" declined to "take the initiative," government would have to become "involved very extensively in state housing, ownership and control for many years to come."[33]

But Housing Enterprises Limited could not get off the ground quickly enough to save the government from the political pressures that developed when housing shortages increased with an influx of veterans after the war's termination. The Wartime Prices and Trade Board attempted to solve the problem of housing returned veterans by drafting orders (never put into effect) that would have allowed "honourably discharged service personnel" and their parents to evict tenants from the homes they owned within three months of giving notice. This attempt to grasp at an easy solution to the problem was rejected when it was realized that many of the evicted tenants themselves would be families of veterans.[34]

Pending evictions clearly spelled a crisis situation to the government. By 23 July 1945 some 8,391 notices to vacate had been filed in the twelve largest Canadian cities. The WPTB lacked equally detailed figures for smaller communities, but estimates ranged from 15,000 to 20,000.[35]

The falling due of notices to vacate set the stage for political protests. Fully 60 per cent of the notices involved the families "of men in the Armed Forces, many of whom are still overseas." Emergency Shelter administrator Eric Gold warned that soldiers who had volunteered for the continuing war against Japan had indicated they would not serve until they were "assured their dependents are properly and permanently housed." He argued that in such circumstances the government could not "permit families to be evicted and left on the street."[36]

The pending evictions led to a bitter struggle between landlords and tenants. Gold reported that Toronto landlords were "descending during the night and physically forcing the tenants out." In Vancouver a tenants' league was formed to picket, prevent evictions, and campaign for low-rental housing. Large meetings of up to three hundred persons were held in Vancouver to protest evictions. The *Vancouver Daily Province* warned that the evictions had been "creating ammunition for Communist agitators." Movers refused to cross picket lines set up by a Vancouver branch of the Labour-Progressive Party. The Vancouver Labour Council wired WPTB chairman Donald Gordon to protest the "deplorable housing situation in Vancouver." Three hundred Chinese protestors attended one LPP rally against evictions. LPP youth club leader Anna Lew denounced discrimination against Chinese renting in certain Vancouver districts.[37]

On 23 July 1945 Gordon convened an emergency meeting of the WPTB to approve a memorandum composed to finance minister Ilsley on the eviction situation. The meeting approved the memorandum, which urged an end to evictions of well-behaved tenants. Gordon warned Ilsley that "soldiers' organizations and others are now busying themselves to protect soldiers' families against evictions." If the "present indicated volume" of evictions were "permitted to mature ... serious trouble" could be expected.[38]

The day after Gordon's memorandum was approved, the WPTB issued order 537. This suspended all notices to vacate given by landlords of self-contained dwellings. All eviction proceedings were stayed and no further notices to vacate permitted. Landlords who had already given such notices to vacate could apply to a court of rental appeals for a hearing. The WPTB internal history of federal rent controls notes that while this freeze "was protested vigorously by the property owners' associations and many real estate firms, it was a relief to thousands of individuals." The government "received almost equal numbers of letters and wires of commendation and protest." A Montreal firm of barristers and solicitors considered challenging order 537's constitutionality. This challenge was dropped after the firm was told that the federal justice minister considered the regulation to be "a valid exercise of the [WPTB's] powers."[39]

The decision to freeze evictions was coupled with another to increase the production of wartime rental housing units. In its 23 July 1945 memorandum to Ilsley, the WPTB argued that "low cost, low-rental housing is urgently needed and is the only solution to the problem." The board noted the "serious delays in getting such houses." These were caused by material shortages and "protracted negotiations with lending institutions."[40]

Indeed, on the same day evictions were frozen, 24 July 1945, C.D. Howe decided that wartime housing production targets would be increased from 3,000 to 10,000 units. Howe sent a telegram to B.K. Boulton, who had replaced Pigott as president of Wartime Housing, directing that this building be done "immediately." Howe asked for a copy of Wartime Housing's proposed allocation to municipalities based on need. Pigott had compiled this estimate several months earlier. Howe believed that "speed must be the primary consideration as the greatest possible number of houses must be built this year" and urged that Vancouver receive "at least a thousand additional houses" as "the most urgent of several situations called to my attention." On the same day Howe also wired Boulton to contact former Wartime Housing general manager Victor Goggin, who had been instrumental in co-ordinating WHL's heavy-construction program during 1941 and 1942. Goggin's sudden rehabilitation as an important person to be consulted points to a major change in direction towards a positive program of veterans' rental housing.[41]

On 25 July Goggin replied to Howe with an outline of a program for 7,000 additional veterans' rental houses. Projects were indicated for every province but Prince Edward Island. A project for 100 houses for Kamloops that had been rejected was now approved. Saskatoon city council representatives phoned and received approval of another 200 homes. Wartime Housing pressed reluctant municipalities such as Hamilton, which had expressed a preference for an NHA home-ownership project, to accept housing projects in accordance with WHL's calculations of need. At its meeting on 8 August the Interdepartmental Housing Committee approved the construction of 1,000 wartime housing units in Vancouver. Donald Gordon informed the IHC that the government was now committed to "taking energetic action to get low-rental housing built."[42]

After Howe's decision to embark on 10,000 units as an annual target of Wartime Housing, IHC's supervision of its operations quickly ended. Municipal co-operation replaced the Department of Finance as the main break on Wartime Housing's building program. Eric Gold explained the new situation to Mitchell Sharp in response to a request to approve seven wartime housing projects on 4 September 1945. Gold told Sharp that there was not a city in Canada that did not need "the number of houses proposed by these projects." Also, by the time the proposals reached the IHC, municipalities had already indicated their need for these homes "by making a substantial financial commitment in order to secure them." Gold was only concerned that the requests received from municipalities would not "total the number of houses that Wartime Housing is prepared to build." The reasons for this gap differed. Toronto had difficulty finding the land for the four hundred houses it was offered. Hamilton refused to sign an agreement for more than one hundred homes.[43]

With social-housing projects being passed through by Howe's reconstruction ministry with surprising speed, the long-time antagonists of this approach in the Department of Finance were also devising their own plans for future housing. Shortly after the decision to freeze evictions, the department took the first steps towards the creation of the Central Mortgage and Housing Corporation. At the beginning of 1945 D.B. Mansur had felt the creation of a central mortgage bank was premature, but he changed his mind during the eviction crisis. At a meeting held in Ilsley's office on 24 July 1944 it was agreed by the minister of finance, acting deputy minister of finance W.A. Mackintosh, Gordon, Anderson, Sharp, and Mansur that such a bank be revived. Mansur was instructed to prepare "a very rough draft bill to incorporate a corporation *to act for the Minister in housing matters*" (emphasis Mansur's). The main purpose of the new corporation would be "the development of private investment in real property and marshalling our rather meagre facilities to that end." Indeed, the CMHC was to measure its success "by the amount of activity not undertaken" by government agencies "in the public housing field." The "primary duty" of the new Crown corporation would be "one of finding ways and means for private enterprise to look after needs in the economic field."[44]

Much of Mansur's draft bill was taken from the Central Mortgage Bank legislation of 1939. Like this institution, the new CMHC would have the power to discount mortgages. Its capital assets were fixed at $25 million, which would be used for government's share of the NHA joint-loan scheme. Mansur also proposed that the buying and selling of joint mortgages, direct lending in areas poorly serviced by financial institutions, and assistance in the establishment of housing corporations should be assigned to CMHC. As a key objective the new Crown corporation would seek to stimulate "private enterprise to serve as much of the need as it possibly can and to serve an ever-increasing portion of the need as time goes on, thus reducing the need to be covered by direct public lending and private participation in housing schemes."[45]

Although investing CMHC with an explicit function to promote private enterprise, Mansur did not do so out of a dogmatic assumption that social housing was unnecessary. Indeed, Mansur felt such efforts should be concentrated in the reconstruction ministry. He believed that schemes such as limited-dividend housing that attempted to "introduce private enterprise into uneconomic activities" would lead to "endless confusion." To clarify the distinction Mansur urged that CMHC be prohibited from making limited-dividend loans, which should remain a ministerial matter. He was pessimistic about the potential of Anderson's efforts. Mansur's proposals were accepted by the federal Cabinet on 2 October 1945 and included in the subsequent Central Mortgage and Housing Act legislation. The new corporation would carry on the work of the National Housing Administration of the Department of Finance, except for its functions in regard to limited-dividend housing.[46]

The protests against evictions also pushed the WPTB to take action against landlords who held housing speculatively, declining to put their accommodation on the market. Emergency-shelter regulations were extended to all parts of the country. It was recognized that "in effect all Canada became a congested area." Each regional Emergency Shelter administrator was instructed to "actively operate to see that all vacant houses in his Region are put to use." Administrators were told to make arrangements "with the light companies, water companies, telephone companies etc," which would ensure a "continuing source of information regarding houses that may become vacant." If landlords would not voluntarily put their properties on the market for sale or rent, they would "be given an order to ... sell the building to a purchaser who will immediately occupy it or to rent it within a certain limited time." Failing either of these two courses, the property would "be placed in the hands of a trust company of rental." Shelter administrators would also maintain "close contact with the Canadian Legion, other servicemen's organizations and labour unions as a means of knowing what families are in distress and as a means for securing local suggestions as to property that might be utilized as housing accommodation." As a result of the government's acknowledgment of the national character of the housing shortage, restrictions on movement of persons into communities seeking family accommodation ceased.[47]

The instructions issued by the Emergency Shelter administrators reveal the painfully acute nature of the immediate post-war housing shortage. The administrators were told to concentrate exclusively on cases of homelessness. Unless a family was homeless or was soon to be "without shelter of any kind," the administrators were to "frankly and honestly express your inability to be of assistance." There might be "the odd case" where a family was housed in "pathetic circumstances." Here the administrator might feel "compelled to do something for it." These "extenuating circumstances" were to be of "a most unusual nature." Once the "restricted class of homeless" was departed from, Emergency Shelter co-ordinator Eric Gold was convinced, the government would be "confronted with countless families suffering inconvenience, inadequacy, discomfort, etc., whom we are totally unable to help and whom we must recognize can only be assisted by the construction of more houses."[48]

The severity of the shortage was also evident from the variety of measures employed by Emergency Shelter officials when faced with the problem of "getting a family off the street" while shelter administrators were looking for accommodation. These included having children placed temporarily in a children's home and housing the family in community and fire halls or police stations and armed forces barracks. Municipalities could make requests to Brigadier Colin Campbell, director general of the Department of Reconstruction's real estate division, to have federal buildings converted into emergency shelters. Temporary buildings were sold to municipalities for this purpose at a scrap value, estimated at 8 to 12 per cent of the original construction cost. Permanent buildings surplus to federal needs were rented at the cost of one dollar a year.[49]

Although CMHC was originally conceived as an adjunct of the Department of Finance's activities in the housing field, continuing housing pressures and debates within government encouraged a broader role than originally anticipated. The co-ordinating body, the Interdepartmental Housing Committee, concluded that all federal housing responsibility should be centralized in one department. On 22 November 1945 David Mansur, Wartime Housing and Trade Board chairman Donald Gordon, Mitchell Sharp of the Department of Finance, and J.G. Godsoe of Construction Control all appealed for housing to be centralized in one agency.[50]

The joint memorandum they issued suggested assigning housing responsibility to an existing federal ministry such as reconstruction or finance. However, Privy Council clerk A.J.P. Heeney took this opportunity to urge Prime Minister King to create a separate federal ministry of housing. Heeney drew King's attention to the "increasing volume of criticism within and without Parliament" on housing issues. He believed the provision of "low-rental housing" had become "one of the most critical problems" facing the government.[51] Heeney's proposal for a separate ministry was rejected, with important long-term consequences. Although CMHC was placed in the Ministry of Reconstruction under C.D. Howe, the Department of Finance would retain control, as it had the majority of seats on CMHC's executive board.[52]

Although CMHC was given the additional duties of responsibility for federal activity in social housing, it remained in the original mould of a Crown corporation whose object was to aid private business in the housing field. The agency would have a peculiar dual role, acting as a ministry in the social-housing field but, until 1964, in a basically token capacity, while also serving as an ancillary service to private mortgage companies, along the model of a Crown corporation. Humphrey Carver has recalled how Mansur's vision of CMHC as a private insurance corporation was reflected in the design of its Ottawa headquarters. Also, to manifest its independence from political influence and Cabinet ministers, CMHC's offices were located two miles from Parliament Hill.[53]

In the months between the passage of the National Housing Act in the summer of 1944 and the Central Mortgage and Housing Act in the fall of 1945, dramatic changes had taken place in Canadian housing policy. The transition to peace was more turbulent than W.C. Clark envisaged when drafting the NHA of 1944 on the assumption that the federal government could avoid social housing as it had done rather masterfully under his leadership in the Great Depression. In little over a year, leases had been frozen, Wartime

Housing's construction programs had surged forward, and tough measures had been taken against withholding shelter from the market.

Although the necessities of politics had forced an expansion of federal activity towards the socialization of the housing market, this shift was the result of short-term measures, well symbolized by Wartime Housing's rental units, designed to be easily sold in the future. By centralizing housing authority within CMHC, the Department of Finance effectively consolidated its influence over federal policy. CMHC would be staffed by the personnel of both the Department of Finance's National Housing Administration and Reconstruction's Wartime Housing Limited, but through their control of CMHC's executive board it would be the former who would be in control, responsible for advising the federal Cabinet on housing policy.[3] Social housing would not be an important priority, but merely an incidental frill to deck out a business enterprise in a garb more appropriate to a public agency, achieving the desired appearance of change in the face of a rigid commitment to the market ethos.[4]

7 The Return to Privatism, 1949–1954

The six years from 1949 to 1954 saw federal housing policy return to private-market stimulation, the preferred approach of the Department of Finance. The vigorous program of direct federal rental-housing construction undertaken by Wartime Housing Limited was replaced by the tokenist approach of finally passing legislation for public housing without encouraging any to be built.

The drying up of mortgage money that had flowed under the 1935 joint-loan system caused CMHC president David Mansur to devise a system of guaranteed loans to revive the market with a fresh infusion of funds from chartered banks. This change was embodied in the National Housing Act of 1954, which still forms the basis of current legislation. Its passage marked the end of the series of housing acts of 1935, 1938, and 1944, and clearly signified that public policy would be committed to assisting private finance.

One of the strangest ironies in the evolution of Canadian housing policy was that the final acceptance of the principle of subsidized low-rental housing, long fought for by Canadian social reformers, should be the occasion for the withdrawal of a strong federal effort in the public rental field. This led social-housing advocate and progressive civil servant Humphrey Carver to conclude that the move was a "shabby trick" by a Liberal government that "would do almost anything to avoid getting into a policy of public housing."[1]

Born of the economic need to house skilled munition-plant workers and sustained by the political need to provide adequate shelter for returning war veterans, Wartime Housing had, from 1941 to

1949, constructed 40,100 family rental dwelling-units. By 1949 the Crown corporation was fast becoming a victim of its own success. More attractive designs were now available to the returning veterans, and improved arrangements extended to municipalities to increase their support of the program. As Carver, also a landscape architect, recalled, the projects were now "excellently designed under the direction of Sam Gitterman, the architect of CMHC, and represented the accumulated know-how of several years in site-planning and house-grouping." They were "vastly superior to what was currently being produced through the private enterprise system, forced into production by the incentives of the integrated plan and by rental insurance." This success encouraged the National Welfare Council, a body of social-work professionals, to urge that wartime housing projects be given over to the management of municipal governments. They would then form the nucleus of the housing stock of new municipal housing authorities. Future building would be encouraged along the lines suggested by the 1944 Curtis Report: low-interest federal loans for financing and annual subsidy payments to ensure affordability for low-income tenants.[2]

The National Welfare Council's proposals for wartime housing were exactly what, as Carver recalled, so "horrified" the federal government. They amounted to the same recommendations that had been rejected in the passage of the National Housing Act of 1944. The need to continue the wartime housing experiment did not reduce the desire of senior CMHC and Department of Finance civil servants to wind it down. This decision was made by C.D. Howe on the advice of the Finance-dominated CMHC executive board in October 1948. It was realized at the time that this could not be done abruptly. The waiting lists of veterans wishing to get into wartime projects would have to diminish. Public opinion must also be prepared for the transition.[3]

By the end of 1948 Howe and Mansur had taken steps that they believed would discourage the construction of new wartime housing by neglecting to promote it. CMHC's negotiators were told to "only talk to a municipality after a letter has been received from them." "No letters" were to be sent to mayors to inform them "that the 1948 program has been extended to 1949."[4] Mansur estimated that this would reduce the building of new wartime units by about half, from 6,000 to 3,000 units. However, despite the end of promotional activities, the demand of municipalities for new units was still so strong that production actually rose to 7,804.[5]

The continuing demand for wartime housing by municipalities increased the federal government's suspicion of these bodies as highly susceptible to political demands for more social housing. Consequently, Mansur would frame the National Housing Act amendments of 1949 with the goal of increasing provincial political responsibilities and municipal costs for public rental housing. This accomplished its desired result; a transfer of political pressure for housing away from the federal government, and a rapid decrease in the level of social-housing production. Only 140 units would be built in one Newfoundland project in 1950. By 1954 only 1,832 low-rental units were constructed nationally. As late as 1964, when the 1949 legislation was finally amended to encourage more public-housing construction, only 11,000 units – about a quarter of Wartime Housing's production – had been built.[6]

The low levels of social housing built under the 1949 NHA amendments were not the product of incompetence but of the intentions of the civil servants and ministers who drafted the legislation. Mansur set forth the basic premise of the legislation in a memorandum of 3 June 1949. He repeated his opposition to federal financing of municipal projects, the approach urged by social workers, trade unions, and the CCF. In its place he outlined the basis of a federal-provincial partnership in which 75 per cent of costs would be assumed by the federal government and 25 per cent by the provinces. Mansur explained to the CMHC board of directors that "the 25% participation by the provinces will tend to keep down the number of public housing projects." This would be accomplished in a number of ways. Mansur knew that at least some provinces would pass a portion of their share to municipalities, which would favourably prove "to be a very real limiting factor" in the amount of public housing built. Burdened with costs, municipalities would also be discouraged by being shorn of responsibility for the management of public housing. Mansur wanted to avoid the creation of municipal housing authorities that would zealously support further socialization of housing. Consequently, the local authorities responsible for the management of housing would all be appointed directly by the province. Ownership would reside with the federal and provincial governments. Mansur emphasized the need "to avoid the establishment of long term vested interests and rights in local authorities." The CMHC directors approved his outline of new legislation at their 25 June 1949 meeting.[7]

During the summer of 1949, after the approval of his proposals by reconstruction minister Robert Winters, Mansur went about securing provincial support for his new public-housing policy. He correctly assumed that Saskatchewan's CCF government would take the view of social-housing advocates and make "a very strong stand that all the money and all the rent reduction fund should come from the Dominion." Indeed, T.C. Douglas, in a 2 March 1949 letter to Prime Minister Louis St Laurent, argued that the low-interest loans available to private limited-dividend projects should be extended to the Saskatchewan Reconstruction Corporation (a provincial Crown corporation) and to municipalities. Mansur did not bother to communicate this proposal to the CMHC board. He also anticipated opposition from the Union Nationale government of Quebec, which preferred to subsidize home ownership. Mansur believed that Nova Scotia, Prince Edward Island, and Alberta were indifferent to low-income housing. The provinces wanting to negotiate would be British Columbia and Ontario.[8]

Mansur chose to obtain a public-housing agreement first with Ontario. Premier Leslie Frost had objectives in this area that were identical to Mansur's. Frost had not extended any assistance to Toronto's pioneering Regent Park project, which had received assistance in land-acquisition funds from CMHC. He was suspicious of its board of directors and of municipally appointed housing authorities in general, believing they would discourage the "ultimate disposal" of their "units to home ownership." Mansur explained to Winters that he would "take advantage of the Province's reluctance towards subsidized housing by the adoption of a plan under which subsidized housing may be provided, but only with the concurrence of the Province." Frost liked Mansur's suggestion that the new legislation would mean the "public housers" would no longer be "able to play off the Dominion against the Province and vice versa," and welcomed the proposed legislation as a politically "effective answer" to "demands for public housing with subsidized rentals."[9]

Mansur also found support in Quebec for his proposals to stall public housing through the federal-provincial agreement. A CMHC memo of the time favourably noted that Premier Maurice Duplessis was a "firm believer in home ownership" and an opponent of "subsidized housing." He believed that "Government-built projects would have the effect of increasing the undesirable trend of individuals relying more and more on the state." Despite his championing of provincial rights on other issues, Duplessis agreed with Mansur that "the Dominion and the Province should work very closely together" in the housing field. Mansur admired Duplessis's views on housing policy and felt he had "more real knowledge of the situation and its needs than any other provincial Premier I have met." Duplessis supported the new NHA section 35, which provided for public housing, since the new NHA amendments also made provision for joint federal-provincial ventures in "providing and servicing new land" for home-ownership projects. Mansur felt that Quebec's own program of subsidizing interest rates on home mortgages complemented NHA joint loans.[10]

The first public-housing project built under the joint federal-provincial program was the Estbary Estates project in St John's, Newfoundland. It was planned according to very minimal housing standards. The pioneer project was not serviced with heat or hot water: tenants were expected to purchase space heaters. The estate was built over three years, and its design was subject to complicated negotiations between the Newfoundland and Canadian governments. After the original design was rejected by the provincial Cabinet, CMHC mollified Cabinet members with the explanation that "any further lowering of standards would not comply with National Housing Act regulations."[11]

CMHC took the attitude that the passage of section 35 had achieved the intended purpose of reducing the political vulnerability of the federal government with regard to low-income housing after the ending of Wartime Housing's extensive programs of direct public rental construction. Little effort was made to promote use of the new legislation, especially in contrast to CMHC's cultivation of private entrepreneurs in the housing industry. No targets were set to meet social needs for improved affordable shelter. On 20 February 1950 CMHC's board observed that it was "not anticipated that there would be a great deal of housing started under Dominion-Provincial plans during 1950." This was not considered a cause for alarm nor for consideration of more promotional efforts. CMHC's board was content that the political objectives of the legislation had been met. The board favourably noted that it was "already" evident that section 35 had accomplished the task of "removing pressures for public housing projects from the Dominion to Provincial governments."[12]

The only other project mentioned in CMHC's first annual report for the fiscal year 1950 was located in Vancouver. This Little Mountain project was not constructed until 1954. Its delays exemplified the tendency of section 35 rental projects to become bogged down in red tape and controversy. The eventual completion of the project was a tribute to the Vancouver Housing Association, which had been campaigning continually since 1937 for subsidized low-rental accommodation. The project was approved in 1950 by the city council over the usual anti-public-housing lobbies of the property industry. These included the British Columbia Mortgage and Trust Association, property-owner groups, and home-builder lobbies. In order to placate such interests, the city council attempted to turn the proposed two-hundred-unit project into an exclusively senior citizens' project. This plan was overturned after a heated confrontation between the council and Vancouver Housing Association spokesman Leonard Marsh, who was now a professor at the University of British Columbia. However, difficulties around tendering costs caused the whole project to be postponed for two years. In 1954 it was reapproved by the Vancouver city council. Public opinion was again polarized. The Vancouver Real Estate Board, the city board of trade, and property owners' associations opposed the project. The Vancouver Housing Association organized a counter-lobby composed of the city Trades and Labour Council, the West End Tenants' Association, the Community Planning Association, the Canadian Legion, the British Columbia Architectural Institute, the Greater Vancouver Health League, the IODE, and numerous other church, labour, and social-service organizations.[13]

When it finally opened in 1954, the long waiting-list of the Little Mountain project attested to the social need for improved affordable housing that existed in the city. Only a third of the families considered eligible were admitted to the 224-unit project. Some 55 per cent of the applicants lived in overcrowded conditions. Others resided in shelters with marked physical defects. These included basement dwellings and homes that lacked water and sinks.[14]

Social-housing problems in Vancouver went beyond the calculations of the waiting-list for the Little Mountain project. Many persons faced with severe housing problems were not eligible for admission. These included families on social assistance, except for a limited quota of pensioners. A contemporary study of Vancouver relief recipients found that they commonly faced severe shelter problems. On average, 37 per cent of their income went to pay shelter costs. Most lived in dwellings with severe physical defects. The Vancouver Housing Association also found marked housing problems among single women who were ineligible for public housing. The survey found that accommodation in lodging-houses without heat or hot water was common. Most rooms had only a hotplate for food preparation. Forty city rooming-houses either lacked basic facilities or were overcrowded.[15]

The narrow definitions of eligibility for public housing in Vancouver did not serve to diminish the waiting-list for the Little Mountain project, which was 402 names long three years after its opening.[16] Further, by 1957 the Vancouver Housing Authority had still been unable to secure funding from the province or CMHC for the same kind of community centre that had proven so successful with wartime housing. The numerous pleas for such a facility made in the city housing authority's annual report brought no action. As late as 1971 press reports commented on tenants' dissatisfaction with the lack of such services.[17]

Public-housing reformers in the period of continuing housing shortages during the first decade after the Second World War were placed in a difficult position. Although the provincial governments' role was not what they had desired, the 1949 NHA amendments achieved essentially what they had advocated in opposition to the secret positions of W.C. Clark. However, they had envisaged a program on the scale recommended by the Curtis Report, which would have had social housing account for the majority of new housing starts.

The social-housing advocates of the fifties strove to make public housing available to the broad range of working-class families, who, their studies indicated, needed it. This led to some unfortunate consequences. A prohibition of relief recipients in public-housing projects was common in the era. The first Bayers Road project in Halifax in 1953 did not admit families with incomes below $1,500 per year. When family incomes rose above $3,200, rent increases were applied to encourage rapid relocation, and some thirty-eight tenants were affected by such increases in the project's first year. Such strict criteria were applied in response to the great demand for public low-rental accommodation: one thousand families had applied for the 161 dwelling-units of the project. This problem was aggravated by the tendency of some provincial governments to encourage public-housing projects that did not provide subsidized rent geared to low-income units. The preoccupation with housing aimed at middle-income groups was a reflection of the continued housing shortage and lagging rental construction. Such unsubsidized units accounted for half of the section 35 rental projects in the first decade of the legislation. Even the subsidized projects of the era had minimum income ceilings. In Ontario, tenants in the projects of the federal-provincial partnership had to have a family income of at least $1,800 a year during the 1950s.[18]

As the post-war housing shortage receded, the spartan accommodation of emergency shelters built from 1944 to 1949 took on a new function. Increasingly, they became refuges for the poor whose incomes were too low for them to be considered eligible for public housing. So devoid of social purpose were the policies of CMHC that the replacement of such shelters with permanent buildings was never given consideration. Emergency-shelter residents were simply ignored. The attitudes of the era recall nineteenth-century distinctions between the "deserving" and the "undeserving" poor.

Although Toronto had the most progressive social-housing policies of any Canadian city, in 1949 it turned the management of its emergency shelters over to a private trust company. It took an attitude similar to that of many private landlords of the era. The city also used emergency shelter as a dumping-ground for tenants of public housing who proved to be unsuitable. The shelters were frequently overcrowded. Buildings would house ten to twelve families with fifty to sixty children. The low rentals of thirty-six dollars a month insured their continued full occupancy, as such accommodation, which included bathrooms, was better than what could be purchased at equivalent prices on the private market. The sensation created by a 1955 United Church *Observer* article, "City Owned Slum," led to the transfer of Toronto's shelters from a trust company to the city's housing authority, responsible for the management of public housing. In Halifax the closure of emergency shelters led to tenants' being evicted by court orders. Families were divided up among the city home, Children's Aid, and the St Joseph Orphanage. The neglect of the plight of long-time emergency-shelter residents exemplifies the truth of Humphrey Carver's quip that the only "party in the housing scene which didn't seem to get much attention was the Canadian family which couldn't afford home-ownership."[19]

The pursuit of home ownership was the only visible social-housing goal of the federal government. It assumed, in Carver's words, the character of a quest for the Holy Grail. Policies in other areas, such as the decontrol of rents, were tied to an overall objective of encouraging home ownership. In its first annual report CMHC justified the Wartime Prices and Trade Board's changes in rental regulations, which had "the effect of permitting an increase in ceiling rents by 18 per cent and 22 per cent," on the grounds of the "increased interest in home ownership" it stimulated.[20]

The only other social-housing program, apart from the minuscule public-housing operation of CMHC in the early 1950s, was limited-dividend housing. This had been the sole, but unused, social-housing program of the NHA of 1944. The amendments of 1948 finally created a workable although small-scale limited-dividend program. The new amendment permitted companies to retain the full value of their properties after paying back their federal loan, which was given at a subsidized 3 per cent interest rate by the federal government. Mansur encountered some difficulty in getting this amendment through, despite what he termed the "completely ineffective" terms of the 1944 NHA. This opposition came from Mitchell Sharp and W.A. Mackintosh of the Department of Finance. They wished to spare C.D. Howe a "prolonged and difficult debate in the House" on what they felt was "a relatively minor matter."[21]

Apart from limited-dividend housing, the only federal effort to encourage private rental housing was the Rental Insurance scheme, another product of Mansur's 1948 NHA amendments. Private rental housing in the decade after the Second World War proved very difficult to establish, even after the decontrol of rents on all new buildings and the awarding of double depreciation for rental construction. Potential investors held back since they believed there would soon be a substantial drop in the cost of residential-construction supplies. In the event, this would not be achieved until the end of the Korean War because of the sustained building boom. Mansur believed that even the "most favourable financial terms" could not "suddenly create rental housing entrepreneurs in English Canada." Apart from luxury housing, the main sources of private rental housing in English Canada had vanished since the Depression. Much of Canada's rental housing in the post-war period were homes of "unwilling" landlords who had repossessed homes in the Depression and were unable to sell them because of rent-control laws.[22]

Mansur rejected the "spectacular approach" to the private rental-housing problem – providing long-term loans at deeply subsidized interest rates – because this would go against the financial market and have the effect of "disturbing the credit structure." Instead, the Rental Insurance scheme was developed, which guaranteed the owners of rental housing sufficient income to pay taxes, debt service charges, operating expenses, and repair and replacement costs. In addition it would provide a "reasonable return on equity investment." Mansur correctly predicted that it would produce only a modest amount of housing, about three thousand units annually.[23]

As well as guaranteeing profits, CMHC provided mortgage money for rental insurance projects if an entrepreneur could not obtain it from a private lending institution. This amounted to providing loans for the construction of unserviced dwellings, "cold-water flats" lacking central heating and hot water. That such poor units would have zero vacancy rates until 1955 is illustrative of the severe housing shortage in the decade after the Second World War.

Not only was much of the Rental Insurance housing unheated, it was frequently of poor design. CMHC confessed to a "general criticism" in Montreal, where most units under the program were built, that those units were "below par" in design. A confidential CMHC report complained of a project's resemblance to a "penitentiary row."[24]

In spite of their frequently low standards of design and servicing, Rental Insurance projects could only be afforded by a middle-class market. In the late 1940s the average level of rent that could be paid by wage and salary earners in nine leading Canadian industries was $43.72. The level of Rental Insurance projects ranged from $74.43 to $83.88. At its 18 April 1949 meeting, CMHC's board of directors concluded that the rents charged under the scheme "appeared to be high for the majority of the tenants." One CMHC study found that, since tenants were being stretched "up to the limit of their resources" to pay rents, it was not necessary to install garages in Rental Insurance projects. The tenants who currently owned cars would soon have to sell them to be able to pay heavy rents.[25]

CMHC rejected the proposals of the Curtis Report and the Canadian Co-operative Union (CCU) to encourage continuing rental co-operatives by lending at the same 3 per cent interest rate extended to limited-dividend corporations. Mansur argued that such assistance would make co-operatives more attractive than home ownership.[26] Newly appointed CMHC research director Humphrey Carver unsuccessfully attempted to persuade the Crown corporation to assist the co-operative housing efforts of the Nova Scotia Housing Commission and the Quebec Housing Co-operative Federation. His modest proposals were for financial assistance to allow these organizations to hire more paid staff.[27]

While deciding to wind down Wartime Housing and protecting the national credit structure at the expense of private rental and co-operative housing, CMHC concentrated on stimulating the mortgage market for home ownership. Such concerns also had an impact on the corporation's policy of rental decontrol. CMHC justified the WPTB's increase of rent ceilings by between 18 and 22 per cent on the grounds of "the increased interest in home ownership" it stimulated.[28]

The federal government wished to return to the private rent market but hoped to escape the political odium of sanctioning massive rent increases. Adopting the clever strategy of David Mansur, it announced that provinces could rescind increases in their jurisdiction for taking over the politically unappealing responsibility for rent controls. Mansur's tactics succeeded because the CCF government of Saskatchewan felt compelled by the sympathy for tenants in its base of party activists to take action to prevent rent increases. Other provinces were forced to follow suit. Provinces could no longer hide behind the convenient constitutional argument that the housing shortage, having been caused by the war, was a national responsibility.

The first steps towards rental decontrol came in December 1946. WPTB chairman Donald Gordon, after "consulting representatives of commercial properties" and senior CMHC personnel, made a number of recommendations for decontrol to the minister of finance. These were to allow a 25 per cent increase in maximum rentals for all commercial space, a general increase in housing rentals of 15 per cent, and an increase of 10 per cent where heat was not provided. For the first time rooming-houses were defined and allowed a 10 per cent increase. Rent controls were removed from seasonal hotels and boarding-houses. Provision was also made to adjust rentals upwards in "anomalous cases."[29]

Gordon's recommendations for changes in rent control were accepted by the Cabinet in December 1946. In recognition of the politically sensitive nature of the issue, it was agreed that "no publicity" would be given to the changes until the minister of finance made a statement in the House of Commons. Before this happened, the Cabinet had some second thoughts on the matter and removed the additional 5 per cent increase for heated accommodation.[30]

The Cabinet was also careful to introduce its package in stages. In February 1947, WPTB order 294 established the mechanism for handling "anomalous cases" of landlords requesting higher rents on properties purchased between 31 October and 25 July 1945. At the same time, WPTB order 294 exempted seasonal boarding-houses serving tourists and vacationers from controls. The 10 per cent general increase in rents was applied on 7 April 1947. To make it more palatable to tenants, landlords had to offer a two-year extension of a lease to obtain this increase.[31]

The alterations made to rent controls over the first six months of 1947 did not fundamentally alter the "total freeze" on the eviction of well-behaved tenants that had been imposed by the WPTB on 24 July 1945. On 9 May 1947 rental administrator Owen Lobley complained that although this had been "announced as a temporary measure," it had now "subsisted for nearly two years." He saw its continuation as a disincentive for "people to own property." Tenants, he believed, had begun to regard their security of tenure "as a constitutional right." Lobley urged the "orderly thawing-out of frozen leases." The first move would be to permit a landlord "to get rid of his tenant if alternative accommodation is available." The second would accommodate the desire of a landlord to occupy his own house in situations "of grievous hardship or moral deterioration."

He recognized that "by their very nature these rules imply discretionary judgement." Consequently, he urged the new WPTB chairman, Kenneth Taylor, to provide that cases be determined by "a persona designata who would exercise discretionary judgement within the orbit of the rules." "These men," he believed, would be "of standing in their communities ... anxious and willing to be patient mediators."[32]

Lobley's recommendations were accepted by Taylor and the government. Forty rental commissioners were appointed to hear these extraordinary cases. Also, in cases of "acute and intolerable hardship" landlords could evict tenants without demonstrating that alternate accommodation could be found. At the same time further decontrols were made in commercial rents. In a move that was to start a pattern of coexisting uncontrolled and controlled dwellings, controls were removed from all new construction completed after 1 January 1947. Taylor also called for measures to begin the transfer of rental controls to the provinces. Again, these changes were carefully phased. WPTB order 742, on 18 June 1947, decontrolled newly built dwellings. This was done to defuse arguments "that rental regulations were discouraging the production of new rental dwellings."[33]

In order to enable the federal government, in David Mansur's words, to "get out of a very uncomfortable field," generous terms were offered to the provinces in exchange for their assumption of rent controls. The federal government offered to pay the costs of provincial administration for the first three years. In addition, it would hand over its rental-control administration and "provide such secretariat as necessary." Mansur pointed out that while the Department of Finance was disturbed by the complaints of landlords, CMHC had "continually met with the criticism that the Rental Regulations are abused and are not protective enough to tenants." Mansur correctly predicted that in such circumstances of political conflict it would "take just about five minutes" for the provincial premiers to reject the federal offer, despite what difficulties they might have in reconciling this "with their views on provincial rights."[34]

On 17 September 1947 Taylor, Sharp, Lobley, Mansur, F.S. Grisdale, and W.L.A. Pope met to discuss means to achieve rental decontrol. The immediate removal of controls, the six agreed, would "result in serious social disturbance and possibly disorder." The supply of rental dwellings was "as tight or tighter than ever before." The lifting of controls without a transitional period would cause the courts to "be swamped with eviction suits; sheriffs would be physically unable to cope with the volume of evictions and veterans and others would picket the dwellings of evictees." At the same time, the six senior officials wanted the provinces to assume responsibility for this difficult transitional process. They maintained that the "time has come to place the responsibility boldly and squarely where it constitutionally belongs, namely with the provinces." It was proposed that the federal government announce withdrawal as of 31 March 1949. The provinces were to be reminded "in no uncertain terms of their powers and responsibilities to deal with and legislate for the situation when the Dominion vacates the field," and to be offered federal payment of the costs of rental administration for at least one year.[35]

Even as they called for the provinces to assume responsibility for rent-control measures, however, the federal civil servants took steps that would make controls difficult for the provinces to administer equitably unless major changes were made. This decreased the popularity of federal regulation and made provincial assumption of controls more politically feasible.

On 1 November 1948 federal rental regulations underwent some major revisions. All rental accommodation would now be decontrolled as soon as it became vacant. This created a situation where identical accommodation would be subject to different prices set by the time the last tenant vacated the premises. Also, landlords were now permitted to raise rents by 10 per cent on unheated and 15 per cent on heated accommodation without having to renew a lease for two years.[36]

In a letter of 23 October 1948 the minister of finance wrote to provincial premiers advising them of the federal government's offer to give them the rent-control field and pay their costs for one year. The only province to enact its own rent-control was Newfoundland. It simply retained its old pre-Confederation rent controls.[37]

The removal of controls on vacated dwellings created a situation where, as Owen Lobley put it, "a premium" was placed "on the eviction of the present tenant so that the accommodation may be let to a new tenant free of control." The WPTB estimated that in Toronto, Hamilton, and Ottawa, "uncontrolled rents exceed the old rentals by about 100%." In Montreal such increases averaged 75 per cent; in the Maritimes, between 50 and 70 per cent; and in Saskatchewan, around 50 per cent.[38]

The high increases of decontrolled rents showed federal officials that chaos would still result from the removal of controls. Lobley estimated that their removal would cause rents to increase by "at least one half" of the level in the decontrolled sector. This meant that rents would go up immediately about 50 per cent in major Ontario cities, 37 per cent in Montreal, 30 per cent in the Maritimes, and 25 per cent in Saskatchewan. In such circumstances, Lobley predicted:

> Every landlord will strive to put himself in a position to reap this golden harvest by sale or rental at uncontrolled rental. Each will terminate current leases as soon as possible, refusing to re-rent houses and holding apartments at rates which, at any rate in the beginning will approximate the present uncontrolled levels. In this wide open market citizens in the low income brackets would not be able to compete and would be pushed to the wall. Many of them would be evicted from their homes and find it impossible to obtain any shelter even remotely suitable. The state of uncertainty, dreadful anxiety and dire hardship would result in a major emergency.[39]

Mitchell Sharp, now director of the Department of Finance's Economic Policy Division, shared Lobley's concerns about the immediate abolition of the remaining rent controls. He saw a situation where "a political emergency would result" from too hasty decontrol, where "landlords would have the first opportunity since 1930 of disposing of an unwanted asset at an attractive price and in many cases tax-free capital gain. They would give their tenants a month's notice and would refuse to re-rent to anyone."[40]

Minister of finance D.C. Abbott accepted Sharp's and Lobley's advice in a memorandum to Cabinet on 1 November 1949 outlining future federal rent-control policy. Abbott supported their view that the continuation of federal controls for another year was a political necessity. He believed the immediate lifting of controls would cause the government to "encounter serious opposition in Parliament, not only from the other side of the House but also from its own supporters."[41]

Although conceding the political necessity for an extension of controls, Abbott outlined to Cabinet several measures for a "further relaxation" of them. Abbott's plan did not attempt to provide landlords with an income that was deemed to be a fair and reasonable return on their operating costs. In fact, no such calculations of costs was prepared during this stage of the decontrol exercise. Instead, the further relaxation of control would serve two related purposes. The first was to demonstrate to tenants that total decontrol was coming. This would show, as Abbott told the Cabinet, "by action that we mean to decontrol." Secondly, the relaxation of controls would "provide some stimulus to the demand for new houses which is desirable having in mind probable economic developments in 1950."[42]

In setting the level of rent increases, the tenant's ability to pay rather than the rate of return to landlords was considered by the Department of Finance to be the critical question. The studies examined not what the just price was but how much tenants and landlords had benefited from wartime and post-war prosperity. The Department of Finance noted that while both groups were better off, tenants had made greater gains. On average tenants earned twice as much as before the war, but the "average rent was up only about 20 per cent." Landlords had the advantage of having to provide fewer services and the favourable experience of zero vacancy rates.[43]

Based on what it was felt tenants could afford to pay, Abbott obtained Cabinet approval for a series of increases in rent ceilings. These would bring rents to approximately half the level the WPTB believed would be established by an uncontrolled market. Rental ceilings for tenants in shared accommodation rose by 20 per cent. Landlords of "heated self-contained dwellings" (generally apartments) were allowed to increase rents by 25 per cent. For "unheated accommodation" (generally houses) the increase was about 20 per cent. Also, steps were taken by the Cabinet to reduce tenants' security of tenure. Restrictions on evictions in the winter months were removed. Landlords could obtain possession of their properties for their own use as a residence after six months' notice, as long as they had owned their properties prior to 1 November 1949. Landlords of shared accommodation were given an unrestricted right to terminate leases after six months' notice.[44]

The Cabinet's further decontrol measures caused enough dissatisfaction among tenants that the continuing presence of federal controls began to become a political liability. Previously, provinces could escape pressures from both tenants and landlords simply by encouraging the federal government to stay in the rent-control field. Now effective pressure was put on the provinces to take concrete steps to reverse the federal increase. This was most pronounced in Manitoba, Quebec, and Saskatchewan.

Manitoba questioned the need for the rental increase. When confronted with the choice between acceptance of the increase or the administration of rent controls, however, it backed off. The Liberal premier of the province, Douglas Campbell, sent Abbott specific questions to obtain more information about the motivation behind the increase and its likely impact on the province. Faced with Abbott's ambiguous reply, Campbell decided to decline the offer of the federal government to vacate the field in favour of provincial control. Instead, the premier reiterated his position of 1948 that "the whole subject of rent control should be left for the attention of the Dominion of Canada authorities."[45]

Quebec, despite the nationalist posturing of its premier, Maurice Duplessis, reacted in a similar fashion. Duplessis complained on 16 December 1949 that the Cabinet's increase was "ill-timed and unjust for all concerned." Although Abbott immediately cabled back to inform the premier that the federal government was willing to "withdraw federal controls" in Quebec at any time the province was willing to fix, the premier declined to assume responsibility for rent control. He claimed Abbott's offer promoted confusion.[46]

Duplessis, like Campbell, had hoped through telegrams to have the federal government reverse its politically unpopular decision to push through a major rent increase. However, both dropped their opposition when faced with the responsibility of administering the program, which would have exposed them to the political protests of the property industry. Duplessis was placed in the awkward position of supporting the retention of federal controls in the interests of political expediency despite his government's nominal provincial-rights orientation.

Although Quebec swallowed its constitutional theories and accepted continued federal controls, the nominally centralist CCF government of Saskatchewan accepted provincial responsibility for rent controls in order to invalidate the federal increase in its province. Premier T.C. Douglas accepted the regulatory role reluctantly. He stipulated that "the recent order permitting heavy increases in rents, straight across the board, is so inequitable and iniquitous that we must submit to your terms to have it annulled in Saskatchewan, no matter how arbitrary we think those terms to be."[47]

Douglas's decision to assume responsibility for controls made all existing federal rental regulations in the province inoperative after 1 April 1950. The federal government agreed to pay Saskatchewan's administrative costs for one year. It would apply federal controls to the province if a court decision ruled Saskatchewan's legislation "invalid due to continued federal controls in other provinces."[48]

Despite the certainty that Saskatchewan would now break the pattern of a return to free-market rents, their success in finally manipulating a province into using its constitutional powers over rent control was a relief to federal politicians and civil servants anxious to be rid of a contentious issue. The Department of Finance told the federal Cabinet that Saskatchewan's actions would "no doubt bring great pressure on other provinces" to "accept responsibility for rent controls."[49]

Douglas was prompted to take action by his province's previous experience with controls and by general sentiment in the CCF and the Canadian labour movements. The Montreal Boot and Shoe

Workers' Union called for a nation-wide one-day general strike against the proposed federal increase. The CCF attacked the government in a four-hour House of Commons debate on the issue and was joined by representatives of the Social Credit and Conservative parties. Douglas confided to Ontario CCF leader E.B. Jolliffe that he felt "certain the CCF people would expect us to tackle the problem no matter how difficult it might be." Both Jolliffe and Douglas believed that with so many dwellings being decontrolled by federal authorities "to some extent," it was preferable "to have the federal government out of the field." Saskatchewan's Mediation Board, which had the power to lower rents, prevent evictions, and protect farmers subject to foreclosure, had some success in protecting tenants in dwellings previously decontrolled by federal regulations.[50]

The Saskatchewan government also used its assumption of rent controls as an opportunity to call on the federal government to introduce a major program of subsidized rental housing. Provincial attorney-general J.W. Corman issued a press release denouncing "the adamant refusal" of the federal government to provide such assistance while "at the same time finding money to subsidize the big steel and gold mining industries."[51]

The partial decontrol measures set in motion federally created hardships somewhat similar to those that Lobley and Sharp had predicted would ensue from total decontrol. In Toronto 1,500 families were ordered to vacate. Some 2,880 people applied for entry to the city's emergency shelters; fewer than one in five were admitted. Three or four tenants a day were evicted through a loophole in regulations that permitted termination of leases if tenants failed to inform their landlords of acceptance of a rent increase within thirty days. Using data provided by social-service agencies, Albert Rose, in a study of 1,058 Canadian households who were renting, found that 27.2 per cent of this sample of families had been evicted. In 72.5 per cent of cases evictions took place because the home had been sold. Subtenancy in shared accommodation accounted for most of the remaining evictions. Half of those evicted paid a disproportionate share of their income in rent. Rose believed his study showed "a real emergency" existed "among middle and low income tenants."[52]

The increasing unpopularity of federal controls among tenants and the example of Saskatchewan's separate controls made pressure for the provinces to take over rent control irresistible. Quebec awkwardly and unsuccessfully attempted to persuade the Supreme Court that federal rent controls were unconstitutional after the situation of wartime emergency had passed. This performance made it more difficult for the province to refuse to accept responsibility for controls. The federal government was aided in this test of its authority by separate interventions from the Canadian Congress of Labour, the Canadian Legion, and the Tenants of Canada. Landlord associations argued that even in wartime federal rent controls were outside of federal powers. The Supreme Court upheld the federal government's power to maintain rental controls in peacetime on the grounds of "peace, order and good government."

Mansur's strategy, however, had secured the federal government relief from this unwanted power. With the example of already functioning provincial regulations in Saskatchewan, the resistance of the other provinces became untenable, and the federal government was able to vacate the field of rent control on 31 March 1951. The same financial assistance and staff services given to Saskatchewan were awarded to other provinces that chose to maintain some form of rent control.[53]

The legacy of rent controls inherited by the provinces was not the best. An uneven crazy-quilt of regulations made it impossible for rent controls to suggest fairness unless the provinces had the political will to make major revisions in favour of tenants who were excluded during the years of partial decontrol from 1947 to 1950. In British Columbia this differential system proved so inequitable and efforts to consider reforms so prickly that the Vancouver Housing Association, a social-housing advocacy group, welcomed the end of rent controls in the province when they were announced in 1955.[54]

In the Maritimes, only Nova Scotia adopted rent-control regulations. These were permissive, allowing municipalities to devise such legislation as they deemed fit to regulate rents. The key city to use them was Halifax. The bitter class struggles stimulated by the issue in the city during the 1950s suggest why the federal government was so eager to award jurisdiction to the provinces. Labour unions and the Nova Scotia Association for the Advancement of Coloured People sustained the city's rent-control program over continued opposition from the city's property industry.[55]

The purpose behind the policy of raising rents to stimulate construction for home ownership would not be tested immediately because of the demands of rearmament occasioned by the Korean War and the general rise in Cold War tensions. Federal economic planners again relegated housing to a lower priority than munitions production. Market forces, as in the immediate post-war period, were permitted to give commercial construction a greater command over scarce materials. This resulted in continued housing shortages for another five years, but without the protection provided to tenants of comprehensive federal rental regulations. After reaching an all-time Canadian record of 91,754 units in 1950, the production of new housing units declined to 84,810 in 1951 and to 73,087 by 1952, the lowest level achieved since 1946.[56]

In addition to slowly restricting new housing production by allowing construction supply to be diminished, CMHC undertook a number of measures to reduce housing demand. This included the suspension of promotional activities, a freeze on CMHC lending values despite home price inflation, and the abandonment of the one-sixth additional loan scheme assumed from the Ontario government. Premier Frost agreed to the latter move only after it was explained to him that it was needed to conserve materials for the Korean War effort.[57]

Bank of Canada governor James Coyne felt that CMHC's efforts to "retard the demand for new housing" did not go far enough. Coyne advocated the end of the NHA joint-loan scheme, which he felt was contrary to an effort "to create housing economically where it is most needed." He urged that all government housing assistance be directed to building rental shelter that would "facilitate an expansion in defence industries."[58]

Coyne's proposals to eliminate joint loans were referred to a committee of Mansur, Coyne, and deputy minister of finance W.C. Clark, where they were defeated primarily by Clark, the original drafter of the joint-loan program. He successfully argued that joint loans had increased construction standards and had protected the government from "the charge of disregarding the small man."[59]

CMHC also restricted its direct loans to builders as a result of the Korean War. At its 9 April 1951 board of directors meeting CMHC decided that all direct loans "should be suspended entirely in metropolitan areas, including a five mile radius from such areas." Later this ban on direct loans was taken further to include all "speculative builders and builders of rental property."[60]

From the inception of direct lending, CMHC took extraordinary pains to avoid confrontations with private lending institutions. Its entry into the field had only come after the concern over controversial Depression-era provincial mortgage-moratoria laws had evaporated in the radiance of post-war prosperity. Negotiations were conducted out between CMHC and private lenders over which would cover various territories. To facilitate such agreements, CMHC representatives met with those of the Dominion Mortgage and Investment Association on 26 May 1950. They agreed that "private informal gatherings" of this kind would "be held at regular intervals and on a formal basis." Monthly luncheons between managers of lending institutions and CMHC staff would be held at the CMHC office. The DMIA agreed to provide "their circulated list pointing out that for a stipulated period there will be areas that they will cover and that they are not interested in any applications outside these areas."[61]

Direct lending was seen as a necessary adjunct to the heavy volume of mortgage loans that life insurance companies were making. In such circumstances, it was seen as inevitable that the less lucrative areas would be abandoned by private lending institutions. An official of Montreal Life expressed concern from "the point of view of public relations" that "there might be the odd inquiry just outside the area designated by him." In such instances CMHC would phone lending institutions before a direct loan was made. However, DMIA officials told CMHC they were frankly not interested "in taking any of the direct loans" that the government was "likely to get."[62]

CMHC worked closely with the insurance companies on a number of areas, especially the encouragement of zoning. Insurance companies told Sudbury municipal officials that they "would not operate unless zoning regulations are enforced." Such pressures in 1950 brought zoning to the northern Ontario communities of Kapuskasing, Capreol, Chapleau, Burkes Falls, Ansonville, Englehart, and Blind River.[63]

Public housing contracted during the Korean War in a manner similar to direct lending. By 1951 public housing had gone beyond the confines of St John's and the battleground of the Vancouver city council chamber. Ontario had built 1,330 unsubsidized rental units; New Brunswick began an 88-unit project in St John; and in Moose Jaw a 75-unit project was started. Robert Winters wrote to provincial premiers to encourage them to keep public-housing projects to a minimum since "defence demands in the construction field are very great and have priority over everything else." He suggested to Saskatchewan minister of reconstruction and supply John Sturdy that section 35 developments should be postponed "except in those areas where there is an immediate need for housing for defence workers." He urged Sturdy to "tell municipalities" that "plans for development under Section 35 should be held in abeyance." Sturdy was not inclined to pass this message on, especially to municipalities where he felt the housing situation was "desperate." As a compromise, by June of 1951 Winters agreed to "an allocation" of 500 public-housing units to Saskatchewan.[64]

In preparing CMHC's defence housing projects, Mansur was determined to avoid the pattern of direct federal financing, construction, and management that had been followed by Wartime Housing in the Second World War. This was in spite of Mansur's own convictions that such an approach was more efficient. It was also contrary to the wishes of the manufacturers of munitions, who wished to repeat the Wartime Housing experiment.[65]

Mansur blocked proposals for a revival of Wartime Housing because of considerations that involved "the long-term point of view." He pointed out that CMHC, for the last four years, had "with some success but much trouble" been attempting to "get out" of its "position as a landlord."[66]

Housing industrial workers Mansur considered a simple matter of extending the physical plant of a manufacturing enterprise. Housing for munitions workers, he believed, should be viewed in the same manner, as "warehouses, garages and other buildings ancillary to the operations of a defence manufacturer." A CMHC memo of the period indicated that industries should no more be provided with homes for one hundred workers than given "the garaging of one hundred automobiles." Manufacturers would be helped to build rental housing for their workers in the same manner they received help to build garages. Both would be done through "capital assistance grants."[67]

Mansur wished to have the housing needs of war industries met as much as possible through home ownership. Consequently, "certified defence workers" were declared eligible for 90 per cent mortgage loans, repayable over twenty years. These loans would first be made to builders who would sell homes to certified defence workers only for two months after the homes were completed. One-ninth of the loan would be held back until the house was actually purchased by a certified defence worker. CMHC agreed to purchase the properties if buyers could not be found. Builders who gave priority to defence workers in apartments would be given loans under the Rental Insurance scheme, on the terms that existed before the Korean War.[68]

The volume of housing built under all aspects of the defence program was not great. By the end of 1952, 167 units for home ownership had been constructed under its auspices. By this time three manufacturers had taken advantage of the federal government's program allowing them capital grants to build rental housing. This program constructed 115 units. One project was established with easy terms for home ownership under a lease-option scheme.[69]

CMHC retained an immense capacity for direct construction, which had flourished so impressively in Wartime Housing Limited, but during the early stages of the Cold War it would not be used for residential construction for civilian needs. Wartime Housing's building arm became Defence Construction Limited. In addition to building for the armed forces, DCL constructed airfield hangars and northern radar installations.[70]

Although the Korean War continued, by 1952 the supply problem eased as the initial phase of Canadian rearmament was completed. In response, CMHC resumed its promotional activities, permitting direct lending by the corporation to builders. It soon became evident that a shortage of private mortgage money had replaced supplies as the key constraint in housing construction. CMHC found that to "all intents and purposes" lending institutions were "not operative" for mortgage loans in northern Ontario. In Hamilton all persons applying for loans to build their own homes were turned down. In Windsor mortgage lending had come "to a virtual standstill."[71]

In order to stimulate the housing market, largely abandoned by private lenders, CMHC resorted to an expansion of its direct-lending activities. Direct loans by CMHC were now the major source of moderate-sized communities from 5,000 to 55,000, such as Granby, Drummondville, Trail, and Prince Rupert, which had been "abandoned by the lending institutions."[72]

The problem of obtaining more private investment in housing was one of the critical situations facing W.C. Clark at the time of his death on 28 December 1952, all the more so because, as one of his obituary tributes noted, the existing joint-loan scheme was "his own invention." It had, the tribute noted, "set up a pattern for combined government and public lending which is different from that existing in any country." Clark took particular pride in the joint-loan scheme in his declining years, in contrast to his attitude at the beginning of the Second World War, when he was inclined to terminate it. An *Ottawa Journal* press account of his death recorded Prime Minister St Laurent's observation that "Dr Clark had taken more pride in his part in shaping the National Housing Act than in any of his wartime acts."[73]

The nature of Clark's all-pervasive influence was well captured by finance minister Abbott's remarks on his marriage of the tasks of "administration and public policy." Abbott recalled that Clark could be "forthright and vigorous" on "an extraordinarily wide range of policy matters." With his "capacity for work" and "tireless devotion to duty," he was indeed a formidable power in the Canadian government. The *Ottawa Journal* account actually equated "Canadianism" with "Clarkism." Unlike Marx's "groping in the nebula of theory," his precepts were said "to be as provable as the working of a V-8 engine."[74]

The tributes well reflected Clark's influence on the Canadian state, especially its housing policies. They also illustrate, however, the limitations imposed by the mechanistic nature of his world-view. He was the embodiment of Frank Underhill's quip that an economist is the garage mechanic of capitalism; any fundamental redesign of a system was beyond his vision. In housing, his repairs had reached the breaking point, for not enough money for private investment could be drawn from the well of insurance companies to sustain what the government viewed as an optimal level of home construction.

After Clark's death, what CMHC termed the task of developing "a wider, deeper basis for investment in housing" was left to David Mansur. Like a true disciple, Mansur was not bound by the exact formula of his mentor's credo but sought the same goal of maintaining housing finance as an outlet for private investment. Mansur consequently was able to break away from Clark's eighteen-year-old recipe of joint loans and developed the new federally guaranteed mortgage as the instrument for securing private investment in housing.

The exhaustion of mortgage money from the private sector in the old joint-mortgage scheme encouraged Mansur to try a new approach to lure investment into residential construction. He had long been aware of the drawbacks to the joint-loan scheme's heavy reliance on insurance companies. For Mansur as for Clark, the entry of the federal government into direct lending beyond the most isolated rural communities posed the threat of the socialization of the housing finance sector. He therefore set out on the course that would result in the drafting of the National Housing Act of 1954.

Although frequently amended, a new NHA or other national housing legislation had not been introduced because Mansur was more flexible in response to the supporters of social housing than Clark had been. Starting with the NHA amendments for limited-dividend housing in 1948 and public housing in 1949, Mansur permitted social-housing projects to be undertaken with federal support but ensured that these were on so small a scale that no dramatic impact would be felt on the private housing market. With a functioning – although essentially tokenist – social-housing sector as a complement, Mansur's new federal insurance of residential mortgages would prove more durable than the joint-loan scheme.

In addition to ensuring traditional NHA lenders, such as insurance companies, Mansur's scheme would tap the previously ignored source of housing investments, the funds of the Canadian chartered banks. Instead of waiting, as Clark had, for suggestions from the financial community, Mansur moved boldly to present his ideas to bank executives. A contemporary article in the *Canadian Banker* revealed that until approached by the federal government, Canadian banks "had neither thought of nor requested to participate in mortgage lending until they received an unexpected letter from the Minister of Finance on October 1, 1953."[75]

For the banks, residential mortgage lending was "an entirely new experience." Since Confederation, bank loans on real estate had been prohibited. Abbott's letter to the banks pointed out that federal mortgage insurance eliminated the reasons for such prohibitions. He also noted that, "in practically all other countries," savings banks were permitted to invest in mortgages.[76]

Mansur believed that the participation of chartered banks would solve the problem of limited geographical distribution that had plagued the joint-loan scheme and led to direct federal lending. He saw the four thousand branches of chartered banks as "providing a mortgage service unobtainable" by other means in reaching to "small and distant communities."[77]

Two weeks after Abbott's letter, a committee of bank representatives met with Mansur and the inspector general of banks, C.F. Elderkin, to hear a more detailed description of the scheme. Later, on 22 October 1953, the general managers of the Canadian chartered banks met with Abbott. This was followed up with a meeting between Abbott and senior representatives of the Canadian Bankers' Association in early December 1953.[78]

The meetings between government and banking representatives produced what the latter felt made for more "workable" legislation than had been originally conceived. The banks did approve of the basic thrust of the government initiative. They felt the proposed scheme combined "some of the better features of the FHA in the United States" with "improvements and alterations to fit the operation into the Canadian system." The banks keenly approved of having the cost of the mortgage insurance paid by mortgage borrowers, through premiums fixed by orders-in-council. Charging borrowers for CMHC appraisals and inspections also won the banks' approval, as did the new NHA provisions for encouraging mortgage liquidity by allowing CMHC's approved lenders to buy mortgages.[79]

In other areas, however, contentious issues arose between the banks and the federal government. The banks rejected CMHC proposals to have the interest rate on the new guaranteed mortgages move down to the current interest rate for NHA joint loans. As a compromise, the banks agreed that the initial interest rate of a mortgage loan would continue during the life of the mortgage. The banks also turned down a proposal to have the government insurance scheme pay bad debts on mortgages by "government guaranteed debentures bearing interest at the current rate on 10–20 year government bonds." Instead, it was agreed that CMHC would pay 98 per cent of the principal and 98 per cent of the interest in cash on defaulted loans. CMHC's offer of a flat $50 for the legal costs of foreclosure was raised, through the negotiations, to an allowance of up to $125.[80]

The most heated area of dispute between the federal government and the banks took place over which partner would have to bear the wrath of the public over mortgage foreclosures. Neither party disputed the desirability of removing former homeowners from the premises in preference to allowing them to rent the property, which became federally owned in the event of foreclosure. Neither wanted CMHC to get into the rental housing business. But the banks objected to the requirement that they would be responsible for evicting tenants before the federal government acquired the property. This odium of evictions, they argued, would "prejudice the good public relations" they felt they enjoyed. However, the federal government remained "adamant" in its position. CMHC pointed out that, since the government provided an "almost full guarantee" with a commercially set basis of return to the banks, "the lenders would have to assume some of the commercial unpleasantness involved." The banks were so displeased by this requirement that they unsuccessfully attempted to have the legislation amended by the House of Commons committee on banking and commerce.[81]

The National House Builders' Association and the Canadian Construction Association both made presentations on the National Housing Act of 1954 before the Commons banking and commerce committee. The philosophical divisions that had characterized these two organizations had diminished since the Depression and war years. The NHBA had lost its consuming anti-public-housing zeal. In turn, the CCA had become more typical of a business body as the liberal-corporatist flavour of its briefs to government disappeared.

In their presentations, both organizations stressed that the basis flaw in the new legislation was that the housing built under NHA could only be afforded by the top 20 per cent of income earners in the country. The NHBA applauded the act's new provisions for an extension of the maximum mortgage amortization period from 20 to 25 to 30 years and the increase in the NHA loan ceiling. It warned, however, that past experience had shown that such increases to encourage lower down payments had not necessarily been applied by lending institutions. Both groups expressed concern over the withdrawal of federal funds from the mortgage market by the transition from joint to guaranteed loans. It was feared that the entry of the banks in some years would not suffice to make up the mortgage money that would be lost by this change. Illustrating its claim that, in cities throughout Canada, the new legislation would "not cater to a very wide band of our population," the NHBA calculated that, in Hamilton, the $75-a-week carrying charges on an NHA home could only be afforded by 13 per cent of the male wage-earners and 30 per cent of the male salaried employees of the city.[82]

The basic reform that the CCA and the NHBA urged to encourage home ownership for lower-income groups was increased provision for the purchase, repair, and extension of existing housing. The CCA urged the immediate proclamation of the sections of the NHA of 1944 that provided for a continuation of the Depression-era home-improvement loan legislation and wartime loans for home extension. New legislation for "open-ended" loans to finance mortgages for older housing and the building of additional rooms was urged. This would provide more money for rehabilitation than was available under the small loans provided under the home-improvement sections of NHA.[83]

Both the NHBA and CCA stressed that financing for older homes was the best way to make the NHA useful to lower-income families and to make home ownership more accessible to the Canadian population. The NHBA pointed out that most working-class home ownership in the nation was in such older housing. These lower-income groups, however, did not have the advantage of the better financing terms of NHA mortgages. In many cases families would have to "assume an existing first mortgage and carry a heavy second mortgage which has been discounted at perhaps 25 per cent or 30 per cent and which carries an interest rate of 7 per cent or over." Further, what limited use working-class families made of NHA took place when they sold their old homes to buy new ones. This, observed R.K. Fraser, president of the NHBA, did "not help the young married couple who do not at present own an older home and have only a limited opportunity to accumulate savings so they will not have to pay exorbitant rents and have to purchase their own homes."[84]

The CCF members of the banking and commerce committee were favourably impressed by the testimony of the NHBA and CCA. Their parliamentary caucus devised an amendment to have the guaranteed loans also provided "for the purchase of an existing residential dwelling." CCF members Ross Thatcher and Claude Ellis stressed that working-class families could afford to purchase used homes at prices around $6,000 to $7,000, but not the typical new NHA homes, priced at $10,000 to $14,000. Robert Winters spoke for all the members of other parties in opposing this amendment on the grounds that "we should concentrate on the construction of new accommodation." This had been "policy" of the government "dating back to the earliest days" of government legislation.[85]

The CCF also directed their criticism of the bill to the continuing tokenism of the 1949 joint federal-provincial partnership scheme for public housing. They noted that the partnership's provisions for allowing provinces to pass on much of the costs of public housing had delayed the development of subsidized rental shelter in many parts of the country. Walter Dinsdale and Stanley Knowles charged that the terms of the partnership prevented any public housing from being built in Manitoba. There, the province required that half of its costs be borne by municipal governments. These in turn could only participate if their expenses were approved by a vote of property-owning ratepayers. As long as it was "possible for a province like Manitoba to crawl out from under its share of the burden and try to pass it on to the municipalities," Knowles argued, it was evident that public housing would only be constructed if the federal government bore "a larger portion of the capital cost" than its current 75 per cent.[86]

Public-housing supporters testifying before the committee charged that the federal government had done little to promote public housing. Longtime social-housing advocate George Mooney, executive assistant to the Canadian Federation of Mayors and Municipalities, said that the federal government had been "reticent" in encouraging public housing. He noted that the handful of public-housing projects in the nation owed little to federal promotion. Instead, they were the work of community social activists such as "welfare people, church people, city council people" who frequently had to endure long campaigns to build a project.[87]

In sharp contrast to the federal government's solicitude for the reluctant chartered banks as it drew them into mortgage lending was the treatment accorded the housing proposals of the Canadian co-operative movement. By 1954 co-operatives in Canada had 1.3 million members and total assets of $496 million. To garner support for their housing goals, the Canadian Co-operative Union and Le Conseil Coopératif Canadien had held a national conference on 14 and 15 December 1953, where a common strategy had been developed for the French- and English-speaking sections of the Canadian co-operative movement and representatives of the Canadian Congress of Labour, among them research director Eugene

Forsey. One of the recommendations to emerge from the conference was "that every effort should be made to have co-operative housing associations recognized as limited-dividend companies for the purposes of the National Housing Act."[88]

After the conference, about the same time that the federal government was negotiating with the chartered banks over the terms of the new act, the co-operative representatives went to meet with David Mansur. Mansur refused their proposals outright. He wrote to CCU president R.S. Staples that CMHC would not "recognize a co-operative as a limited-dividend company unless of course the co-operative undertakes to rent the units to people who have no co-operative interest in the project."[89]

Rejected by the federal government, the CCU felt its "only recourse was to appear before the Standing Committee on Banking and Commerce," which was considering the new NHA. The CCU noted that the likely higher NHA interest rates under the guaranteed-loan scheme would result in a noticeable reduction of those who could afford housing in co-operative home-ownership projects. They also argued that giving low interest loans to limited-dividend companies but not the co-operatives was a form of "discrimination." The CCU brief pointed out that "the idea of a limitation on dividends is inherent in co-operative effort." Union president Staples observed, "Those of us here at the head table could organize a corporation to build houses for all in this room and if we would accept certain provisions the government would provide most of the money over many years at a low rate of interest. On the other hand, if those of us at all these tables wished to provide houses for ourselves, we cannot get that sort of assistance."[90]

Worried by the sympathy of the banking and commerce committee for the CCU's position, Winters and Mansur fostered the impression that they were considering a modification in their position. They implied that limited-dividend loans would be made available to co-operatives that would use the funding for accommodation of low-income families who did "not obtain individual ownership of the houses." By creating confusion, Winters and Mansur prevented the introduction of an amendment that would have obtained the support of some Liberal MPs as well as CCF members and Conservatives such as Donald Fleming. Winters "narrowed objections" to one point. This was the matter of "independent administration" required in the text of NHA section 16 for limited-dividend projects. Mansur claimed that CMHC would lay itself "open to continual controversy if administration of the housing project were not in some respects independent of the tenants." But before the legislation was introduced for third reading in the Commons, Mansur moved to quiet the co-op supporters before they met with the parliamentary committee. He wrongly intimated that co-operatives might qualify if they were willing "to turn over administration in certain aspects to some independent central body such as a provincial union." This devious advice persuaded the CCU to make "no issue" of their desire for an amendment on the floor of Parliament. The government was saved from considerable embarrassment over the introduction of an amendment that might have been supported by members of its own party.[91]

After the NHA was safely through the House of Commons, the government made its position quite clear to the CCU. Staples noted that the government's position was not "very reassuring" since "a housing co-operative not controlled by its members would not be a co-operative at all." Mrs Anne Rivkin of the Saskatchewan Economic Advisory and Planning Board wrote to Allan Armstrong, director of CMHC's public-housing division, about the "problems of a group of working women." They wished "to construct a building containing 16 bachelor apartments," using funds available under the limited-dividend section of NHA. When Rivkin asked if there was any reason why the "prospective tenants" could not "form a Limited Dividend Company to rent apartments to themselves," Armstrong's marginal notes on the letter supplied the answer. This was prohibited because section 16(3) of the new NHA required "independent administration." Likewise, the British Columbia Teachers' Federation was told, when it requested a limited-dividend loan for a co-operative project, that CMHC was "unwilling to finance under the limited-dividend section of the Act a project where a prospective tenant held or had shares or a beneficiary interest in the shares of the borrowing company."[92]

The CCU, in its desire to obtain limited-dividend loans for co-operative housing, was quite flexible. It was on the verge of forming a unique form of co-operative housing well adapted to Canadian circumstances. The union estimated that homes under its limited-dividend co-operative formula could be available for families of an annual income of $1,750, close to the ceiling for the period's public-housing projects. The CCU was willing to accept that such homes would never be transferred to market ownership; indeed, this was part of its philosophical commitment against speculation in housing as a commodity. But of course, this would have run counter to the whole direction of the National Housing Act, which was designed to maintain the private mortgage market.[93]

The National Housing Act of 1954 was the last time a new housing bill was introduced by the federal government. Largely, the legislation simply copied the NHA of 1944 as amended, apart from replacing the joint-loan scheme with a program of mortgage-loan guarantees. The fact that a new housing act has never since been introduced indicates how firmly Mansur's version of the assisted-mortgage approach has been fixed in Canadian housing policy.

The National Housing Act of 1954 confirmed, in final legislative form, the conservative tenor of Canadian housing policy ever since the federal government had assumed a housing responsibility in 1935. The legislation stabilized the basic thrust of federal policy simply to stimulate residential construction by disturbing the existing private market as little as possible. This had the socially inequitable result of channelling most federal assistance to the top 20 per cent of Canadian earners, those who could afford to purchase a newly built, NHA-financed home. Social-housing programs, whether limited-dividend or public housing, would be residual in nature, applied at the whim of the mortgage market or born of the vigour of the campaigns of local activists. After 1956 even public housing would be distorted by its association with the rehousing of families displaced by urban renewal. With suburban housing being generated for an affluent minority by NHA-guaranteed loans and NHA-financed servicing and land assembly, at the same time as the homes of the urban poor were being razed for commercial redevelopment financed by NHA, federal housing policy by the mid-1950s had become a cruel caricature of the worst inequities of Canadian society.

8 Drift within Close Confines, 1955–1992

The National Housing Act of 1954 set the basic parameters of federal housing policy for the next three decades. Unlike W.C. Clark's relatively rapid-fire succession of three housing acts in seven years, David Mansur's housing legislation, although frequently amended, has endured for thirty-six years. Mansur's framework, while sharing Clark's adherence to the private market, had greater flexibility and hence more longevity. More easily than Clark's rigid determination never to allow a single unit of subsidized shelter, it could be stretched to accommodate demands for socially sensitive policies. Starting with his own NHA amendments in 1948 and 1949, Mansur had shown a willingness to concede to the demands of social-housing supporters, but he had cleverly combined this with the adoption of programs that would ensure that such innovations would be on a small scale. In this spirit federal policies would evolve in subsequent years.

After passage of the first effective program for any permanent low-rental housing in 1948, the basic pattern Mansur had established – federal efforts geared to encouragement of the market sector while sustaining a residual social sector for low-income groups, which could not develop at a rapid pace – went through three basic changes in three decades. First, from 1954 to 1964, a booming market sector contrasted with a painfully slow-moving social sector, whose basic goals were often subordinated to the big-city boosterism associated with urban renewal. In the next decade, public housing encouraged by the 1964 NHA amendments was rapidly constructed for low-income families. However, in the frantic pace to catch up with past unmet unit targets, the high-rise developments, marginal locations, insensitive designs, and ghettoized accommodation associated with such projects caused public housing to be brought to a sudden halt in the 1970s. The third shift in federal policy was embodied in the NHA amendments of 1973. This legislation, passed in response to demands for a "comprehensive" rather than "social" housing policy, facilitated the transition from building segregated public housing to support for non-profit and co-operative housing projects designed to be attractive to a wide range of income groups.

In spite of these shifts, Mansur's basic policy framework remains unaltered. In response to concerns over bulldozers destroying cherished residential communities and skyrocketing home prices and rents, enormous pressures have at times been generated to scrap Mansur's approach and adopt a comprehensive policy modelled on the housing achievements of Scandinavian nations and other social-democratic countries. This period of "turbulence" saw such major federal studies as the Hellyer and the Dennis-Fish reports repeat essentially the same message, which was underscored by the resignation of the minister responsible for CMHC, Paul Hellyer.

Despite considerable sound and fury, the results achieved, even after the significant boosts for social housing of 1964 and 1973, did not signify the major transformation urged by critics. Like the NHA amendments of 1949, which produced public housing but only on a minuscule scale, those of 1973 did sanction the long-awaited non-profit and co-operative projects. But, as in the earlier generation of social housing, the new, more appealingly designed projects for integrated communities would be built on a limited scale. After a short-lived burst of reformist energy unleashed by the political necessities of an NDP-supported minority government, barriers were quickly developed to discourage the growth of a new "third sector." Federal assistance to land banking ended abruptly in 1978. A strict allocation list was developed of the number of co-op and non-profit housing units that could be constructed in a given year, invariably below the capacity of the third sector. This was part of a major delegation of housing programs to the provinces. Its failure was enhanced by provincial disinclination to add on separate subsidies in addition to federal payments. Pressures from co-operatives, native organizations, and private non-profit projects barely sustained a federal housing role, whose scope diminished under the Conservative government of Brian Mulroney. So narrow had the scope of federal intervention become that, by 1986, the reformist-minded government of the Ontario Liberals took an unprecedented step in launching its own Homes Now program to boost the sagging production of new social-housing units.

THE TRAGIC GAMBLE OF URBAN RENEWAL

The first dent in David Mansur's housing framework was the NHA amedments of 1956. These for the first time allowed urban land purchased by federal funds under NHA to be used for commercial redevelopment. This clause, which opened a veritable Pandora's box of local greed and boosterism, had complex and contradictory impacts. While it brought social housing to such previously impenetrable fortresses of hostile real estate reaction such as Winnipeg and Calgary, it accomplished this worthy social aim at a heavy price. The local business elites who were lured into providing subsidized rental housing for low-income families did not magically undergo a change of heart as a result of federal largesse, although they met the letter of the law requiring rehousing of dispossessed tenants. Such elites frequently combined their urban-renewal schemes with the razing of historic neighbourhoods, underpaying expropriated homeowners, and the construction of over-large public-housing projects, devoid of needed complementary recreational services. Public opposition to arrogant urban-renewal projects helped to nurture support for the major shift to the delivery of third-sector social housing by non-profit associations, co-operatives, and municipalities that would be embedded in the 1973 NHA amendments.

The ambiguous legacy of the 1956 NHA amendments can in part be traced to Cabinet satisfaction with the framework Mansur devised in 1954. Mansur recognized his success in imprinting his image on federal housing policy, by resigning shortly after passage of the NHA of 1954. The appointment of Stewart Bates as his replacement as president of CMHC was a sign that innovations would not be welcome. Formerly deputy minister of fisheries, Bates was inexperienced in housing matters and brought no bold new agendas with him to CMHC. Although Bates later backed the socially minded supporters of public housing within CMHC and pushed through the important 1964 NHA amendments as a final achievement, he had enormous difficulty contending with the conservative CMHC board. Dominated by the Department of Finance, the board would make it difficult to present new approaches before Cabinet, a difficulty frequently compounded by the conservative character of certain ministers responsible for CMHC, notably Bay Street business magnate Robert Winters.[1]

The business-oriented push to federal policy that would continue after Mansur's retirement was exacerbated by the 1956 NHA urban-renewal amendments. Later problems would have been reduced had the recommendations of CMHC's policy-planning body, the Advisory Group, been more fully implemented. Established by Bates, chaired by Humphrey Carver, the Advisory Group made suggestions for urban renewal that were targeted at helping the poor. These included the provision of 100 per cent loans and grants and start-up funds for non-profit housing associations in order to reduce the red tape strangling public housing by requiring the federal government to assume a greater share of costs and to provide subsidies for housing rehabilitation. Despite Bates's support, all these policies would be rejected.[2]

Bates had important exchanges with Winters and the CMHC board that boded ill for urban renewal. Supporting the ideas of the Advisory Group, Bates tried unsuccessfully to persuade Winters of the need for subsidized home repair. Becaused it lacked the commercial allure of big redevelopment projects, Bates's view that "in social terms the need for decent, safe and sanitary accommodation has no necessary relationship with demolition" would not sway Winters. Attempts to change the Clark-Mansur framework were quickly detected as heresy by the CMHC board. On 12 February 1957 the board gave Bates a lesson in its philosophy of social housing. It informed him that his attitude that public housing should be "primarily an instrument of social policy to remedy the conditions of the poor who live in bad housing" was wrong. Instead, "the needs of individual tenants should be secondary" to "economic and urban development considerations." Public housing would provide only "a bare minimum of housing for the occupants," while being used to improve the overall appearance of the community. Spartan shelter would make it "clear" that CMHC was not "competing with private enterprise." Private enterprise, the board felt, would build "a more attractive product for those who can afford it."[3]

The urban-renewal game offered millions to those who would remake Canadian cities along the lines of the latest skyscraper projects promoted by the land-development industry. It would be most effectively used for this purpose by local commercial and real estate elites, who were able to seize on it and push ambitious schemes forward quickly. Many cities, such as Halifax, were well prepared for such an approach, having earlier lobbied for the legislation. The cost of remodelling the central core of every Canadian city along "superblock" lines similar to those dreamt of by Clark in the 1920s would play a major role in Paul Hellyer's termination of the program in 1968.

The story of urban renewal in Halifax illustrates the heavy social costs that would later make it politically impossible for a populist-minded minister like Hellyer to support such programs – especially after being confronted with demonstrations of residents about to be evicted. The city had pushed for the legislation, quickly prepared the required planning studies, and was prepared to face concerted protests from residents angered by the threat of displacement. While Halifax mayor and land developer Leonard Kitz immediately praised the 1956 NHA amendments as a "godsend" to the city, residents to be affected protested that "a small group in authority has decided to confiscate an area of the city [and] force the people out despite their unanimous disapproval."[4] Along such lines the urban-renewal debate would rage in Halifax and most major Canadian cities for the next dozen years.

Halifax commissioned University of Toronto planning professor George Stephenson to undertake a study of urban renewal. Stephenson provided the city's tightly interlocked media, business, and political elites with what they wanted: a rationale for the clearance of the city's poor from areas they wished to have redeveloped for more remunerative use. In one sweeping statement Stephenson urged that the low-income housing of central Halifax be demolished since it was "repellant to good commercial development." Likewise, the blacks of Africville must be moved since "the land they now occupy will be required for the future development of the city." Stephenson admitted he had "no accurate record of conditions" in Africville, although he knew many residents had become "home owners after many decades of struggle to obtain some security."[5] He did detail the housing problems of welfare recipients who were housed in the central part of the city, between Jacob and Market Streets. This was used as the basis for the "sweeping away of the worst housing in the City" in this district.[6]

In total, Stephenson estimated that 7,000 persons would require rehousing as a result of his slum-clearance recommendations. His report recommended that only 4,500 persons be housed in new limited-dividend and public-housing projects. Admitting the "serious difference" in these figures, Stephenson optimistically assumed that "some of these families and persons will obtain accommodation within the city." Such a prediction was contrary to his own description of existing overcrowding, which was being used to justify the report's redevelopment proposals.[7]

Halifax city council did not trouble itself with the net housing deficit of Stephenson's report. It approved his proposals forty-eight hours after they were released. The council's implementation of the report was harsher on low-income groups than Stephenson intended. The twenty-six-acre city prison lands were not used for low-income limited-dividend housing, as Stephenson recommended. They were used instead for luxury condominiums. Although the terrible housing conditions of welfare recipients were used to justify redevelopment, little concern was given to the plight of such poor persons dislocated by the scheme. The same report that adopted Stephenson's plans also concluded that the "four hundred roomers and boarders are not required to be looked after ... there is no reason why they should not fend for themselves, particularly as there is no financial assistance available from NHA for this purpose."[8]

The 267 family residents to be displaced by redevelopment had some protection because of the NHA's requirement that provision be made for their rehousing. Consequently, the Halifax council approved the 349 Mulgrave Park public-housing project on 11.5 acres of land already owned by CMHC. This target was small in relation to the combined needs of the relocated families, which amounted to a 35-family waiting-list for the existing Bayers Road public-housing project and the 198 families in Halifax still living in emergency shelter.[9]

Although CMHC architects devoted considerable care to the award-winning design of Mulgrave Park, local authorities paid scant attention to the social needs of the new tenants. They took little care, for instance, to examine the affordability of public housing for the displaced. No significant problem was expected since, with "one additional working member over and above the head of family for each of two families," income would be only "slightly less" than the level required. Even worse, after the approval of the Jacob Street slum-clearance project on 20 February 1958, the city immediately began the associated demolitions, expropriations, and evictions in spite of the fact that public housing at Mulgrave Park would not open until October 1960. Distress began to be deeply felt after two months. The city manager estimated that evictions had reached "five or six hundred" persons during that brief period, with 110 persons "put on the streets" in a single evening to permit the demolition of fifteen buildings. In addition, according to the commissioner of public works, "a great number" had not come to "attention."[10] A study of Halifax blacks by the Dalhousie University Institute of Public Affairs found that even after the construction of Mulgrave Park, most displaced blacks had to double up with families in the immediate vicinity of cleared areas or occupy houses soon slated for demolition. Likewise, the city welfare department reported the spread of slums to other areas of Halifax "because of overcrowding and owners converting single family houses into rooming houses."[11] Upon entering Mulgrave Park, the displaced often found the subsidized rents too expensive to pay. Public-housing authorities temporarily allowed rental arrears to accumulate. This ended in May 1962, and evictions took place at the rate of two per day. Higher-income tenants admitted to Mulgrave Park solely on account of being displaced by urban renewal were all evicted after a year's residence.[12]

Halifax social-welfare officer H.B. Jones stressed the negative impact of slum clearance on the city's housing conditions in an 11 July 1961 statement to the city council's housing policy review committee. Jones estimated that one thousand families lived in one room and asked if anyone had "an idea how many families have been disrupted so far by the redevelopment scheme?" He told the committee:

> We haven't gotten within at least 3,000 apartments which we need. I would think that would be a bare minimum. You drive along the road, you see a house is coming down, you say "Isn't that nice the slum is being demolished," and we know the buildings of low standards should be demolished; but how many times have you heard anyone say, "What happened to the people?" The week before last, we had a man and his wife and three or four children sleeping in a car in Halifax. They are now living in Springhill. The man found an old house up there. We have pushed hundreds of families all over Halifax County. We have them living in Digby, many in Cape Breton; we have them up as far as Windsor and beyond. We are driving these people out.

Jones estimated that two hundred families a month were leaving Halifax because of the pressures of redevelopment.[13] Housing committee chairman Lane responded, "If you find a family cannot live on a man's reasonably decent income it is because the woman is not educated to administer that income properly."[14] On 8 August 1961 the council resolved to "aggressively continue the established policy of slum clearance," ordering monthly reports of violations of minimum standards with "the details of all the violations discovered and the action taken" to compel compliance with city laws.[15] Faced with such demands for improvements without financial aid, landlords would demolish their properties or raise rents, and low-income homeowners would be forced to decide between paying money they could ill afford on repairs or losing their homes altogether. A housing study conducted by H.S. Coblentz for the city acknowledged conflict between "planners educated to lead the community towards a middle class ideal and ... the wishes of the residents of slum and blighted areas," and noted that housing often "appears obsolete only so long as it is inhabited by members of a lower class."[16]

The saga of urban renewal in Halifax would epitomize the bitter class conflict it stimulated across the nation. After lying vacant for five years, the land of the city's impoverished blacks was finally converted into a giant office complex in the same league as Place Ville Marie in Montreal or the Toronto Dominion Centre. Much of the land belonged to Mayor Leonard Kitz, who became a member of Halifax Development Limited. This company consisted of most of the corporate elite of Nova Scotia and was formed after Major General K.C. Appleyard strolled down to the Halifax Club to develop a syndicate. As well as Kitz and Appleyard its members included Roy Jodrey, director of fifty-six companies, including the Bank of Nova Scotia and the Nova Scotia Light and Power Company, Russell Harrington, another principal in Light and Power and in Eastern Trust, and Harold C. Conner of National Sea Products, the largest Canadian fish-processing firm. These investors received the benefits of CMHC's $4-million expenditure for land acquisition and the city of Halifax's payment of $5 million for new roads and sewers, for a purchase price only 10 per cent of these costs. By 1980 HDL owned over a million square feet of retail and office space, 444 apartments, a Canadian Pacific Hotel, and the Nova Scotia Savings and Loan Company.[17]

Removal of a racial minority, pursued so openly in the earlier Halifax urban-renewal project, was repeated in another pioneer undertaking in Vancouver. Here the group targeted for displacement was the city's Chinese community. According to press surveys, urban renewal was resented by both the Chinese and the Italian residents of the disrupted communities. Besides their ethnic communities, the neighbourhood's residents also appreciated its closeness to employment and its cheap rents. The clearance was regarded by the Chinese residents as racially motivated because of their or their parents' forced displacement by the torching of their homes prior to the First World War. Chinese community spokesperson Foon Sun observed that "the Chinese seem to be goats of the whole thing."[18]

Like Halifax's blacks, Vancouver Chinese were cleared out of their neighbourhoods to encourage industrial and commercial developments on the lands in which they had formerly lived. Humphrey Carver has described how Vancouver's director of planning, Sutton Brown, "mounted his main attack on the Strathcona neighbourhood." Although two public-housing projects were built to house the displaced, Vancouver city council had too little concern for the victims of the attacks to accept the requests of the Vancouver Housing Association that they ensure "that persons displaced by slum clearance, who do not move into public housing, find suitable accommodation." The failure of authorities to assist single displaced persons was particularly onerous in Vancouver since it was the older hotels and rooming-houses, where low-income singles lived, that were being demolished rapidly by redevelopment.[19]

In Montreal, the Canadian city most marked by the inequalities of the national vertical mosaic, urban renewal was also characterized by particular hardships for low-income ethnic minorities. In his 1965 commentary on urban renewal Carver condemned the failure of the city to rehouse persons displaced by the city's ambitious projects. He observed that in Montreal most clearly among Canadian cities, "the dynamic energies of capitalism are displayed." Here, "magnificent and beautiful office towers" stood amidst "the surrounding areas that are being laid waste." By 1965, 1,779 families in the province of Quebec had been displaced by clearance but only 796 units of subsidized housing had been completed. The actions of the federal government, not formally part of the urban renewal, exacerbated the problems of displacement in Montreal. The CBC's French-language headquarters involved the destruction of 1,250 homes. The building of Expo '67 helped to reduce Griffintown from a community of four thousand persons to a few hundred. It also contributed to the complete disappearance of Victoriatown. Montreal's Chinatown disappeared under the federal governments's Place Guy Favreau. By 1968, two thousand homes a year were being demolished by urban renewal in Montreal. The 305 former families of Victoriatown had their shelter costs increased by 36 per cent as a result of their relocation. Another 109 families displaced in the Côte des Neiges district faced a 62 per cent increase. At one point Deputy Mayor Lucien Saulnier defended his city's housing record with the observation that "we have already over 400 dwellings built and occupied and are engaged in the clearance of 4,500 others."[20]

An undisguised zest for demolition was characteristic of most urban-renewal studies funded by CMHC. One early such examination of Saint John, New Brunswick, by Albert Potvin, concluded that "a total of approximately 4,000 (homes) are in such a state of deterioration as to require immediate demolition and replacement." Opposition to the project, from residents affected by its clearance of fifty-seven acres, was so intense that protest leaders complained that "so many people in the area have died because of the strain, we won't have any members." Neighbouring Moncton, like Halifax, quickly destroyed eight hundred homes following the release of its urban-renewal report, which recommended turning over "all obsolete and non-conforming lands" in central Moncton to commercial development. In Sarnia a 170-acre residential subdivision was turned into an industrial park so the residents' complaints about air pollution would not interfere with the workings of a neighbouring chemical plant.[21]

In promoting its urban-renewal scheme, CMHC would sanctify the marketplace ethos generating pressures for redevelopment. Planning traditions stressing the stabilization of residential areas, conservation of older housing, and containment of commercial shadowing were swept away. The credo of the real estate trader became equated with high national purpose. In 1964 Potvin wrote in CMHC's publication *Urban Renewal and Public Housing*, "trying to dress up older buildings that have served their intended purpose for the best part of their expected life span, retarding the demolition of age-worn homes, slows down the natural development and expansion of communities." In "many cases," Potvin argued, it would be better to "hasten the end through major surgery." Canadians would have to come to accept "the idea that a spent and outmoded neighbourhood can be profitably amputated through a single operation." Potvin went so far in this line of thinking as to lament that "homes in Quebec's walled city," because of their appeal to tourists, escaped "the fate of many newer structures standing in the way of progress." He fully accepted as a sign of a neighbourhood's decline the fact that it was no longer buoyant "by real estate standards." While Potvin noted that his views were not shared by most Canadians, he correctly observed that they were accepted by "civic authorities and hard-headed business men alike." They recognized "that preservation ... only leads to stagnation."[22]

Towns and small cities followed the pattern that major metropolitan centres set for urban renewal. A 1965 study of Amherst, Nova Scotia, recommended razing 190 dwellings covering eighty acres. Similar steps were called for in an examination of Nova Scotia's Pictou County, which urged the establishment of a "Critical Improvement Area." Here "all the houses" would "be cleared." In Trenton, Ontario, "a colony of marginal small and poor housing" was slated for removal from "a strategic waterfront site." Another section of Trenton's working-class homes "if cleared would be appropriate for commercial and light industrial uses."[23]

The trap that urban-renewal schemes meant for low-income Canadians who could not afford to participate in any existing federal housing program is well illustrated by the CMHC-funded urban-renewal study of Glace Bay. Completed in 1964, the study was undertaken by the long-time president of the Canadian Institute of Planners, Norman Pearson. At the time Pearson was identified with the most socially responsible members of his profession. He noted that Glace Bay's housing problems had been intensified by the refusal of lending institutions to make loans in the community, not just to coal miners but also to persons with secure jobs in neighbouring communities.[24] Since most homes were sold at low prices to employees of the Dominion Coal Company, some 89 per cent of Glace Bay's housing was owned by its residents. Although substantial improvements had been made from the time when the community's homes were owned by Dominion Coal, 40 per cent were in need of repairs. Another 19 per cent were overcrowded and 23 per cent lacked hot water. Some 30 per cent were without baths or showers, and 57 per cent were heated by stoves and space heaters.[25]

On average, only twenty homes a year had been built in Glace Bay since the end of the Second World War. Homes built according to NHA specifications cost $10,000 to $14,000 a year and so could only be afforded by 10 per cent of the community's households. Many residents had been able to build homes in the post-war years only because of the subsidies provided under the Veterans' Land Act. The co-operative housing program of the Nova Scotia Housing Commission had become too expensive for most Glace Bay residents. From 1945 to 1954 only 18 per cent of the participants had had to take out a second mortgage, but by 1964 this figure had risen to 47 per cent. Such increasing costs contributed to a decline in the co-operative program from 289 units in 1959 to 49 in 1964.[26]

In Glace Bay the NHA's home-improvement loans (which finally came into being after 1954) were in amounts far too small to deal with the need for major repairs. Through conversations "with the various locals of the United Mineworkers of America," Pearson found "that new foundations and minor repairs and internal work such as replacing furnacing and wiring might typically average $3,000 per dwelling." Dealing with problems of subsidence would bring the cost up to $3,500, and complete reshingling and a new roof, to $4,000. Unless such costs could be covered by personal or family savings, they had to be obtained by finance-company or personal loans, both of which carried "very high rates of interest." Miners naturally feared to undertake such borrowing, which would be equivalent to a year's wages. The debt burden could cause the loss of their homes in case of an interruption of their regular earnings by injury or unemployment.[27]

Despite the hardships that discouraged rehabilitation of housing, Pearson noted that the miners were still able to come up with the funds to make very practical, innovative, and decorative improvements to their homes. He observed that "even in the poorest areas there was evidence ... of attempts to paint and patch, even if major work is beyond family means." Homes had entrances relocated for greater privacy and additions were also frequently made.[28]

Since existing provincial and federal programs offered no solutions (even public housing could not be afforded by 59 per cent of the community's residents), Pearson recommended the application of regulatory techniques. He admitted that this approach "was not locally popular." A minimum-standards by-law would be set according to NHA specifications. This would be a requirement for the receipt of any federal or provincial aid. The by-law would be "enforced on the basis of annual inspections and an annual register checking progress or deterioration." "Well-qualified inspectors" would examine all homes in Glace Bay "and identify those needing clearance." The homes that violated by-laws would be "acquired and removed."[29]

Pearson's report called for the demolition of 902 homes, approximately a third of Glace Bay's housing. He felt it "did not matter" that many of the town's homes were "still good places to live," although below NHA standards. The standards would give "a basis for subsequent inspection and enforcement action, encouragement to individuals to improve when they see the municipality determined to act with vigour." Pearson also urged that all housing adjacent to the new Deuterium heavy-water plant be cleared, assuming "that all those affected will be able to make their own adjustments." He maintained that "a home is a home, but a rehabilitation shack is still a shack." He told the city council that the only alternative to his recommendations was to "tear the whole town down" after fifteen years.[30]

Pearson's plans illustrate how federal housing policies could do nothing to improve the shelter conditions of many low-income Canadian families. In this community of 24,186 persons, only 41 per cent of households had incomes high enough to be eligible for public housing. The Glace Bay urban-renewal study proposed to rebuild the town to a standard that 90 per cent of its residents could not afford. The report recommended the demolition of a third of the town's housing stock. If actually carried out, this would have been accomplished by having one section of the community send inspectors to close and later demolish the homes of the rest. More likely the recommendations would have served as a basis for coercing families to "improve" their homes at all costs, with a few examples sufficing to make the "needed" impression. The displaced residents would have nowhere to go as they could not afford the new NHA homes.

Pearson's recommendations could not be carried out because there was no local business elite interested in using the cleared land produced by demolitions. The exception was the powerful Crown corporation Deuterium, which wished to have homes that disrupted its operations removed.[31] Given the state of existing legislation, the only alternative Pearson could propose to his recommendations was to point to the need for new policies, such as assisted home ownership based on the local success of the Veterans' Land Act, or subsidized home repairs. No such recommendations in urban-renewal reports would be forthcoming.

Displacement of low-income groups was also encouraged by CMHC through its promotion of building-standards by-laws on existing homes. Like the regulations of public-health officers, these rules had the impact of making it a legal offence to be poor. This was quite directly spelled out in the city of Ottawa building-standards by-law, promoted as a model by CMHC. The by-law prohibited overcrowding in "excess of one person per hundred square feet of habitable accommodation." Each dwelling in the city was legally required to have "a bathtub or shower and hot and cold running water." No financial assistance for improvement was offered. As proof of this program's success, CMHC offered a photograph of a home whose owner, being "financially unable to keep his property in good repair," decided instead to "demolish the building and sell the vacant Land." In eleven years, from 1952 to 1963, 1,383 dwellings were demolished for violations to Ottawa's buildings by-laws. Full-time inspectors not only responded to complaints but reviewed assessment rolls. All homes with a low assessed value were presumed to merit inspection as likely by-law violators. The Ottawa program was directed by Peter G. Burns, who served as redevelopment and standards officer for the city of Ottawa while on leave from his post with CMHC.[32]

Urban renewal's essentially commercial and speculative motivation and the hardships it imposed on low-income persons have been especially well documented in the case of the city of Hamilton. Its mayor, Victor Copps, in a 1968 commentary on the city's experience with the program, did not claim any success in providing better housing for the city's low-income residents. Indeed, he cautioned that urban renewal did not "provide instant housing for those lacking decent accommodation for their families." Instead, he stressed that "urban renewal is profitable and one of the most important side effects on finance is this profit." The degree of profit depended on "the degree of upgrading of the land in the redevelopment area." Copps told his fellow mayors that the "excitement and the profit" of urban renewal made it "well worth the difficulties and dangers" it presented. Each city could have a development as "proportionally exciting" as Montreal's Place Ville Marie. Urban renewal allowed municipal politicians an escape from such "humdrum matters as"filling potholes, collecting garbage, building sidewalks."[33]

After the completion of its 1958 urban-renewal study, the city embarked on the immediate clearance of residents from the Van Wagner's beach community. This seventy-five-acre working-class neighbourhood was turned into a municipal park.[34]

The second phase of urban renewal, which took place in Hamilton's North End, was announced as a balanced combination of rehabilitation and demolition. A small pilot project in rehabilitation was undertaken here, which came to the conclusion that NHA standards were too high for affordable older housing. The impact of the demolitions in the North End project were examined by two McMaster sociology professors, Franklin J. Henry and Peter C. Pineo.[35]

Henry and Pineo's study documented the heavy financial costs that urban renewal imposed upon the displaced. The average rent paid by relocatees rose from $55.80 a month before displacement to $94.40 afterwards. Mortgage payments for relocatees jumped an average of $25.80, to $58.90 a month. This amounted to an 87 per cent increase. Taxes also increased. The average price paid to the displaced for their homes was $10,700; the new houses they purchased averaged $15,500 in price. An average of a mile and a half was added to the relocatees' trips to work. In comparison to the control groups, the average family income was reduced. Relocation reduced parents' contact with their grown sons and daughters "from daily or almost daily contact to once a week contact." The average number of close relatives seen weekly or more frequently also declined. In comparison to the study's control group, relocatees experienced a higher rate of serious illness and more colds and sweats.[36]

In the section of their study based on interviews with relocatees, Pineo and Franklin recorded a virtually uniform negative response to the impact of urban renewal. Some complained of being forced to rent, not having obtained enough from the sale of their homes to buy a house. Others lost employment for taking time off to move. Renters who had made improvements to their homes received no compensation, despite previous promises of public officials. Families were sometimes forced to move in "zero weather" conditions. This resulted in "hardship and sickness." Relocatees frequently complained of being shown "dumps" by urban-renewal officials, "as if these places were good enough for people in the North end." Often three moves had to be made within the space of one year. Higher purchase price of new homes reduced families' savings to the levels of a decade earlier. Stores were sold to the city at a price a third of that needed to establish a new business.[37]

Little concern was expressed by Hamilton officials to improve the lives of people displaced by urban renewal. Demolition of homes was carried out in the summer of 1965, but public housing was not made available until 1969. The demolition of three hundred working-class homes in the summer of 1965 created a "seller's market" for Hamilton real estate companies. A 59 per cent rise in the price of homes took place in the city, compared to the national average of 14 per cent.[38]

A frequent complaint reported by Henry and Pineo was that the displaced had "lost their belief in the integrity of city officials." Promises of compensation were made that were never kept. Complaints of being treated as "second-class citizens" were common. Individuals who had earlier experienced hardship from relocation were not permitted to address public meetings. The displaced felt compelled to accept the price offered by civic officials since "the threat of going to court was more than a poor man could pay." Some relocatees were hospitalized with nervous breakdowns. Homes temporarily rented by urban renewal were sometimes so poorly heated that pipes froze. Furniture had to be sold because of the smaller homes that had to be purchased in other cases. Typically, one relocatee lamented the loss of his hobby shop and wine cellar. The compensation given was not adequate to purchase a home with similar facilities. In another case, a relocatee told the surveyors she would "always hate urban renewal for upsetting her health, her marriage and her family."[39]

What to business-oriented municipal politicians was the adventuresome and profitable game of urban renewal was stopped by citizens' protests, rather than by academic disclosure of the hardships imposed by the program on working-class families and small businessmen. The real estate groups supporting urban renewal had seen the forced displacement into higher-cost shelter it involved as a positive alternative to public housing. Although Humphrey Carver claimed credit for the inspirational name of urban renewal, the basic concept originated in the United States. In 1942 the National Association of Real Estate Boards had called for the federal government to expropriate slum areas, clear the land, install public services, then give "the building sites to private redevelopment companies or individual builders for a long period of years." This proposal was denounced at the time by United States Housing Authority director Nathan Strauss as "a speculator's dream come true!" He accurately predicted that such a scheme would increase overcrowding and make the difficulties of low-income families "more acute." Strauss found it "strange" that "respected economists and city planners" should support programs advocated by "a wealthy aggregation of real estate interests" and "the powerful, reactionary lobby of the United States Savings and Loan League." By contrast, no Canadian urban expert until the late 1960s would denounce this alliance that promoted urban renewal. Consequently, Carver learned, much to his surprise, that "private developers, the advocates of public housing and CMHC" would all "find themselves strangely linked together as a joint establishment to be looked on with suspicion and distrust."[40]

The effective force of citizen protest in stopping urban renewal is best demonstrated in the city of Toronto. The city's plans for clearance were on a truly massive scale and called for the removal of all working-class communities from the city's downtown area. The city's 1956 urban-renewal study called for the demolition of a twenty-seven block area bounded by Queen, Dundas, River, and Church streets. At the same time, an Industrial Leaseholds' scheme envisaged office towers and industrial redevelopment on 142 acres of residential land between Queen and King and the east side of Parliament. Some 252 homeowners in the Don Mount area (expropriated for urban renewal) had to pay an average of 56 per cent more for their new homes than what they received in compensation. This amounted to $5,370, the equivalent of an average year's income for the displaced.[41]

Toronto broke the pattern of urban renewal because those affected by redevelopment schemes were able to organize both publicity-friendly confrontations and complex litigation, aided by social workers trained in the tactics of Saul Alinsky and such community-linked professionals as recent law graduate John Sewell and Karl Jaffary, son of a pioneer Canadian public-housing advocate. An urban-renewal scheme at Trefann Court was blocked by such tactics as a three-hour occupation of the Toronto Board of Control, picketing the homes of municipal politicians, and the use of lawsuits to block redevelopment action. The publicity such protests engendered caused Paul Hellyer, then minister of transport with responsibility for CMHC, and the federal task force on urban renewal to visit Trefann Court on 1 October 1968. Here they were greeted by placard-waving protestors from the residents' association.[42]

Hellyer's personal encounter with the victims of urban renewal could only help to increase his bold determination to abolish it, despite continued support for the scheme within CMHC. He has recalled how in their "bulldozer approach to urban redevelopment" CMHC officials were "as wrong as they were stubborn." Hellyer's decision was part of a recognition that the disruption of urban renewal, in increasingly well-organized working-class communities, was too costly. Hellyer would come to regard this cancellation as the beginning of the "greening of Toronto" and to mock urban renewal, observing that, "in order to eradicate the 20 to 30 per cent of buildings that were rotting beyond repair, whole blocks were demolished. Thousands of sound houses capable of being rehabilitated at reasonable cost, together with thousands of others in perfectly good condition were destroyed. The economic waste was enormous. But far more importantly, the sense of community, that certain intangible something that gives a district life and meaning, was eradicated. An atomic bomb could have scarcely produced greater dislocation."[43]

THE SURGE OF THE THIRD SECTOR

Although hyperbolic, Hellyer's comparison of the impact of urban renewal to an atomic bomb also captured some of its explosive impact on Canadian housing policies. The very disruptive force of its destruction fostered an angry public reaction that dramatically changed social housing in Canada. The federal bulldozer was replaced by start-up grants that empowered community groups across the country to build living environments according to their own dreams. A more dramatic departure from the arrogance of the boosterism of ruthless commercial elites could not be imagined.

The shift to the "third sector" of municipalities and private non-profit housing bodies did not come overnight. Services clubs and municipalities had built non-profit senior citizens' apartments since Mansur's pioneering 1948 NHA amendments, which established social housing in Canada on a permanent basis. But such accommodation was seldom found for familiy housing, championed by social-activist groups, in contrast to the more genteel backers of better shelter for seniors. Moreover, the surge of the third sector would greatly improve existing seniors' housing built since 1948, as attitudes that stigmatized the poor, epitomized by the spartan and unpopular bachelorette apartment, collapsed with the greater competition among social-housing providers. In vivid contrast to the bleak institutional-style, high-rise apartments towering over cleared land that characterized public housing under urban renewal, the new third-sector developers would compete with entrepreneurial developers for the middle-class housing market, assuring higher-quality accommodation to low-income residents of mixed-income projects.[c]

Hellyer's sudden termination of urban renewal was part of the turbulence that had begun to shake public confidence in federal housing policies. This was reflected in his unusual housing task force and the CMHC task force on low-income housing. Both had the goal of replacing the NHA of 1954 with new federal housing legislation geared to a comprehensive housing policy. Both would encourage the land-banking, rehabilitation, and third sector achievements of the 1973 NHA amendments.

Concern about the quality of public housing was closely linked to urban renewal. In major urban centres since 1956, the two had tended to go hand in hand, with the welfare of tenants secondary to that of commercially motivated redevelopment. Even where public housing was built in suburban locations, new projects, such as those of the Ontario Housing Corporation in metropolitan Toronto, tended to be on the high-rise model. Such projects invariably caused difficulties in projects heavily populated by tenants with young children, who would often use elevators, stairwells, and corridors as their recreational areas. Suburban projects were often relatively inaccessible by public transit, located in less desirable areas, and suffer from the wind-tunnel effect caused by high-rise design.[44]

Criticism of public housing in the mass media and by the Hellyer task force tended to focus on projects in the metropolitan centres of Halifax, Quebec City, Ottawa, Toronto, and Vancouver. In these cities alone, average project sizes were in excess of 150 units. Earlier large-scale projects associated with urban renewal were no longer being duplicated by public-housing authorities at the time the criticism of their alienating, high-rise, ghetto-like concentration of the poor peaked. Also, CMHC funding criteria had been changed to incorporate community recreation centres into public housing, absent from earlier projects.[45]

Avoiding "monster"-sized public-housing projects and providing recreational services in social housing were the most easily implemented reforms. The magnitude of the challenge lay in the failure of both public and private sectors to develop a capacity for managing family rental housing. Most major apartment developers refused to accept families with children. Tenants interviewed expressed dissatisfaction over delays in repairs and unresponsive attitudes, which, if similar to problems encountered in the private rental sector, collided with their own belief that government should provide a higher level of service.[46]

Concern was also expressed in the early 1970s over the quality of life in senior citizens' projects. These first emerged out of the 1948 limited-dividend housing NHA amendments, which permitted municipalities to establish their own housing companies. Lacking rent-geared-to-income subsidies, many tenants in these projects lived in poverty, paying over half their income in rent. This was eased after 1964 with the new pattern of building rent-geared-to-income housing for seniors, on the same basis as the public-housing program.[47]

Affordability was the most easily resolved problem in senior citizens' housing. Although the lack of suburban resistance to seniors' projects had prevented the severe problems of waiting-lists characteristic of the family public-housing sector, similar problems arose in terms of poor project design, management, and amenities. A similar desire to avoid competing with the private sector delayed the incorporation of recreational facilities, such as craft rooms, until groundbreaking assistance was provided by the private Atkinson Foundation in an innovative R.J. Smith project of the Metropolitan Toronto Housing Company (MTHC).[48]

Results of an inquiry by the Canadian Council on Social Development, published in 1973 in a volume entitled *Beyond Shelter*, revealed that senior tenants continued to suffer from the strength of the principle of non-competition, or residual housing. One of the most poignant testimonials to the humiliating impact on tenants of the residual-housing approach was the council's study of the MTHC's high-rise Downsview Acre project. The study found an isolation of tenants, typical of large projects located in suburban areas. A tiny recreation room could hold few of the residents, and residents complained that their apartments had not been repainted in six years. Tenants requiring special assistance were given no help in matters such as cleaning blinds, painting, or replacing tap washers. When one elderly single woman complained of a cold draft, the superintendent relied "You're not senior citizens, you're just in this building because you couldn't afford anything better."[49]

Beyond Shelter compiled a stark list of basic inadequacies in senior citizens' apartments. Only 54 per cent of residents surveyed found access to services such as shopping centres and medical offices easy, a problem more severe in suburban project locations. The report noted considerable dissatisfaction over the sharing of toilets and baths in hostel developments, and that many developments were not within walking distance of community facilities. Also, an often poorly working buddy system was more prevalent in government-run developments than in private non-profits, unsatisfactorily compensating for lack of staff. Only 5 per cent of seniors' projects had tenants represented on governing bodies, and these were exclusively in private non-profits. Community groups also had more contact with non-profit projects than with public ones, and the former had a higher staff ratio, often including volunteers from the sponsoring body. The greater social orientation and availability of services offered by non-profits meant that they tended to serve a more physically incapacitated population than did the public-housing sector. They also served a higher income group. As a result of the introduction after 1964 of rent-geared-to-income housing in the public sector, senior citizens' apartments tended towards two classes of social housing.[50]

Beyond Shelter made several recommendations to increase the quality of senior citizens' apartments. It urged that the cramped bachelorette units be replaced with one-bedroom apartments. Housing projects were to be designed to facilitate the provision of social services. More recreational programming was recommended. Its emphasis would be on assisting residents to plan their own activities. Similarly, the craft and hobby rooms found in only a quarter of projects should become universal features. Access to community facilities should be a precondition for CMHC approval. More volunteer activity in the public-housing sector was called for. Resident participation in management was urged. *Beyond Shelter* stressed the need to give the same financial support to private non-profits as public housing. It also envisaged a broad income mix in all future senior citizens' apartments. Fixed maximum and minimum income limits would be eliminated and subsidies given to tenants who could not afford economic rents through shelter allowances.[51]

While less militant seniors had their advocacy performed for them in the pages of *Beyond Shelter*, pressure by public-housing tenants came from the grass roots. This included the formation of a national public-housing tenants' association, government-paid public-housing tenant organizers, and briefly lived protest-oriented publications such as *Ontario Tenant*. One model of tenant self-management, in Vancouver's Little Mountain public-housing project in 1969, involved the establishment of a co-operative store, a recreation program, visits to senior citizens' study clubs, health programs, and a nursery school. The project's tenant association also developed a management course in conjunction with the University of British Columbia Extension Department.[52]

The criticism of the authoritarian and residual approach of past public-housing efforts found powerful political expression in the reports of both the Hellyer task force and a CMHC low-income housing task force known through two of its report's principal authors, lawyer Michael Dennis and researcher Susan Fish. Both reports urged replacement of public housing with new forms of social housing that did not stigmatize its residents. They recommended that shelter subsidies be given universally instead of being restricted to low-income families. To prevent such funds being wasted in rent gouging or property speculations, both reports recommended bold federal action in the areas of land banking and production of new housing by co-operatives, non-profit associations, and municipalities. Reversing previous CMHC injunctions, the Dennis-Fish Report urged the federal government to make "a conscious decision to compete with private builders in the provision of rental housing for the majority of Canadian tenants." It predicted that by "involving an articulate middle-income constituency" that was "better able to bring pressure to bear, paying their own way and unwilling to accept second best," project planning and operation of social housing in Canada would have to improve. In place of provincial bureaucracies such as the Ontario Housing Corporation, social housing in the future should be built by more responsive "co-operatives, non-profit institutions, service clubs, community groups, municipalities."[53]

The Dennis-Fish Report and Hellyer task force faced the same opposition within the federal government that supporters of municipally administered non-profit housing had in the 1940s. The Department of Finance had always opposed a strong municipal housing role for the same reason the reformers wanted it: the former feared and the latter desired any move towards the socialization of the housing market.

The ringing endorsements of co-operatives by the Hellyer and Dennis-Fish reports challenged a long period of hostility to them on the part of the CMHC board of directors. In 1960 CMHC did sponsor the Midmore Report on co-operative housing – not surprisingly, considering that Humphrey Carver was in charge of such decisions. Despite such support, a memorandum prepared by the corporation after its release suggested, "We have no special mission to encourage co-operatives ... The question of the 'next' steps in connection with the Midmore Report is entirely a question for the co-operative movement itself." Further CMHC discussion of the Midmore recommendations took the view that the "harmonious relationships" between residents in communities would be jeopardized by "ventures ... in co-operative housing." That "king of stability," the Canadian "home," being a private place, would be undermined should co-operative housing flourish in the county.[54]

While the federal government could disrupt neighbourhood stability through urban renewal, it declined to introduce any possible turmoil in neighbourhood life through support for new forms of housing tenure. However, when co-operatives actually got under way, CMHC would be forced politically to assist in their financing. This would take place through the first project, the Willow Park co-operative in Winnipeg.

The pioneering Willow Park project was the product of eleven years of work, from 1959 to 1967. It stemmed from the interest and commitment of a group of staff at the University of Manitoba whose earlier attempts to start a co-operative development geared to home ownership had failed because of the lack of suitable land. Their efforts led to the formation of a study group and provided the impetus for the formation of the Co-operative Housing Association of Manitoba on 23 January 1960. The associations's first board of directors included representatives of the province's leading labour and co-operative groups. Among these were the Winnipeg and District Labour Council, the Manitoba Pool Elevators, United Grain Growers, the Co-operative Life Insurance Company, and the Co-operative Fire and Casualty Company.[55]

Willow Park was received coolly by both conventional private lending companies and CMHC. Its success depended on the united efforts of the province's labour and co-operative movements. This helped to sustain the project past such initial setbacks as the decision of the Winnipeg city council not to provide land for the project. That decision was later reversed, and the city agreed to provide a sixty-year lease, set at a nominal price of $50,000 for land adjacent to its public-housing project. The by-laws of the new co-operative were taken from the earlier study group's examination of Swedish housing co-ops. CMHC's assessment of the lending value of the project forced the removal of many items originally planned. Even after this decrease in standards was met, CMHC indicated it would not forward any loans until the project was 80 per cent occupied by co-operative members. In such circumstances, the project was funded initially by a $330,000 loan from the Manitoba Federated Co-operatives and the province's grain pool co-ops. After the CMHC decision to invoke the 80 per cent rule, $2 million was lent by the Co-operative Credit Society of Manitoba and guaranteed by the Manitoba Federated Co-operatives.[56]

The Willow Park model of housing co-operation between labour and co-op groups was applied across Canada. The Canadian Labour Congress and the Co-operative Union of Canada organized a National Labour Co-operative Committee in 1964. A second housing co-op started in Abbotsford, British Columbia. Here, the directors of a community credit union took the view that "their members had housing problems which could not be resolved by simply lending money for ever-increasing down payments." A subsequent credit-union general membership meeting voted to give $2,100 of surplus earnings towards start-up funds to sponsor a co-op. This project involved thirty family and fifty-four senior citizens' homes and a two-storey recreational hall. It also encouraged senior citizens to design their own housing, which led to such unique features as the inclusion of a second bedroom for visitors or as an aid in times of ill health.

Both Abbotsford and Willow Park underwent expansions. By 1978 three separate co-op housing projects had evolved at Willow Park. They involved 450 homes and shared a common community centre and co-op store. A few miles from Abbotsford, a local of the International Woodworkers of America initiated the first housing co-op exclusively for senior citizens. This involved 84 homes. The British Columbia Carpenters' Union also helped to promote co-op housing in the province. It set aside money for a staff person to promote the concept and used its pension funds for investment in housing co-operatives. In Windsor the United Autoworkers took the initiative in the early Solidarity Towers project, built in 1968. Its 293 units rose along the waterfront on some of the choicest residential land in the city. The project was regarded by some UAW members who became co-op tenants as equivalent to a 50 per cent wage increase.[57]

One of the first steps of the newly formed Co-operative Housing Federation of Canada, which replaced the National Labour Co-operative Committee, was to make a presentation to the Hellyer housing task force in 1968. It was well received, for the task force viewed co-operatives as an attractive alternative to public housing in providing shelter for low-income persons. More explicit recommendations for the strengthening of the co-operative sector came from the Dennis-Fish Report in 1973. It shared the Hellyer Report's basic assumptions about replacing public housing with shelter allowances. However, it put more emphasis on the need to expand non-profit and co-operative housing, where its predecessor had focused on making entrepreneurially produced housing affordable to low-income persons.[58]

In addition to dissatisfaction with urban renewal and public housing, a major source of public controversy in the late 1960s and early 1970s was the escalating costs of home ownership. Policy conflict over this issue was at the heart of the dramatic events surrounding the resignation of Paul Hellyer. The possibility of a ministerial crisis over housing policy had been present since the creation of the Canadian Construction Association in 1919. Had Joseph Pigott, T.H.

Anglin, or Hughes Cleaver been senior Cabinet ministers, a similar crisis could have occurred. The same personalities could have advanced policy recommendations acceptable to the building industry, which would similarly have been opposed by the Department of Finance, financial institutions, and real estate interests. Hellyer, himself a builder, was a kindred spirit to the commercial builders of the 1930s and 1940s, who had favoured bold interventionist measures by government. His ideas would face even more difficulties than those of these predecessors. The realty and financial interests that had earlier opposed measures towardss the socialization of the housing market were even stronger now due to the emergence of large land-development corporations.

Despite Hellyer's air of contempt for CMHC, many of his proposals for solutions to housing problems bore a surprising similarity to those advocated by progressive civil servants within the agency. Carver has recalled how the CMHC Advisory Group developed the New Communities program. Under this proposal, it was envisaged that the "federal government would provide long-term loans for assembling land and the debt would be discharged out of land sales." The Advisory Group believed that "the eventual increase in market price caused by the growth of the community would accrue to the community itself rather than to the private speculator, and so help to meet the costs of public housing and other social services."[59]

The Advisory Group's proposals received the support of the then prime minister, Lester Pearson, who made them the focal point of the 1967 federal-provincial-municipal conference. At once, however, this plan was "seriously undermined," as Carver's memoirs relate, when "a senior official of the Department of Finance managed to delay cabinet approval of the conference papers to such an extent that the legislative proposals could not be sent out to the provincial premiers in advance of the meeting." Carver viewed this "lordly attitude" of the Department of Finance as typical of its relationship to CMHC.[60]

The Finance Department's sabotage of CMHC's proposals for a major federal program of public land assembly was repeated two years later when it helped to scuttle similar recommendations emerging from the Hellyer task force, which had arrived independently at the same emphasis on the need for public land banking. Hellyer has recalled how his own experience as a builder convinced him of the desirability of public land banking. He found that land was developed better "in large parcels ... than 50 or 100 acre subdivisions." Some private developers involved in large-scale land assembly made similar comments to the task force. Teron's Kanata land assembly, near Ottawa, was regarded as particularly impressive, "a breath of planned air amid the unreasoned sprawl of urbanization." Teron told the task force that there was "no use dreaming about planning a city unless you own the land."[61]

The Hellyer task force was also favourably impressed by the impact of land banking on controlling land prices. It observed the relatively low price of serviced land in Saskatoon, a result of land banking, where lots were "among the cheapest in the country" although priced "to yield a modest profit." Here and in Red Deer, Alberta, revenues from land banking encouraged a high quality of municipal services, encouraged generous provisions for parks, open space, and school sites, discouraged urban sprawl, and gave a good supply of lots to small builders. Both cities used the NHA federal-provincial land-banking program established in 1949 to expand their holdings of land acquired for tax purposes in the Depression.[62]

Hellyer's proposals soon encouraged the opposition of the Department of Finance. This quickly emerged at a high-level dinner party held between task force members and civil servants after the completion of the final report. The Department of Finance representative was a former close associate of W.C. Clark, the current deputy minister of finance, Robert Bryce. Bryce expressed his suspicion of municipal interference in real estate markets, a suspicion remarkably similar to his mentor's. He also indicated opposition to the task force's call for 100 per cent federal loans for land banking and servicing. Bryce and other senior officials, such as future deputy minister Tom Shoyama, maintained that 90 per cent federal loans would encourage greater municipal integrity. The remaining 10 per cent would be financed on the private market. Bryce also expressed concern over the scope of the program and wanted to know its limits.[63]

The Hellyer Report's proposal for large-scale municipal land banking posed a potential threat to the profitability of the vast land assemblies of private developers. A few years later the Dennis-Fish task force found that six leading developers controlled most of the acreage expected to be used for residential development in the metropolitan areas of Edmonton, London, Ottawa, Hull, Toronto, Vancouver, and Winnipeg. In Toronto three firms associated with the Bronfman family owned 5,000 acres on the western fringes of the city. Power Corporation controlled 6,000 acres in Ottawa. Genstar Limited, a Canadian subsidiary of a Belgian conglomerate, owned 2,500 acres in Winnipeg alone. By 1975 it would own 17,300 acres throughout western Canada. In 1973 CMHC researcher Peter Spurr discovered that corporate land assemblies involved 103,092 acres around twenty-two Canadian cities.[64]

Despite the dramatic political battle that ensued over the land-banking recommendations, Hellyer's proposals were modest, cautious, and politically safe. Unlike CMHC's ambitious and potentially risky proposals for new communities, Hellyer's proposals had the undoubted political merit of leaving the municipalities to deal with the thorny question of purchasing and servicing land quickly enough to protect homeowners from being hurt by inflated land prices. If they had the political will to accept generous federal financial assistance and engage in an ambitious land-banking and -servicing program, the opportunity was there for municipalities to secure more effective land-use planning and reduce housing-price inflation. Centres in which real estate interests hostile to such a course prevailed were free to reject such programs, but at the risk of facing the wrath of potential first-time home buyers and small builders squeezed out by developers. The federal government would be insulated from criticism of home-price increases, which its measures to stimulate the mortgage market would inevitable create.

By focusing on the decisions of municipal governments, Hellyer's proposals also safely avoided grandiose schemes to channel urban growth into new cities. Instead they provided flexible tools for municipalities to plan their expected urban growth and, by doing so, to escape the heavy additional servicing costs of urban sprawl. The lure of having municipalities assume the profits of being their own land developers was thought by the task force to be sufficient to secure their enthusiastic co-operation.

It might appear, given the already-existing land banks controlled by giant land-development companies, that the job given to municipalities was akin to Thomas Adams' earlier efforts to lock the barn door after the horse had fled. But as the development-industry critics of the Dennis-Fish study and a report undertaken by CMHC researcher Peter Spurr were quick to point out, the fact that these land banks contained all the land municipalities expected to need for development did not mean that they had cornered the market on land that could be developed. There was still such land to be purchased, as the limited efforts at government land banking that were carried out in the boom indicated. In 1969, through agents working for a law firm, the Alberta Housing and Urban Renewal Commission had been able quietly to assemble 4,864 acres held by 38 owners. The commission adeptly bought this land acreage, without resorting to expropriation, at prices averaging $2,094 an acre. After the announcement was made of the new assembly, Edmonton-area lot prices declined by about $1,000. The commission had shown some ingenuity, buying land that had previously been zoned for industrial development. A study of land assemblies during the early 1970s found that the Saskatchewan Housing Corporation was generally able to purchase land at or below prices on the private market, while the Manitoba Housing and Renewal Corporation was able to purchase at the market price in most instances. The provincial agency in these years having the most difficulty securing land at low prices was the Ontario Housing Corporation. The prices it obtained, however, were even lower: it purchased land at $1,692 an acre in its most criticized expense for assembly of 5,000 acres near Ottawa in 1973.

The success of land banking where it was pursued with a political will, despite the unfavourable conditions of a real estate boom, indicates that the Hellyer task force was at least promoting a realistic possibility when it proposed having municipalities to break the real estate market. Land banking is more easily undertaken as a silver lining of tax defaults, as was the case in Saskatoon in the 1920s, or by municipal foresight in times of depressed land values, as was the case in the purchase of land at deflated prices by German cities in 1919. In any event, boom conditions necessitated the kind of streamlining of the existing federal-provincial land-banking program that the Hellyer task force had in mind. Favourable opportunities had to be seized upon quickly. They could not be delayed by the complex three-party negotiations favoured by the existing program. This would lead to cities being held to ransom by the inflated prices of landowners, or forced into the politically unpalatable course of expropriation.

For all the sensation it produced, the Hellyer task force's controversial land-banking proposal was nothing more than a modification of an existing federal program to make it more effective. Concerns for provincial rights could easily have been incorporated by the precedent of the NHA amendments of 1964, which proved effective in expanding public housing. As with those amendments, so under Hellyer's proposal, provinces would have been able to direct any federal loans to themselves if they wished, due to their constitutional authority over municipal government.

The critics of the Hellyer proposals within the Department of Finance feared large-scale support for land banking. Their concerns centred on the effects of such a comprehensive program being carried out by municipalities rather than on the political criticism that would ensue from a small-scale effort that proved a failure in controlling housing-price escalation. Although the still-secret nature of the documents evaluating the recommendations leaves final answers to the future, the basic pattern of social conflict involved in the decision gives some hints about the outline of the deliberations. If Hellyer's proposals had proved effective, they would have meant a substantial monetary loss to financial institutions, land developers, and real estate interests. These were not inconsequential economic groups. In 1975, 2 per cent of the Canadian GNP involved real estate transactions. This amounted to $1.3 billion. In 1968 mortgage holdings accounted for 51.5 per cent of the investments of life insurance companies and 54.7 per cent of trust-company holdings.[65]

Much of the political debate around Hellyer's land-banking recommendations side-stepped questions of the merits of this proposal. Such evasion was especially true in the case of the constitutionally focused arguments of Prime Minister Pierre Trudeau. Although all the evidence is not in, his taking refuge in an appeal to provincial rights, despite his reputation as an advocate of a strong federal government, conveys the impression that he was seeking to hide conflicts among economic interest groups behind the wallpaper of constitutional abstractions. The task of openly disputing the merits of the land-banking suggestion fell to the back bencher Robert Kaplan. He stressed that, since developers and speculators already owned all the land needed for urban development in most Canadian cities, the land-banking recommendations of the report were superfluous. They would, he argued, "only involve the transfer of huge sums of money from the Treasury, without bringing the land purchased any closer to development."[66]

Kaplan's attack on a land-banking program designed to lower residential lot prices was highly compatible with his own self-interest. If successfully implemented, land banking by municipalities would have reduced the profitability of the land assemblies of corporate developers. Kaplan himself was part of the property industry, in partnership with a major Toronto developer with interests on the suburban fringe, his father-in-law, Max Tannenbaum.[67]

That Trudeau should have to rely on members of the property industry to justify the omission of land-banking improvements from the 1969 NHA amendments did not augur well for the legislation and its consequences. In his speech introducing the legislation Trudeau defined federal jurisdiction in housing matters as constitutionally restricted to "[increasing] the flow of private lending." Hellyer later affirmed that the 1969 amendments were essentially the same as those he had proposed; the only exception was that "the crucially important sections" dealing with land development were simply "cut out."[68]

The amendments of 1969 repeated the pattern of legislation in 1935, 1938, and 1944. In all instances the new laws were formulated after an extensive and wide-ranging public debate. Numerous options were considered, but in the end the federal government simply reinforced the private mortgage market. The amendments of 1969 introduced numerous measures to encourage private lending for home ownership. The maximum period of federally guaranteed mortgages was increased from thirty-five to forty years. A five-year rollover mortgage was introduced. The federal Cabinet was empowered to alter loan-value ratios and limits. In his speech introducing these new features Trudeau pointed out that they complemented other government policy initiatives designed to strengthen the private mortgage market. These included the automatic adjustment of the NHA rate to the general interest rate and changes to the Bank Act. Both encouraged the return of chartered banks into the residential market.[69]

Critics of the land-banking recommendations of the Hellyer and Dennis-Fish reports have argued that it was unrealistic to expect that the relatively small proportion of the housing stock represented by newly built homes could set standards for the home-ownership market, which was heavily influenced by the resale value of older homes. Such a perspective, however, ignores the recommendations for inner-city land acquisition and housing rehabilitation intended to discourage the gentrification stimulated by more sensitive approaches to urban renewal. To encourage the acquisition of the older housing stock by co-operatives and non-profit organizations and dampen real estate speculation, the Dennis-Fish Report urged that 100 per cent grants be given for housing rehabilitation by "municipalities or non-profit groups purchasing existing larger rental projects and operating them on a non-profit basis."[70]

One of the ironies of the impact of the 1973 NHA amendments was that their provisions accomplished what the foes of urban renewal wanted in terms of better services while abetting the displacement of the very persons the program was designed to help. Albert Rose has observed how, in the Don Vale district of Toronto, the 1960s foes of urban renewal achieved "improved civic housekeeping, rational street patterns, additional social and recreational facilities, housing rehabilitation, more adequate urban planning." Such benefits, however, were not "delivered to those who apparently fought for them against very severe opposition a few years ago." These persons were largely squeezed out of their neighbourhoods by higher housing costs. In the absence of measures to control the real estate market or transfer income to lower economic groups, improved community amenities, funded by the Neighbourhood Improvement Program (NIP), became the means by which a neighbourhood lost its working-class residents.[71]

The federal government refused to allow NIP funds to be used for land acquisition in inner-city areas. Although such proposals were supported by the city of Toronto, they were rejected by the federal government. Attempts by the residents of the South Carleton community of Toronto to use NIP to encourage the acquisition of housing by non-profit groups met with the response that they violated the "integrity" of the program. Since NIP could not be used to retain the community's stock of low-income housing, the South Carleton residents' committee pulled out of the project. They decided that the scheme's funds for beautification would increase property values and so heighten the pace of the displacement of working-class families.[72]

Land speculation could have been curbed if housing had been bought up for control by non-profit and co-operative groups. In the absence of such curbs, the most effective programs of neighbourhood improvement would cause the displacement of working-class residents by stimulating an increase in land values. In Vancouver, a Strathcona-area program based on the rehabilitation and upgrading of existing facilities achieved such dramatic improvement that land values increased until many of the original inhabitants could not afford to live there.[73]

The combined impact of NIP and RAP (Rehabilitation Assistance Program) was too tepid to counter trends towards the displacement of low-income families from their neighbourhoods, especially after the shrinkage of the federal housing role after 1978. Individual low-income homeowners would invariably be helped by better terms for housing-rehabilitation loans than they could get on the private market, but such improvements could not stem the tide of commercial redevelopment and gentrification. Nation-wide, decline from 31 to 27 per cent in the population of inner-city areas took place from 1971 to 1976. Once encouraged by urban renewal, the "loss of access to central area housing stock from the resurgence of commercial rebuilding in downtown areas" became recognized by federal housing planners as a socially disruptive trend. In Vancouver, from 1973 to 1976, some 4,400 residential units were demolished.[74]

The impact of the government response to the Hellyer task force was to fuel housing-price inflation. Its recommendations for stimulating home ownership were adopted, while proposals for curbing costs by streamlining the land-banking program were rejected. From 1979 to 1975, residential lot prices would increase on average by 40 per cent above the Canadian general inflation index. In Alberta and British Columbia the increase would be 129.4 per cent. In Ontario it was 95.2 per cent.[75]

The volume of mortgage money dramatically increased as a result of the 1969 NHA amendments. In 1970, $2 billion in residential mortgage approvals were made. By 1976 this had increased fivefold. This surge was accompanied by a 6 per cent decrease in mortgage rates. Demand for home ownership was also stimulated by exemptions from capital gains taxes, which gave homeowners preferential treatment. Increases in real income and falling rates of return on other financial investments also contributed to demand pressures.[76]

The boom in residential lot prices created opportunities to earn returns far above any normal rate of profit from other legal forms of investment. A typical Calgary suburban home in these years worth $63,000 had $20,00 of its value accounted for by the developer's profit on the sale of the lot. In Winnipeg, such developer's profits on land amounted to $7,100 to $10,000 on a new $50,000 home. For Winnipeg homes in the $60,000- to $65,000- price area, this figure ranged from $15,000 to $24,000. In Vancouver, vacant residential lots selling for $15,957 were resold for $30,804. In Erin Mills, near Toronto, lots selling for $33,394 cost the developer only $12,668 to produce.[77]

During the early 1970s the average rate of profit in the fifteen largest public-land-development corporations was between 30 to 40 per cent. This inflation of residential lot prices also proved advantageous to financial institutions. The inflation of lot sales above prices gained by normal returns on investment added $17.4 billion to the value of residential mortgages on new homes constructed from 1970 to 1978.[78]

Among the 1973 NHA amendments adopted under the pressure of an NDP-influenced minority government was an expanded land-banking program. Land-banking increased from $75 to $185 million annually in federal expenditure. The government pledged to develop a $500-million program of advanced municipal land acquisition over several years.[79]

The new or finally developed public land banks of the mid-1970s were able to achieve the goals of greater public revenues and the provision of lower-priced housing. Edmonton's Mill Woods assembly, purchased by government in the summer of 1969, sold lots in 1976 for $12,510. Private developers' similarly situated residential lots cost $24,000 to $29,000. In Halifax, lots in public land banks sold for $6,830 while comparable lots of private developers sold for $11,000. Such bargains inevitably proved quite popular. Some 3,000 applications were made for the 725 serviced Mill Woods lots. In 1976 Thunder Bay city lots sold for $11,500; similar lots of private developers cost $20,000. The Malvern assembly in North York sold lots at prices $24,000 below market value. Land-bank administrators developed a number of administrative techniques to restrict speculative resales by homeowners.[80]

With the return of the Liberals to majority government, the demise of the land-banking program quickly followed. David Greenspan, a Liberal Party supporter and lawyer, was appointed to head a federal-provincial task force on the cost of serviced residential building lots. The committee's recommendations served as the basis of the termination of the entire land-banking program, which had been in operation since 1949. On September 1978, the same day as the release of the Greenspan Report, the minister of urban affairs, André Ouellette, announced the program's termination.[81]

Greenspan's report maintained that high lot prices were an inevitable consequence of any attempt at land-use planning to secure a better urban environment. It concluded that "legitimate concerns over environmental issues, service standards, development patterns and densities have led to restriction which underlay the lot supply shortage during the boom." Such a conclusion ignored previous realities of Canadian history; frequently, as in the years 1900 to 1913, high lot prices arose in the absence of any planning regulations. Greenspan also overlooked the experience of other nations with comprehensive programs of advanced public land acquisition. The report simply assumed that lot prices would collapse if building permits could be approved on any of the Toronto region's 700,000 vacant acres.[82]

Even more manipulative was the Greenspan Report's attack on municipal planning departments for failing to keep inventories of vacant lots. The report actually ignored these very studies when they conflicted with its own conclusions. Despite the alleged "shocking lack of data" of these authorities, municipal planning studies were numerous and invariably drew conclusions at variance with Greenspan. They documented the fact that, in a given year, more lots were available for building-permit approval than developers were interested in building on. A substantial lot surplus consequently existed in the midst of a rapid lot-price inflation. Civic planners in London found, in August of 1974, that the city had an inventory of 1,676 approved lots. On 31 December 1976 a similar study in Winnipeg found that the city had 3,768 serviced and approved lots awaiting development. Municipal inventories in Calgary found that, during June of 1973, the city had 15,000 lots on stream. On the average, only 4,000 to 5,000 thousand homes were built annually in the city. At the time of the release of Greenspan's report, some 17,623 lots were available for building in the Hamilton-Wentworth region. Even after using a safety factor of 100 per cent, the estimated need of builders in this district was for only 9,477 lots. In the neighbouring Niagara region, some 30 per cent of the semi-detached lots created from June 1970 to June 1978 remained vacant. Only 62 of the 152 subdivisions in the region created before 1976 were fully occupied with homes. In the twelve subdivisions created in the next two years, applications for building permits were applied for on only 51 per cent of the lots.[83]

The housing-price boom was more complex than the classic situation of oligopoly, where a few sellers conspire to control the market. Such factors as real estate hype and the promise of tax free profits from future resales played a role, as did panic buying, born of the assumption that prices for a home in the foreseeable future would be even steeper. Forces stimulating market demand allowed developers to sell lots at prices that secured windfall profits. The developers were not "forced" to earn such high profits by a shortage of lots and serviced lands, which in all major Canadian cities were far in excess of building needs. Greenspan's report also ignored the servicing studies of CMHC's Land Assembly and New Communities division. This branch of CMHC undertook an infrastructure mapping program that covered 75 per cent of Canada's residential land supply. The studies did not include "infill lands" within existing built-up areas, but even with this conservative factor built into their estimates, CMHC found massive surpluses of serviced, developable residential land in most major Canadian metropolitan centres. Despite soaring lot-price inflation, Calgary and Edmonton had a 3- to 4.9-year supply of serviced land. The greater Toronto region had a 5-year supply of serviced land, Windsor a 6.9-year supply, and Halifax 7. Among the highest were Hamilton with 8.9 years, Sudbury and Thunder Bay with 9, and Winnipeg with 10.9. London and Ottawa had a supply of 11 and 12.9 years respectively. In three years, sewers already planned for construction would create enough serviced land for 10 years of expected residential growth across Canada. Most Canadian cities had a supply of 3 to 5 years of building lots at a given moment, less than the total surpluses of serviced land but still substantial. CMHC concluded that "the land available for development in 16 cities (excluding Montreal) is usually two and one half times the estimated demand." On average this amounted to a 145 per cent surplus of the supply of land available for development.

Also, CMHC found that, in seven metropolitan areas, the combined holdings of the four largest landowners were enough to supply all the projected housing demand.[84]

Greenspan's report pointed to low lot prices in Montreal and urged that it be regarded as the model for the nation. Here, municipalities serviced the land themselves, not relying on private developers to do the job. Greenspan urged that the federal government assist municipalities financially so that they could afford such servicing costs across the country. The solution, he stated simply, was to "attack municipal deficits, not municipal virtues." This recommendation conveniently ignored the findings of the CMHC Land Banking and New Communities division that the pattern of Quebec municipalities was not geared to supplying an adequate supply of serviced lots but to subsidizing massive waste and urban sprawl. In contrast to the 3- to 5-year supply of serviced lots in most Canadian cities, Montreal had a supply for 15 years. In three years Montreal would have a 73.7-year supply of serviced land. Currently, in 1978, it had a 33.9-year supply.[85]

Quebec municipalities were able to maintain an adequate supply of lots for small builders by costly and extravagant servicing schemes paid for by their taxpayers. In the rest of Canada requirements that developers provide services encouraged more compact urban development and the concentration of the supply of future building land in fewer hands. Developers had the opportunity to sell more lots at lower prices to small builders in the subdivisions they owned for which planning approval was granted. This was not done, as market conditions allowed developers to sell lots at conditions that resulted in unusually high rates of investment return. Public land banks had become one of the remaining strongholds for small builders, who increasingly were excluded from developer-owned subdivisions.[86]

Nevertheless, the boost to land banking, undermined by Greenspan, was highly symbolic. Its brief success, from 1973 to 1978, showed that the agenda of those civil servants in CMHC concerned with curbing the inflation of the real estate market, focusing federal housing intervention on the needs of low- and middle-income Canadians, and developing a strong third sector in housing could be suddenly taken out of the world of secret memoranda and put into legislation under the pressure of political survival. Humphrey Carver has recalled how "the original version of the testament written by the Advisory Group in 1965, which had been put into an authorized version for a 1967 conference and then in Lithwick's revised version in 1970, finally reached the statute books in 1973."[87]

The 1973 legislation did provide a major boost to third-sector housing. Within a few years of its passage it led to the situation that the critics of public housing wanted; all new social housing was built by the non-profit corporations of municipalities and private associations, or by co-operatives. The ghettos of low-income groups confined to second-rate projects designed to proclaim the social status of their residents were confined to projects constructed before 1975. Further, the new projects met expectations for higher-quality design, resident participation, and social integration. And often they would be carried out by the critics who had originally urged their adoption. Michael Dennis, for example, would become president of the newly created Cityhome non-profit housing corporation of Toronto. He would oversee such third-sector triumphs as the conversion of the former Bain and Spruce Court units of the 1912 Toronto Housing Company into co-operatives, the remarkable historic preservation and infilling of the Hydro block, and the large-scale integrated new community of social housing in the St Lawrence project.

The NHA amendments of 1973 stimulated the third sector in a variety of ways. For the first time, 100 per cent loans were made available to non-profit and co-operative housing groups. A separate section of NHA was established for co-operatives, who finally shared in the same preferred rates given to limited-dividend housing corporations in the past. Non-profits and co-ops became eligible for loans to purchase and rehabilitate older housing. They were also provided with the same rent supplements given public-housing tenants, permitting them to reserve a quarter to a third of project units for low-income families. Further, the legislation was actively promoted by CMHC. Alexander Laidlaw, a veteran theorist and leader of the Canadian co-operative movement, was employed by CMHC to encourage and guide new co-operative housing groups, and start-up funds were provided for them.

Despite the significant new thrusts towards a comprehensive housing policy, the 1973 NHA amendments moved in contradictory directions. While acting to ease speculation through its improvements to land banking and the third sector, in other ways the legislative program fuelled residential property speculation. In this respect the amendments had an effect similar to the termination of urban renewal, which ended federal subsidies for commercial redevelopment but increased gentrification by improving services and amenities, leaving the real estate market untouched.

The choices facing CMHC after the passage of the amendments of 1973 were well summarized by Alexander Laidlaw in a 31 December 1973 memorandum. Laidlaw attempted to persuade his CMHC colleagues that it was now "especially appropriate for CMHC, as a public corporation, to concentrate on public and social housing, leaving the private sector to carry on as far as possible under its own resources." Since, in its "earlier years," CMHC had provided "the underpinning required by entrepreneurs in the housing industry," Laidlaw felt the time had come for the corporation to devote "its energies and resources" largely to the third sector. The adoption by CMHC of an aggressive condominium-promotion policy meant a rejection of Laidlaw's suggestions for a new course. Of the CMHC booklet "Ten Advantages of Condominiums," he commented that the only advantage listed that co-operative housing could not provide was "the opportunity for capital appreciation normally associated with the ownership of real property." Laidlaw felt that "the difference is real and fundamental," since housing co-ops did "not provide the same opportunity for private gain."[88]

One of the great ironies of federal housing policy was that, in rejecting Laidlaw's advice and employing aspects of the 1973 NHA amendments at variance with the Dennis-Fish recommendations, CMHC would be forced into a situation of technical bankruptcy. The humiliation of a bail-out by the federal treasury was part of the inevitable price of the pursuit of an impossible dream of making home ownership affordable to low-income groups, even in the modified condominium form, in the heated real estate market of major metropolitan centres.

In the long-campaigned-for changes of 1973 was one program that the reformers had actually warned against. This was a new provision for assisted home ownership. Both the Dennis-Fish Report and Carver's Advisory Group had made what would prove to be prophetic predictions concerning the dangers of assisted home ownership in heavily populated urban centres. Both agreed that such subsidies for low-income home buyers would be appropriate for rural or small-town communities such as Glace Bay, districts that were characterized by a high proportion of homeowners and a stable or even depressed real estate market. Both reports predicted, however, that in booming urban centres, giving more dollars to prospective home buyers would result in escalating housing prices, poor-quality construction, and the bankruptcy of many low-income families.[89]

In issuing its warnings about the impact of assisted home ownership, the Dennis-Fish Report pointed to the unhappy experience of such efforts in Canada and the United States. One such example was the experience of projects under the $200- million CMHC Special Innovations program launched in 1970. One Montreal-assisted home-ownership project built under this scheme was surrounded by a quarry and rubbish dump, since the developer wished to reserve his better lands for higher-income groups. Row housing constructed elsewhere was frequently poor in location, materials, and design. Likewise in the United States, assisted home ownership tended to encourage builders to construct second-rate homes. Shortly after purchase, low-income families found themselves stuck with "faulty plumbing, leaky roofs, cracked plaster, faulty or inadequate wiring, rotten wood ... lack of insulation and faulty heating." The rate of arrears in such projects was double the normal rate. Dennis and Fish concluded that assisted home ownership would inevitably encourage shoddy building and would add "fuel to the fire" of housing price inflation by increasing "expectations of future capital gains."[90]

Dissension over issues such as the encouragement of home ownership among low-income families or the merits of condominiums and co-operatives is evidence that the hoped-for new thrust towards a comprehensive housing policy was not achieved by the 1973 NHA amendments. Housing policy remained a battleground for the value conflicts of Canadian society and would respond quickly to the ebb and flow of political trends.

Co-operative housing did get a major boost from the 1973 amendments and Laidlaw's promotional activities. Within four months of the passage of the 1973 amendments, the number of co-op housing groups in the Toronto area grew from three to thirteen. Growth in British Columbia was especially rapid after the passage of the legislation. It was aided by the province's credit unions, labour unions, and the provincial NDP government. Also, as Laidlaw, now CMHC senior adviser, pointed out in a memorandum of 20 February 1974, CMHC personnel in the province proved to be "enthusiastic and helpful." By the time of Laidlaw's observations, three co-operative projects were under construction in British Columbia. Twelve were in preparation. One was a pioneering mobile-home co-operative park. In Saskatchewan, where 42 per cent of the population belonged to credit unions, the central union began to take a leadership role in encouraging co-op housing and acquiring land for projects.[91]

Laidlaw assisted many new co-operative groups in communities across Canada. He was aware of the mix of community activists needed to spark the development of a co-operative housing project in an area. In the case of Saint John, New Brunswick, he saw that the key to establishing a pilot co-op project of fifty to one hundred units was to get together "a good sponsoring group based on labour unions, churches, two consumer co-ops in the city, credit unions, family services." Laidlaw would impress upon such groups how they could realize some of their long-cherished hopes for housing as a non-profit service controlled by its members. Given the bad experience such organizations had had with the federal government in the past, this often proved to be a difficult task. In Halifax, only after an initial meeting were the participants Laidlaw brought together able to go beyond "abstract matters" and "vague plans." Even more difficulty arose in Cape Breton. Here Laidlaw met with "about 20 people from the College of Cape Breton, labour unions, churches, welfare agencies and citizen action groups." From the descriptions these people gave of housing conditions in the Sydney area, Laidlaw concluded that the region must be "among the worst in Canada." The assembled activists complained of being "woefully neglected in the matter of housing by all levels of government." With this focus on the woes of the past, Laidlaw found it difficult to generate interest in participation with the new co-operative program.[92]

Like land banking, the third sector's expansion began to grind to a halt after the return of the Liberals to majority government. In fact the weakening of the third sector and the termination of land banking were closely related because of the opportunistic fashion in which the Greenspan Report's recommendations for abolishing land banking were implemented. The report was used to provide an excuse to kill the federal land-banking program and so gratify the related cost-cutting concerns of the Department of Finance and the interests of land-development corporations in minimizing competition from small builders, municipalities, and co-operatives. The report's call for infusions of federal funds to duplicate the over-servicing of Quebec municipalities was not adopted. The new federal initiatives it inspired were purely negative in character. Federally funded land banking was not only abolished in suburban areas but was also precipitately terminated as a means of acquiring urban sites for the social-housing programs of municipalities. This caused much of the momentum of the innovative non-profit and co-operative projects to be last, for these groups lacked the capital of private developers to build up their own large land assemblies.[93]

The city of Toronto had used the federal land-banking program to provide land for its own non-profit housing corporation and for co-operative groups. A month after the program's termination, urban affairs minister André Ouellette told city officials that CMHC-insured private-sector loans would be made available for municipal land assembly. A year later a city planning report found that "nothing more has been heard of this proposal although the Mayor has pursued this matter with the Federal Minister." The same study found that the federal cut-off of land-banking funds had "cast ... into limbo" the city's plan to encourage the construction of more social housing. Toronto was denied the assistance of the CMHC New Communities program to assemble parkland for its major St Lawrence non-profit and co-operative housing development. This program, so long fought for by supporters of social housing within CMHC like Humphrey Carver, operated on a minuscule level. Only $1.9 million was spent on it between 1973 and 1978. After the draft development of a land-assembly program financed by the private sector, fatal roadblocks were raised by disputes over acceptable regulations for the scheme. Two years after public land banking ended, even representatives of the Housing and Urban Development Association of Canada began to call on the federal government to maintain inventories of land.[94]

The demise of land banking also helped to undermine the always shaky Assisted Home Ownership (AHOP) scheme. Originally, in 1973, AHOP had provided direct loans to low-income families with one or more dependent children for the purchase of a home. An interest-reduction loan provided 8 per cent interest rates (compared to a market rate of 12 to 12.5 per cent), and grants up to a ceiling of $600 per year were designed to keep monthly mortgage payments under 25 per cent of gross income. Escalating land costs and a conservative drift after the return of the Trudeau Liberals to majority government encouraged the transformation of AHOP from a social-housing program to a prop to assist the mortgage market. By 1976, income restrictions for eligibility were lifted. The earlier mortgage subsidies and grants of up to $600 were replaced with an interest-reduction loan and a $750 annual maximum grant. This grant would be phased out over five years.[95]

The new AHOP program had the unfortunate result of enticing low- and moderate-income families into buying homes that became more difficult to pay for as federal subsidies ran out. Six years after the purchase of their homes, families faced increases of 150 to 200 per cent in their initial monthly housing charges. After the end of five-year mortgage renewals, interest rates suddenly jumped from 8 to 16 per cent. Mortgage values became greater than the depressed market value of the homes.[96] Unemployment and unstable real estate prices also contributed to defaults.

The collapse of AHOP brought CMHC to a state of technical bankruptcy. In 1979, defaults forced it to pay out $490 million in claims to private lenders. In 1980 these payments amounted to $450 million. Such payments for insured loans resulted in CMHC's borrowing $212 million from the federal government's consolidated revenue fund. In expanding suburban centres such as Mississauga and Brampton, AHOP repossessions put CMHC into the unwelcome situation of being the biggest landlord in community.[97]

Particularly disastrous experiences tended to befall condominium owners in the AHOP scheme. In the greater Toronto region, with new single detached homes costing a minimum of $60,000, only condominiums could qualify for the maximum $45,000 AHOP ceiling. As a result of AHOP foreclosures CMHC became the owner of 28,500 condominium units. These were worth more than one billion dollars. Also, it became obvious that condominium projects built under AHOP were plagued with structural defects. These habitually included leaky roofs and walls, poor insulation, and the use of cheap materials that deteriorated quickly. Faced with such insoluble problems, developers in the Toronto region had begun by 1980 to build exclusively for the luxury market. Newly built condominium units now sold in the range of $88,000 to $165,000.[98]

The retreat of private investors from the building of housing for medium-income families resulted in CMHC's improvising unusual expedients to lure them back into the housing market. The Canada Mortgage Renewal Plan attempted to cushion the worst effects of interest-rate fluctuations on homeowners. This program was modest in scope compared to the Canadian Home Ownership Stimulation Plan. This involved an outright grant of $3,000 to first-time home buyers of either a new home or a resale property. The sum expended on this short-lived scheme was almost equal to five years of CMHC's constrained social-housing budget. From 1 June 1982 to 1 January 1984, $782.4 million was spent on the special home ownership stimulation scheme. From 1979 to 1983, only $792.1 million had been spent by CMHC on all its varied social-housing programs.[99]

To stimulate private rental construction, CMHC improvised the Canada Rental Supply Plan. This provided owners of new rental-housing projects with interest-free loans for a period of fifteen years. The only guarantee of ensuring modest quality, as opposed to luxury housing, under this scheme was to limit federal loans to $7,500 per unit of housing built. With tax incentives included, total federal subsidies amounted to a third of the capital cost of such units, substantially greater assistance than that provided to non-profit and co-operative groups. CMHC did not respond to difficulties in the housing capital markets by increasing its direct-lending activities, as it had in the late 1950s. In 1983, the corporation made only 130 direct loans for home construction.[100]

Inflated capital and land markets paradoxically meant that the single detached newly built home became affordable in most Canadian metropolitan centres only for pioneering developments of co-operatives. Metropolitan Toronto's first co-op project involving single detached homes opened in 1979. Named the West Humber Co-operative, it comprised thirty-three homes. The project was only possible because it was built on low-priced land acquired from the now defunct federal land-assembly program. Eight of the community's residents earned less than $13,000 annually and so were eligible for subsidized rent supplements. The project was sponsored by the Toronto Labour Council's Housing Development Foundation.[101]

The lack of access to large land banks continued to cripple other legitimate non-profit and co-operative groups. These were placed at a disadvantage in competition with entrepreneurial builders for limited social-housing allocations. The land holdings of developers allowed them to assemble proposals quickly for the hastily announced annual social-housing-unit allocations for a given Canadian region. In the Toronto region, for example, Cimpello Charitable Trust was created by the giant corporate developers the Del Zotto family. The development company's access to massive land assemblies allowed it to garner a substantial per centage of the unit allocations for federally assisted social housing in the region.[102]

Co-operative housing had achieved enough success stories by 1978 to win important allies. Five years after the passage of the 1973 amendments, co-operatives had grown from 1,500 to 22,000 units. The movement had sparked many innovations and struck roots deep enough in Canadian society to resist mounting efforts to pull it out. A variety of adoptions of the co-op principle made it helpful in a striking range of areas.

An International Woodworkers of America credit union in New Westminster purchased a forty-eight-suite apartment block for conversion to a co-operative for the use of its members. Through its sponsorship of co-operatives, a credit union in Campbell River, British Columbia, became the major housing developer in its community. A housing co-op in Regina was formed to meet the housing needs of single parents. In Mission, British Columbia, St Andrew's United Church sponsored a project of ninety senior citizens' homes in a wooded setting, complete with its own community centre and four units designed to accommodate the use of wheelchairs. The "Ideal Village Co-operative" of a suburb of Ottawa emphasized "the natural lifestyle of its members, who are all non-smokers." Another Ottawa-area co-operative, based on a "commitment to Christian principles," stressed the "bond of association between members." The Toronto Building and Construction Trades Council established a Labour Council Development Foundation to sponsor co-operative projects. It lowered construction costs by eliminating the expense of contractors and subcontractors; in their place, tradesmen were hired directly by the foundation for co-operatives. An early foundation project was the Grace MacInnis Co-operative, which saved Victorian homes threatened with demolition by a developer planning a condominium project.[103]

However, in spite of such achievements growing pressure on the entrepreneurial housing sector and the conservative drift after the 1975 return to majority of the Trudeau Liberals condemned the third sector nourished by the 1973 legislation to a perpetual state of siege reminiscent of the continual review of Wartime Housing. Trudeau's close personal friend, CMHC president William Teron, would subject social-housing innovations to a deadening scrutiny that would have won the applause of W.C. Clark.

During his tenure Teron was granted a leave of absence to manage his private business affairs. His developer-oriented views were expressed clearly when the Canadian Co-operative Union sent a journalist to ascertain his position on housing co-ops. Teron criticized co-operatives for failing to incorporate "the idea of capital gain for the individual home-owner into the co-operative model." He maintained that such "greed seems not to miss anybody."[104]

On 5 May 1978 urban affairs minister André Ouellette announced changes to federal policy on co-operatives that appear to have incorporated the attitudes of Teron and other market-oriented persons within CMHC. The proposed changes in policy, released by Ouellette in a document titled *New Directions in Housing*, effectively called for an end to the federal co-operative program. If passed, these changes would have ended the co-operative nature of the twelve thousand housing units built since Willow Park opened in 1964. Ouellette proposed to terminate the formula introduced through the 1973 NHA amendments for encouraging co-operative housing. Long-term fixed subsidies were to be replaced by annually adjustable operating subsidies. Before receiving any allocation for new construction, co-ops would have to buy up any unsold or unrented private-sector housing in their communities. This was announced when CMHC was beginning to dispose of the large number of vacant condominium, AHOP, and Assisted Rental projects that had fallen into its possession due to foreclosures and bankruptcies. The Co-operative Housing Foundation observed that this would amount to dumping co-op projects that could not be sold on the private market because of "poor location, poor design or construction, or projected high operating costs." Further restrictions were also called for on co-operatives' freedom in design and development.[105]

The most sweeping of the proposed "new directions" were those that would have ended the co-operative nature of existing projects. All incentives for co-operatives to lower their rents by such means as resisting speculative opportunities and instituting energy-conservation measures and other improvements were eliminated. Co-op rents would have to reflect housing costs determined on the basis of current interest rates. Any surplus that these rents generated would be given back to CMHC.[106]

These ingenious measures aimed at transforming co-operative housing into a pliable instrument for CMHC to supplement, but not alter, the private housing market were defeated by a broad alliance of co-operative, labour, and religious groups. Support came from the Canadian Labour Congress, the national Anglican and United churches, several municipalities, and two provincial governments. Also, other co-operative groups and provincial credit-union centrals gave support.[107]

The support of many powerful groups in Canadian society strengthened the negotiating team of Canadian housing co-operatives that met with Ouellette in the months following the *New Directions* announcements. Co-operatives announced their acceptance of the federal government's demand that they obtain their capital financing from non-governmental sources. This was made possible when Co-operative Trust of Canada offered to replace CMHC as the source of capital loans for co-operatives. Instead of lending directly, CMHC would play its more acceptable role as a guarantor of these loans. It was expected that Co-op Trust would be able to tap into the funds of credit unions, churches, and trade-union pension funds in making its housing loans. CMHC agreed to provide grants to reduce housing co-ops' interest payments so that tenants' monthly charges would equal rents at the lower end of the market in their communities. CMHC agreed as well to continue to provide subsidies for rent-geared-to-income housing for low-income families. Although such subsidies for low-income residents would be continued, annual capital grants to lower interest rates below market values would be phased out.

With private financing from the co-operative sector playing a stronger role, CMHC agreed to impose fewer conditions on project developments. Co-operatives continued to be eligible for subsidized rehabilitation loans. At least 15 per cent of the units in new projects would be reserved for low-income families obtaining rent-geared-to-income shelter supplements. Provinces could provide further subsidies for low-income families without a corresponding reduction in the federal contribution. New co-operative groups became eligible for start-up grants of up to $75,000. Similar grants were allocated for non-profit resource groups providing technical and co-ordination services. A member of a co-operative negotiating team concluded that their victory in saving their movement "was won through political pressure applied by a small movement with big friends and big clout."[108]

Although the co-operative housing movement was saved, annual battles now loomed over the extent of the unit allocations established by CMHC. Typically, in 1980 the federal government budgeted for an allocation of 2,500 co-op housing units, although the demand was twice as great. By March of 1980 all the co-op allocations established for most regions of the country were used up. These were only increased after a campaign by co-ops, credit unions, churches, labour unions, community groups, tenant associations, and social-service organizations.[109]

Canadian co-operative housing increasingly evolved innovative techniques by which residents developed democratic means to establish control over their residential environment. This was most vividly displayed in the case of Montreal's Milton Park, where it became most apparent that urban renewal and co-operative housing are polar opposites. The project was originally part of a large-scale assembly of the Concordia Estates land-development corporation. In 1972 this company demolished 250 homes for a combined hotel, office, shopping, and luxury apartment skyscraper, a scheme that was forced through in spite of hunger strikes and occupations by residents of the affected community. The development company's collapse prevented the project's extension to the Milton Park community. It consisted of 135 buildings containing 600 rooms. The ensuing rehabilitative project, begun in 1980 and completed in 1983, involved 14 housing co-ops and 6 non-profit groups. It upgraded the community's housing stock without displacing the district's low-income residents. Renovated duplexes charging $180 rental in Milton Park were estimated to cost $600 on the private market elsewhere in the city. Even welfare recipients could still afford to live in the community after rehabilitation. Some units were rehabilitated to meet the special needs of senior citizens and the disabled. One apartment was redone to suit the needs of a tenant confined to a wheelchair. A non-profit rooming-house was developed and community control established over the district's twenty-five commercial properties.[110]

The success of the limited but creative third-sector housing developments in the family sector, fostered by the 1973 NHA amendments, resulted in fierce criticism from the property industry similar to that encountered by Wartime Housing. In 1979 the Toronto Real Estate Board (TREB) urged that home-ownership subsidies replace those given by the federal government to non-profit and co-operative housing. The board argued that "people are poor because they have either little ability, or have not worked hard enough." The TREB also proclaimed that "home owners make better citizens." Such sentiments were close to those of a task force appointed to review CMHC, chaired by Donald J. Mathews, during the short-lived government of Conservative Prime Minister Joe Clark. It recommended an end to all federal third-sector housing subsidies and subsidies for rehabilitation. In their place a shelter-allowance system would aid low-income tenants competing in the private market. This call was repeated in the 1983 CMHC evaluation of the non-profit and co-operative housing program. Despite its call for the restoration of the private market, the report cited many achievements of the program. These included providing affordable housing for low-income groups, maintaining high quality of design and maintenance, strong satisfaction among residents, and the elimination of much managerial expense.[111]

The growth of the non-profit housing sector has been constrained by the neo-conservatism of all provincial governments except Ontario's, and a similar spirit in Ottawa since the 1982 election of the Conservative government of Brian Mulroney. Where in the past the non-profit sector was clearly constrained by a quota of unit allocations, these allocations can no longer be met because of the unrealistic maximum unit price (MUP) ceilings. Municipal non-profit housing companies have difficulties getting contractors to bid within the price ceilings of such contractors, who often do so only out of inexperience with realities of the cost of such construction.[112]

While MUP keeps the non-profit half of the third sector down, the major brake applied to the co-operative sector is the lack of positive federal encouragement. After the death of Alexander Laidlaw in 1983, his position as CMHC adviser on co-operative housing was not filled again. Consequently, CMHC plays a passive role in co-operative housing, limiting its role to providing loan guarantees and rent subsidies to low-income families. The promotion of co-operative housing is solely the responsibility of the Canadian co-operative movement. It has designed some innovative features to accomplish this goal, such as encouraging credit-union depositors to use their RRSPs for a fund to encourage co-operative housing. Although it lacks the often considerable land banks of private developers, the co-operative housing sector shows considerable strength. In 1987 the 40,000 co-operative housing units had a book value of $3.5

billion. There were 4,600 co-op units under construction and 5,500 units in the planning stage. To guide such projects, various co-operative housing federations and resource groups employed over 450 full-time staff.[113]

In many ways 1978 was a pivotal year for federal housing policy. Paradoxically, it underscored how much value-judgments on the relative worth of different segments of the Canadian population would be reflected in housing. This year, which marked a major retreat in the low- and middle-income rental housing sector for families, also represented a glowing success for seniors. The call of *Beyond Shelter* to replace the spartan bachelorette unit with a more roomy one-bedroom apartment was now being forced, by the market, on providers of shelter for seniors. The boost in seniors' housing, encouraged by the 1973 NHA amendments, was now being reflected in the marketplace. A CMHC-funded study of senior citizens' apartments built under the auspices of religious bodies in metropolitan Toronto reported:

> In comparing the various government-developed and -operated seniors' residences to congregational facilities, the researchers could not help but note that the latter appeared less austere and institutional. All the residences operated by religious congregations, for instance, included donated furnishings for social and recreational areas that gave the complex a cheery, welcoming appearance, and were supported by organized volunteers who raised funds and provided "homey" services not readily available in government facilities. Volunteerism provided labour which government could not possibly afford. In contrast, the researchers learned of a recent instance wherein a volunteer group seeking involvement in a recreational program at a publicly operated facility in Metropolitan Toronto was discharged by the recreational director, because they were seen as meddlers and a threat to the director.[114]

As low-income seniors were no longer trapped by the subsidized rents of social-housing projects that now competed with each other for tenants, unpopular, cramped bachelorette units became vacant across the country. So successful had the third sector for seniors' housing become that controlling the oversupply of it became the focus of attention. Calgary, for example, by the 1980s had begun to suffer from a 30 per cent vacancy rate in senior citizens' apartments.[115]

The success of public policy in the seniors' sector serves to highlight the national failure in family housing and its deep roots in social values. Although the federal government's termination of its program of public housing in 1978 was part of developer-turned-

CMHC-president William Teron's drive to wind down federal intervention in a number of areas, opposition – largely from suburban middle-class homeowners and conservative governments – had already brought it to a halt in many areas four years earlier. This pattern was most evident in Toronto, which had been the corner of the nation most receptive to public-housing projects. A metropolitan civil service sympathetic to social housing, with the support of political figures such as the early Metro chairmen Frederick Gardiner and William Allen, had connived with the province to keep the location of Ontario Housing Corporation projects secret from suburban councils and ratepayers.

After a long-time leader of suburban resistance to social housing, Alec Campbell, gained entry into the heart of the enemy's camp through his election as Metro chairman, OHC's project locations became a matter of public knowledge. This caused OHC to be barred from all of Etobicoke north of Highway 401, from the last remaining residentially zoned vacant lands in North York, and brought it into continual harassment on every new proposal in Scarborough.[116]

The third sector, based on voluntary and municipal initiative, removed some of the hostility of distant provincial bureaucracies to social housing for families. But its efforts could not produce the breakthroughs for families that actually created vacancies and forced the dramatic upgrading of older projects for seniors. The city of Toronto was unique in pursuing a bold, comprehensive development strategy, comparable to that of the now moribund provincial housing corporations.

Although 113 municipal non-profit housing companies were established from 1973 to 1987, most played a reactive role, struggling to develop comparatively few units under their own management. The fierce battles of the 1950s over social housing did not end when the income-segregated public-housing projects of the past were replaced with more appealing income-integrated units. Winnipeg, for instance, continued the polarization of housing debate characteristic of its past, with a non-profit corporation being created only by the mayor's tie-breaking vote. After ten years of operation, the Winnipeg Housing Rehabilitation program could boast only a portfolio of fewer than four hundred units.[117]

The relative uniqueness of Toronto's boldness in the family-housing field meant that it was impossible there to repeat the dazzling success of seniors' housing. No problems of vacancies plagued public-housing managers, making it easy for them to resist the reforms urged by tenants and social-housing advocates. While unpopular bachelorette units would be replaced for seniors by their new improved choice of accommodation, this lack of alternatives for families would perpetuate the worst abuses in public housing and embolden provincial opposition to improvements encouraged by the federal government.

CRIPPLING THE THIRD SECTOR

While the 1973 NHA amendments boosted the third sector, a number of steps taken by the federal government in 1978 would prove crippling to it. This would set a pattern, finally cumulating in the end of the directly delivered federal co-operative program in 1992, which would greatly constrain the expansion of social housing in Canada. Only in Ontario would the federal retreat be evenly matched by an expansion of social-housing activity on the part of a provincial government.

In addition to eliminating land banking, the federal government introduced other measures in 1978 that frustrated efforts to produce social housing for families. One policy sometimes blamed, the shift to private financing, was less critical than it might seem. Federal guarantees provided private credit, and the co-operative sector enjoyed close relationships with Co-op Trust. While co-ops would often die from lack of land, the difficulty of securing private credit was of less significance.

What was critical in 1978 was the shift of housing responsibilities to the provinces, which would have largely unfortunate consequences for social-housing development. While this shift was justified by the assumption that provinces would "stack on" their subsidies to those provided by the federal government, this process developed quite unevenly. From 1978 to 1981 provincial assistance dropped to only 9 per cent of federal subsidies, and three years later provinces had further reduced their spending on housing.[118] Such moves by neo-conservative governments, as in British Columbia, made it difficult for municipalities to produce affordable housing for low-income families. Provincial policy prevented Vancouver from achieving such targets despite its generous leasing of land for social housing at below-market prices.[119]

The more distant federal government, less vulnerable to the influence of anti-social-housing groups than the provinces, found itself in the 1980s championed by third-sector advocates. This preference for a strong federal role was sustained by such varied groups as the church-funded Urban Core Support Network, federations of housing co-operatives, native organizations such as the Assembly of First

Nations and the Native Women's Association of Canada, and by numerous social-housing advocacy groups based in cities across the country. The co-operative movement was convinced that at least half the provinces had no active commitment to co-operative housing and that some were actively hostile. As a result, in 1985 the movement successfully mounted an intense lobby of MPs and Cabinet ministers to ensure that co-operative housing remain a program unilaterally delivered by the federal government.

In other non-profit sectors less well-organized groups lost out to the provinces, which often excluded them from planning on social-housing issues. Only the strong federal constitutional position of the Department of Indian Affairs has allowed native groups to be directly represented in the intricate negotiations of the federal-provincial Policy and Monitoring Committee. This obscure body annually sets national housing policy. Other interested parties have difficulty in determining which level of government has the ultimate responsibility for their area of concern.[120]

The conservative policy directions launched after 1978 contributed to the mounting problem of homelessness in the 1980s. The need, detailed since the 1930s, for subsidized, low-rental housing did not go away; rather, it increased as the limited supply of private low-rental housing was eaten away by gentrification and other efforts to upgrade it. Consequently, the series of moves that dampened third-sector housing activity had the effect of allowing more people to fall through the holes in the social safety net, which was in other ways diminished by policies of fiscal restraint.

In the 1980s Canadians rediscovered the problem of homelessness, the reality faced by the worst housed in the nation, a shelter-dependent population unable to secure personal control over living space. The treatment they received from the Canadian welfare state was especially harsh. The state consigned single homeless persons ineligible for general welfare assistance – because of their lack of a permanent address – to the uncertain resources of private charity and the severe accommodation provided by hostels.

The great housing surveys of the Depression set a pattern of ignoring the plight of the single homeless person. The public housing that would finally be constructed after 1949 was entirely geared to families. Hostels for single men would be run on the same lines as during the Depression years. The chaotic situation of social-welfare assistance after the federal government's departure from the field in 1941 improved somewhat after the government again assumed some responsibility for the cost of relief with the passage of the Unemployment Assistance Act of 1956. However, the ineligibility of the hostel-dependent population for relief continued because of such requirements as the need for a permanent residence.

Hostels have continued essentially to be warehouses for the destitute. Characteristically, such facilities for men provide only a bed at night. They lack privacy and the minimal control over personal possessions that could be provided by lockers. Typically, hostel residents must leave the premises at 7 a.m. and begin to line up for a ticket to secure a bed at night at 4 p.m. No resources are provided for counselling and skills development, on the assumption that residents simply need to get out and look for work during the day.[121]

A new attitude towards the homeless was marked by the emergence of women's hostels, founded by the women's movement to provide support functions. Women's hostels had previously been paternalistic institutions designed to rescue women from the economic necessity or intimidation that compelled prostitution. Now such hostels provide care, in homelike surroundings, without a punitive attitude to residents in their counselling, life-skills training, and job-search programs. Their success is stark contrast to the facilities provided for men; in 1985 the Ontario Welfare Assistance Act funded men's hostels at $7.50 per day per resident and women's hostels at $25.[122]

The plight of the homeless was made worse by trends of urban renewal and gentrification, unaided by the massive NHA funds of the past but left unchecked by the abandonment of land banking and other positive moves to retain low-income housing stock. Typically, in Toronto some 13,000 units were lost to rental deconversion as former multiple apartments and rooming-houses became townhouses. In Montreal the stock of rental rooms similarly dropped from 15,000 in 1971 to 6,000 in 1986. Such a situation spelled disaster for programs aimed at seeking the rehabilitation of psychiatric patients. The increased housing shortage forced such persons out of the personal independence of rooming-houses and into dependence shelters unable to provide support services.[123]

Increasing pressures on shelters brought a recognition of the need to provide supportive housing for singles. By 1984 model projects for single homeless people were begun that provided separate rooms for residents in place of dormitory accommodation. Metropolitan Toronto created an innovative Singles Housing Corporation. The Homes First Society developed a project designed to get people out of shelters and into affordable housing. In a departure from normal rules, general welfare assistance applicants in Toronto were given up to half a month's entitlement as an advance to assist in obtaining shelter and given the guidance of a housing counsellor. As a result of such innovations, discrimination against singles seeking social housing finally diminished.[124]

At the beginning of 1988 Ontario became the first province to end discrimination against single persons seeking rent-geared-to-income housing. This innovation was applied by the federal government to co-op housing, but only unevenly by other provinces to their public-housing portfolios.

Making singles eligible for social housing, however, increased the length of already long waiting-lists for subsidized units. The largest single source of such units, the Metropolitan Toronto Housing Authority, in September 1988 had a 10,400-household waiting list, some 38,000 people, although only 8 or 9 units become vacant each day.[125] Although programs in place to address core housing needs had an appropriate program for every conceivable need, their underfunding and restrictions on development perpetuated the problems they were designed to address. Using the classic criteria of affordability and adequacy, a 1985 CMHC study found that 500,000 Canadian households had serious housing problems.[126]

The heightened problem of homelessness falls upon families as well. Even cattle-exhibition dormitories and armouries have been pressed into service as family hostels. As waiting lists for social-housing projects mount, hotels and motels are employed to shelter families dependent on social assistance. Parents lose authority over their children, and frequent moves destroy needed classroom continuity. An increasing number of parents feel compelled to give up their children to relatives, foster homes, or adoption agencies because of an inability to secure appropriate housing. Often, the long strain of searching for shelter imposes great emotional damage on families.[127]

The gap between the institution of a program and the alleviation of the problems it was designed to address becomes most apparent in the area of native housing. As even the most casual visit to a reserve or urban concentration of native people reveals, Indians in Canada have shelter problems far more serious than those of any other ethnic or occupational group. Compared to only 6.7 per cent of the general housing stock, 50 per cent of native housing is in need of repair. While 98 per cent of non-native homes have running water, indoor toilets, baths, and showers, 62 per cent of native homes have such amenities. Despite these bleak statistics, the benefits of a bold third-sector approach were extended to natives in 1974, when non-profit native housing companies became eligible for CMHC grants to build or buy housing for low-income families and seniors. But as for the non-native third sector, funding does not meet the needs of the native non-profit housing corporations. Federal funds provide 1,000 units per year, while the native non-profit sector has the capacity to build or acquire 4,000 units annually.[128] Furthermore, CMHC's bureaucratic regulation of native housing operations frequently exceeds such supervision by provincial governments. A native community that established its own sawmill and designed and built its own homes at half the cost of a government unit was declared ineligible for subsidy assistance out of lack of strict conformity to CMHC rules.[129]

Mounting waiting-lists, homelessness, and the increasing magnitude of unmet core housing needs have drawn attention to the gaps in the safety net of social assistance that allow so many to fall into destitution. Commonly, changes in one area of social policy have been stymied because of the effects of others. For instance, concern is widespread that any increase in social-assistance payments will simply be dissipated through higher rents made inevitable by the tight rent market caused by near-zero vacancy rates in many Canadian cities, especially since roomers and boarders are usually exempt from provincial rent-control regulations. These problems are compounded because social-assistance levels set on the basis of provincial averages fail to take into account the higher shelter costs of major urban centres. In Toronto 87 per cent of single recipients who rent on the private market spend more than 80 per cent of their allowance for shelter. In Ontario a third of those not living with relatives or in subsidized housing spend more than 50 per cent of their allowance on accommodation.

The gap between the real cost of living and the unreality of social-assistance payments has spurred a huge increase in emergency food banks, making them one of the few growth industries in the social-services field. This is a needed closure, by private charity, of a gaping hole in the public social safety net. A Toronto health department study found in 1985 that housing-affordability problems caused malnourishment and child neglect because families reduced spending on basic necessities. Crowded and inadequate dwellings imposed particular health and social stresses on children, including weight loss, a slowing in intellectual development, and retardation of physical and motor skills.[130]

Reports on the housing conditions of social-assistance recipients in the 1980s reveal essentially the same pattern as the housing survey of the 1930s. The remedy urged in the Depression, subsidized low-rental housing, was in place, but it was available to relatively few of those who needed it. In Ontario only one in seven recipients lived in rent-geared-to-income housing, putting those forced to seek private accommodation at a significant disadvantage.[131]

The lack of effective co-ordination of social-welfare and housing policies has led to a recognition of the need for review to achieve the comprehensive system envisaged by the James Report of 1944. Just as the James inquiry evolved in a time of unusual demands for social security, occasioned then by the Second World War, a similar body, the Social Assistance Review Committee of Ontario, was formed in a time of mounting evidence of severe social strain and the political pressures of an NDP-supported Liberal minority government. Its aptly named report, *Transitions*, released in September 1988, made a bold call for a transition to a social-welfare system that would guarantee all a "frugal standard" of comfort, enabling "recipients to integrate into the community, achieve self-reliance and exercise choice."[132]

In its proposals for repair of the social safety net, *Transitions* combined recommendations for income security, assistance eligibility, accessibility, personal development, respect for family life, and housing to achieve its goal of "equal assurance of life opportunities in a society that is based on fairness, shared responsibility and personal dignity for all." Its recommendations were carefully designed to prevent increased social-assistance payments from being eroded by higher shelter and other costs.[133] Rate setting would be established by legislation, including a yearly indexation of rates based on changes in the cost of living and a family-budget approach that would incorporate reference to expert opinion and actual expenditures. In contrast to other expenditure items, shelter costs would be paid to a ceiling that would include heating and hydro costs. Social assistance would be provided to all needy families (a considerable change from current rules for single-parent eligibility) and would be extended fully to the working poor to eliminate financial disincentives to work.[134]

Transitions coupled its recommendations regarding the adequacy of and eligibility for social assistance with a comprehensive set of recommendations regarding housing policy, to ensure that the greater funds provided the poor would not be dissipated by their vulnerable position in a tight housing market. The impact of the MUP ceilings was listed in great detail. They assumed economies of scale "unrealistic for smaller projects," failed to take into account local market conditions, and fuelled neighbourhood opposition to new social housing projects since the cost ceilings made it "impossible" for a housing project to meet expectations regarding its"appropriateness and attractiveness."[135]

In addition to recommending an increase in MUP ceilings, *Transitions* addressed the impact of the lack of land banks for non-profit housing developers. It recommended that the land shortage of third-sector housing be remedied by the deliberate use of government land holdings for this purpose. To assist the third sector, it urged a three-year rather than the current one-year period for land acquisitions and rezonings to be completed.[136]

Whether *Transitions* is implemented or becomes another unmet standard like the Purvis, Rowell-Sirois, and James committee inquiries of the past depends on the value-judgments of Canadian society. *Transitions* itself makes its values quite explicit, stemming from a belief in the inherent worth of each person. All are "presumed capable of reason, choice, self-realization and independence"; further, it is posited "that society has a responsibility to assist its members in their development and integration, with a framework of economic equality and social justice."[137]

Although *Transitions* states that its values "have guided the modern welfare system" in its evolution, events only a few weeks after its publication showed that far different, less benign attitudes were shaping Canadian housing policy. Former Toronto mayor John Sewell was fired from his position as chairman of the Metropolitan Toronto Housing Authority, responsible for managing the Ontario Housing Authority's housing stock in the Toronto region. The furore over the decision crystallized the conflicting values that lie behind opposing positions on the future of Canadian housing. Sewell's firing offers insight into how conditions intolerable in market or third-sector housing could be regarded by governments as satisfactory in public housing reserved for low-income families. Astounding revelations of the use of MTHA properties as dumping-grounds for used cars, the lack of superintendents in even the most massive buildings, the unenviable record of having the worst elevator service in the country, and the lack of security protection for residents all point to a punitive attitude towards low-income groups. This was even more apparent in the reaction to Sewell's plans to redevelop MTHA's "large swaths of purposeless grass and asphalt between buildings," used for the "unimpeded, casual disposal of garbage, litter and derelict vehicles." By commissioning architects such as Jack Diamond, skilled in the artful design of infill housing, Sewell demonstrated that twenty thousand new units could be built for a variety of income groups on MTHA's land.[138]

The prospect of the end of the public-housing ghetto set Sewell on a collision course with entrenched federal and provincial policies. His dismissal was calculated on the assumption that the political costs of alienating low-income voters was affordable. Sewell's departure came during a conflict with a federal-provincial task force, established by CMHC, that was preparing to sell to private developers "locationally obsolete" public-housing projects surrounded by gentrified areas that had become the preserve of upper-income groups. That such a body would contemplate a proposal that explicitly deemed low-income families unworthy of the amenities of the central city reveals the nature of the values that have helped to shape existing policy.[139]

Sewell's dismissal was subsequently to unmask an even less savoury aspect of Canadian politics. The affair revolving around Liberal Party fundraisers Patti Starr and Elvio DelZotto, also a major land developer, illustrates how far marketplace values had penetrated the social-housing sector. It was revealed that Sewell, in his efforts to upgrade public housing, had alienated the powerful Starr and the DelZotto family, especially by refusing to contract MTHA work to DelZotto companies. The DelZottos had previously been accused of taking scarce social-housing allocations away from co-operatives and community-based non-profit organizations, bribing an OHC official, and introducing elements of organized crime into the construction industry. These accusations would now pale beside the suggestion of bribery of the executive assistant of Premier David Peterson, Gordon Ashworth.

The reverberations of the Starr affair had even more significance because it occurred in Ontario, the province with historically the strongest commitment to building a substantial social-housing sector. In other provinces the influence of social-housing opponents among real estate and allied business groups was strong enough to prevent such programs from succeeding, particularly after federal cutbacks in social housing were begun by the federal Liberals in 1978. The Starr affair, in essence a model for entrepreneurial manipulation and distortion of social-housing programs, could therefore have happened only in Ontario, since elsewhere provinces tended to avoid vigorous efforts to provide an alternative to housing produced for profit.

With its "Red Tory" tradition going back to the province's Loyalist founders, Ontario has historically been at the forefront of innovation in government housing policy. Its major innovations can be traced back to the limited-dividend projects of 1914 sponsored by the Toronto Housing Company, through the Bruce Report of 1934, the opening of Canada's first housing project in Regent Park in 1949, the rapid expansion of public housing under OHC from 1964 to 1974, and the strength of municipal non-profit housing thereafter. In 1988 it even launched the ambitious Homes Now program, the first in Canadian history to involve a totally provincially funded social-housing program. It included a component for advanced land acquisition, the absence of which since 1978 had so crippled the third sector.

Other provinces have largely abandoned new housing targets under the guise of restraint programs. This has forced groups concerned with social housing to encourage co-operative housing without the assistance of provincial governments. Cliff Andstein, secretary-treasurer of the British Columbia Federation of Labour, has observed that government housing programs, even if they were available, would not be the best solution since they "tend to create a dependency situation and weaken groups' efforts to provide housing that effectively responds to the needs of the people they're serving." Urging planning within a legislative and regulatory framework, Andstein suggests that in British Columbia alone there are $87 billion in government pension-fund assets and $15 billion in private pension funds that could be directed, in part, towards co-operative housing if the community had some say over their use. To date, he notes, only a few construction unions in British Columbia have invested any share of their pension funds in housing. This, of course, took place where there was an immediate benefit of employment for members.[135]

The uniqueness of Ontario's approach was further highlighted by its moves to foster co-operative conversion as the cheapest way to produce social housing, including a considerable portion of the "flipped" Cadillac-Fairview rental-apartment stock (which had been sold at speculative prices to a chain of buyers before government seizure). This new thrust in both provincial policy and tenant activism demonstrates that, despite the cutbacks in social-housing programs, considerable dynamism was still being shown in the third sector, now assisted by new interest in social investment among trade unions looking for less state-dependent strategies for employment in the wake of the free-trade agreement with the United States.

The three decades after the passage of the National Housing Act of 1944 began with an essentially tokenist public-housing program, produced as a compromise between pressures for reform in Canadian society and fanatical opposition to the principle from the Department of Finance, led by the adroit president of CMHC, David Mansur. This program was later carefully nurtured by CMHC civil servants, such as Humphrey Carver, who closely co-operated with their colleagues outside of government, supporting their efforts with federal research funds and ingenious but dangerous methods, such as urban renewal, that brought public housing into such bastions of real estate reaction as Edmonton, Calgary, and Winnipeg. In the climate of controversy of the 1960s, these essentially paternalistic approaches to social housing were modified, typically by the emergence of new co-operative and native-housing corporations that were directly controlled by tenants, and municipal housing corporations that proudly competed with entrepreneurial housing producers.

Three phases of housing policy shifts did not uniformly alter various social-housing projects over time. High-rise, income-segregated public-housing ghettos continued much as they had when subject to a barrage of criticism in the late 1960s – without the procedures for tenant control over management and without the government funding for organizers that flourished briefly in the early 1970s. The fate of public housing for families contrasted vividly with seniors' housing, where the pressure of vacancies forced rehabilitation of earlier projects, built under the residual-housing philosophy of the past, up to the higher standards of the third sector, aiming for a higher-income market. The 1980s saw the first recognition in public policy of homeless singles, but the massive backlog often meant only the opportunity to be put on a three-year waiting-list for assisted shelter.

Social housing had become infinitely more complex than the early pioneering projects that saw municipalities divided between real estate interests on the one side and trade unions and social-service organizations on the other. Now a great variety of groups integrated firmly into the fabric of Canadian civil society had become deeply enmeshed in the delivery of housing – service clubs, credit unions and co-operative trusts, trade unions, and churches. While NIMBY (not in my backyard) residents' associations reinforced the real estate lobbying of the past, social housing had become and remains a central concern of Canadian public life.

The vision of idealists concerned about shaping a more just society finds expression in developing housing protected from the property speculation of the real estate market in increasingly varied forms: a church or trade union turning its parking lot into a seniors', non-profit, or co-operative housing project; tenants uniting to purchase their apartment building; seniors displaying "Grey Power," as in 1978, to save a city of York project from demolition through co-operative conversion; or shelters for the homeless turning their bleak hostels into apartments for independent living.

9 Between Community and Anomie: The Poles of Canadian Housing Policy

Throughout the twentieth century Canadian housing policy has tended to drift between the poles of community and anomie. Anomie was expressed more frankly in the innocent era of the boom of Laurier "prosperity," when big-city boosters and social reformers debated the signals produced by overcrowding. At a time when the rock pile constituted the welfare state and elderly persons frequently retired to poorhouses, slum conditions could still be equated with prosperity and a skilled artisan's sacrifice of his garden accepted as a price for living in a big city.

The frankness of turn-of-the-century boosters in equating overcrowding with prosperity accorded with the statistical inquiries of the era, which demonstrated that the blight of the housing crisis, once associated comfortably with Europe by even the most zealous of reformers, such as Sir Herbert Ames, had come to Canada. Through the remainder of the century Canadian society would be engaged in a debate on how to remedy this situation, where generally a third of the population could not afford to live in housing that was conducive to their health.

During the first three decades of the emergence of a recognized housing problem in Canada, escape from crowded, unsanitary housing in the urban core was possible on the suburban fringe. This safety valve was eventually closed off for all but the middle class by fire codes, public-health regulations, and planning controls. Even when such escapes were possible, they were purchased at a heavy price, not only in dollars but in terms of the initial hardships to families that tended to replicate the conditions from which they were fleeing. Also, periodic real estate booms closed off this option, and busts also invariably led to massive foreclosures on now overmortgaged properties, a phenomenon that largely discredited the federal government's first brief experiment in housing legislation in 1919.

In the first three decades of the century Canadian and American housing policies differed from those of the democratic states of Europe in essentially the same ways. While Canadian reformers were not infected with the same dogmatic preoccupation with building codes as their American contemporaries, their unsubstantial limited-dividend housing achievement meant that their efforts were concentrated in the same fruitless direction – encouraging the destruction of the limited housing stock affordable to low-income people. Neither the pre–First World War efforts to build non-profit projects for skilled workers nor the rental subsidies adopted after the war, which became common in Europe, were attempted on a significant scale in North America. In Europe the labour movement, in many northern countries organizing three-quarters of the labour force, could elect national governments after the First World War. In North America labour was in rapid retreat, both economically and politically, after a brief upsurge in the immediate post-war period. In Canada the return to prosperity of the construction industry meant that discussion of social housing quickly died, save for the most cautious of proposals from a few social workers.

While the absence of government intervention in social housing from 1900 to 1930 can be ascribed to similar patterns of privatism in Canada and the United States, the Depression saw the two nations diverge surprisingly because of one man, W.C. Clark. The same social forces that generated the New Deal's pioneering public housing were present in Canada. In both countries, construction associations of enlightened large-scale builders led the way, building coalitions that included labour unions, social workers, planners, and architects. Neither Mackenzie King nor R.B. Bennett was particularly hostile to Franklin Roosevelt's zest for public housing. These Depression-era prime ministers were, however, guided by W.C. Clark, who was definitely determined to build a wall against creeping socialism from south of the border. While cleverly tying up the construction contractors in bureaucratic knots during the Depression, Clark had less success when they became dollar-a-year men employed by government during the Second World War. Such frustrated visionaries as J.L. Pigott were able to realize their thwarted

Depression dreams in the activities of Wartime Housing Limited, but would again be defeated by Clark in setting the housing agenda for post-war reconstruction.

One of the many legacies of W.C. Clark, intended as an instrument to foster the private market, was the creation of the Central Mortgage and Housing Corporation. Although still heavily influenced by Clark's Department of Finance, CMHC eventually encouraged a more flexible federal housing policy. While a long-time friend and intimate of W.C. Clark, CMHC's first president, David Mansur, did not share Clark's fanatical determination to block the establishment of a single social-housing project lest it provide a subversive example of an alternative to the private market. Beginning with the first effective federal legislation for limited-dividend housing in 1948 and the even more heretical combination of land banking and public housing through a complicated and therefore limited federal-provincial partnership in 1949, Mansur provided the basis for a small-scale, essentially tokenist, social-housing effort. His absorption of the Department of Finance's National Housing Administration into CMHC removed the most strategic fortresses in the Canadian government for assaults on social-housing innovations.

Mansur's drafting of the National Housing Act of 1954 set the basis for sustained federal assistance to the private mortgage market. Although periodically revised by such measures as a temporary venture into mass direct lending at the end of the 1950s and new stimulus to lure chartered banks back into mortgage lending with the 1969 NHA amendments, the basic framework of government-insured mortgages remained. The longevity of the NHA of 1954, which still forms the basis of federal legislation, reflects the continued thrust of government efforts to prime the pump of the private market.

Mansur's departure with the passage of the National Housing Act of 1954 was highly symbolic. Having inherited a pattern of market thrust and social tokenism, Mansur's successors at CMHC would attempt to tamper with the master craftsman's politically adroit framework. While major dents were made in Mansur's casting of federal policies, they could not shift its central direction.

The complex heritage Mansur left to CMHC was most evident in the new pattern that emerged after his departure. His broad-minded approach was epitomized by his decision to hire one of the most insightful critics of his own housing policies, Humphrey Carver, to the post of CMHC research director. Carver was a close associate – and remained so after his CMHC appointment – of Canadian social-housing activists, foremost among them his long-time friend the crusading W.H. Clark. Carver and Clark would try every possible stratagem to encourage the federal government to adopt social-housing policies, including, in the 1960s, mobilizing the poor to speak from their own perspective.

Mansur's departure and his succession by Stewart Bates, former deputy minister of fisheries with little experience in housing issues, showed that the federal Cabinet of Prime Minister Louis St Laurent believed housing problems had been solved when Mansur turned the tap for a flood of new investment in the mortgage market from private banks. Although Carver won the ear of the open-minded and socially concerned Bates, winning the ear of the federal Cabinet was another matter, particularly when he had to go through a sceptical CMHC board of directors, heavily influenced by the Department of Finance. The department continued, in its lordly way, to exert a conservative influence over Canadian housing policy.

As a way of nurturing support for social housing in difficult circumstances, Carver devised urban renewal. His cleverness in doing so was akin to the mischief of the sorcerer's apprentice. By playing with the devil of real estate greed and big-city boosterism, he unleashed a whirlwind of destruction beyond his control. Urban renewal entailed a clever combination of social conscience and land-capitalization increases and did bring public housing to Montreal, Winnipeg, Edmonton, Calgary, and a host of lesser Canadian cities, but it was all the more needed to shelter those displaced by its own projects. This was a critical development, for the first acceptance of public housing in these hostile political environments would later provide the basis for more extensive and varied methods of shelter subsidies once programs got under way. But they would also bring public housing into disrepute, because of the protests caused by demolitions fostered by urban renewal and the high-rise projects that were often built in conjunction with it.

While Carver was dangerously invoking tempests of greed in urban renewal, which his previous clashes with anti-public-housing lobbies should have warned him against, many subsequent events that unexpectedly fostered more conservative housing bureaucracies would have challenged the most perceptive crystal-ball gazers. This was a peculiar chain of events that occurred after the National Housing Act of 1964, which led to the formation of the Ontario Housing Corporation.

Carver, like most social-housing advocates, had in the 1940s supported the recommendation of the Curtis committee for federally financed and municipally managed public housing. With the 1964 NHA amendments he finally secured what was hoped would be an alternative to Mansur's deliberately cumbersome 1949 federal-provincial partnership. The new amendments provided for federal financing of the housing projects of both provincial and municipal governments.

The success of Carver and his colleagues in 1964 was somewhat eroded by a strange series of deaths that underscores the role of individuals in policy evolution despite the constraints imposed by the situation of their times. The NHA amendments were followed shortly afterwards by the deaths of both Stewart Bates and the director of Ontario's housing branch, Wilfred Scott, who together had viewed the new legislation as fostering a stronger and more effective social-housing partnership between municipalities and the federal government. These deaths coincided with an Ontario Cabinet shuffle after the replacement of Leslie Frost by John Robarts, and a sudden shift in Ontario's views about housing responsibilities. Now Ontario took control over essential decisions on public housing through a surprise volte-face, the creation of the Ontario Housing Corporation.

The creation of OHC was the beginning of a decade-long contraction of both federal and municipal roles in social housing, a process that would only be reversed by public criticism of provincial arrogance, which, like urban renewal, would end by generating further public opposition to social housing. OHC had many major triumphs in boosting backlogged public-housing construction, a particularly impressive record considering that it operated in many areas where the conservatism of municipal governments would have prevented them from attempting any serious social-housing efforts.

OHC's success had, however, a darker side. Its achievement as a social-housing developer was never echoed in the realm of management. Also, it operated everywhere in the province, including areas such as Toronto, Hamilton, and Ottawa, which had to give up their public-housing portfolios while suffering less successful management methods. Criticism of OHC by social activists in Toronto dovetailed with real-estate-motivated barbs from suburban areas, causing OHC's demise and an unexpected reconcentration of social housing within the central city's limits.

OHC's stormy history was perhaps the least serious problem unleashed by the growth of provincial housing powers. Ontario's example was followed by other provinces less committed to a bold social-housing program. In Quebec, for instance, the Ontario model foundered on the rocks of French-English constitutional battles, which aided real estate speculators in their campaign to have the bold land-banking recommendations of the Hellyer task force scrapped. Provinces got into the assisted-home-ownership field, setting the stage for the technical bankruptcy of CMHC through the disastrous AHOP program.

Carver and Clark, through their promotional efforts focused on the Canadian Council for Social Development, were able to restore the framework of federal-municipal co-operation in social housing, aided by the controversies generated by the Hellyer and Dennis-Fish task forces. Anti-urban-renewal activists, municipal reform politicians, anti-poverty activists, and the trade union and Canadian co-operative movements all contributed to the boost given to third-sector housing by the NHA amendments of 1973. These finally achieved the direct federal funding of social housing that reformers had so long been advocating.

The period from 1973 to 1978 would be a short-lived Valhalla for Canadian housing reformers. During that time Canadian housing policy seemed to be moving in the comprehensive direction epitomized by the social-democratic Scandinavian states. But the new balance and expansiveness of housing policy would prove to be delicate and politically vulnerable. In 1978, federal termination of land banking and transfer of responsibilities to provinces would encourage a rapid contraction of the growth of third-sector housing, contributing to mounting homelessness in the 1980s.

Complicating the economics of widespread homelessness was the extraordinary effort forced on the homeless to search for shelter, thus wasting time that otherwise would have been spent in job training, therapy, and employment searches. Considerable sums that had been spent on psychiatric care, addiction rehabilitation, or vocational counselling were dissipated in the all-consuming search for basic shelter – the first necessity in getting one's life in order.[1] Despite the heavy social costs of this vicious circle, only in Ontario was any new life really put into social-housing programs, which continued to wither after the fateful curtailments of 1978.

The value-driven nature of the Canadian response to widespread homelessness can be seen in the contrast between programs aimed at the rehabilitation of veterans after the Second World War and the treatment accorded the former psychiatric patients of the 1980s. Immediately after the end of war in Europe, all levels of the Canadian civil service recognized that the housing crisis was playing havoc with efforts to rehabilitate veterans. They could not attend university courses or necessary therapy sessions because they were unable to find housing in major urban centres such as Toronto, where these services were provided. For the respected war veterans, however, a direct federal program of rental housing construction was undertaken. The response to the plight of the homeless in the 1980s, a less-valued group in society, was far less generous.

The history of Canadian housing policy holds a mirror up to Canadian society, revealing new aspects of the national image usually hidden from historical analysis focused on individual figures or on charged disputations over constitutional amending formulas, national symbols such as the flag, or unification of the armed forces. Behind the glare of publicity showered on more glamorous but often trivial issues, substantive debate continued. Despite changes in Cabinet, all movements in policy would have a certain rhythm; the Department of Finance would oppose bolder social-housing programs, and competing government agencies would help to organize disparate groups in civil society, such as the co-operative movements, residents' associations, churches, and social-service agencies, to support and fight for such programs.

Beyond the interdepartmental battles and their roots in conflict within Canadian society, housing policy can be clearly linked to the values of political culture. One obvious tendency is Red Toryism, a nationalist-minded, democratic corporatism with a stress on social solidarity. The Red Tory tradition cannot be properly understood solely within the context of the Conservative Party. Its values have certainly animated nationalist figures in the Liberal Party, such as Newton Rowell, who participated in Borden's Unionist Cabinet. It can also be found in the social liberalism of Mackenzie King and in the practical programs of the CCF-NDP, despite their occasional rhetorical outbursts in favour of eradicating capitalism.

No matter how broadly defined, however, Red Toryism is a minority, although influential, political current in Canada. In the housing field its impact was originally even more confined to Ontario than in subsequent years. Socially minded Ontario Tories like Fred Beer pushed the first Toronto limited-dividend housing company, while Conservative premier Sir William Hearst spearheaded the drive for the first national housing legislation in 1919. Epitomizing the noblesse oblige of the paternalistic reformer, Conservative lieutenant-governor Herbert Bruce was a key popularizer of the basic principles of social housing in the Great Depression. Another Toronto Red Tory, T.L. Church, used Bruce's report to launch the back-bench revolt that created the Dominion Housing Act. A similar-minded Ontario Tory was Wartime Housing president Joseph Pigott, who concentrated most wartime housing in Ontario, partly because of the relative absence of the heavily real-estate-motivated opposition that so stymied his company's operations in Quebec. Premier Maurice Duplessis brought public housing to that province after his conversion by socially minded reformers in his party. With the return of the Liberals of the Quiet Revolution, social housing languished, only to be revived when the Union Nationale returned to power and in 1967 adopted the bold model of a provincial Crown corporation, borrowed from John Robarts' Ontario Housing Corporation.

While OHC may seem the epitome of the top-down Tory approach, its successor as a successful producer of family social housing, Cityhome, was created by a similar political culture. It was shaped by that contemporary incarnation of Red Toryism, David Crombie, who successfully recruited one of the leading theorists of the new third-sector approach to social housing, Susan Fish, as a provincial Conservative candidate and eventually to secure elevation into the Cabinet. The retreat of Ontario Conservatives away from such interventionist policies since 1985 has left them in the political wilderness.

While anti-American Toryism can hold sway in Ontario, whose memories of the revolution were reinforced by the War of 1812, it has been relatively feeble across the rest of the country. In western Canada, in strong contrast to Ontario, manufacturers never interested themselves in providing better shelter for their workers. Here the preoccupation of business elites with real-estate-fostered class conflict has made western Canada the political bastion of Canadian social democracy. This polarization, however, has not turned western Canada into a North American version of the political culture of Scandinavia. While NDP governments have encouraged social housing, their very strength and success have not had a moderating effect on the opposition but have further intensified political warfare, encouraging successor governments to halt all NDP-sponsored housing innovations. Indeed, such reactionary mentalities, transferred to the burgeoning Reform Party, resulted in the federal government's putting a sudden end to unilaterally sponsored co-op housing in the 1992 budget. While in housing western social democrats behave like Ontario Tories, their opponents are better organized and possessed of a missionary zeal, epitomized by the Alberta Social Credit Party's combination of imported American-style religious fundamentalism with the most pronounced hostility to public housing of any Canadian provincial government.

With a loyalist heritage and British connections similar to those of Ontario, Atlantic Canada has seen a more modest version of pro-social-housing Toryism. Halifax and St John's were among the first to use the 1949 NHA amendments. The Catholic Antigonish movement pioneered a program of self-built co-operative homes for workers, in co-operation with the Nova Scotia Housing Commission and later all Maritime governments. But none of the Maritime provinces has been as bold as Ontario because limited financial resources have compelled governments simply to plug into federal programs.

With western Canada characterized by ideological polarity and the Maritimes by financially induced passivity, Quebec perhaps best illustrates the marginality of Red Toryism outside its fortress in Ontario politics. It was clearly a red tint in the Union Nationale blue that allowed the Quebec government to force an unwilling Montreal city council to accept social housing in the 1950s. Similar motives were behind the Union Nationale's bold launch of the Quebec Housing Corporation after over a decade of inactivity in social housing. But these brief bursts of activity highlight the Whiggery of the norm. Judged by its housing record, the Quiet Revolution seems especially tranquil because the Liberals did not dare to provoke the indignation of real estate interests hostile to public housing, as their Union Nationale predecessors had. Although embarking on determined moves to nationalize hydroelectricity in imitation of an earlier generation of Ontario Tories, Quebec Liberals would not attempt to copy their current counterparts in moving to provide affordable housing for low-income families.

While strains of Toryism have woven the fabric of social support for the varied social-housing projects and subsidy payments across the country, other Tory visions of community have been blocked by Whig opposition. This ranges from the fear of affluent neighbours that a proposed co-operative housing project will affect property values to the ideological hostility of right-wing Cabinets in western Canada, the conversion to American conservatism of the national leadership of the Conservative Party of Canada, the laissez-faire liberalism of the Quebec Liberals, and the usual dogmatism of the federal Department of Finance.

The clash of Whig and Tory values, so recently replayed in the national trauma of the free-trade debate, continues to define the flux of Canadian housing policy, as Tories try to curb the real estate market and Whigs stimulate it to foster real-estate-motivated ends such as greater land capitalization. The fixed nature of land seems to make it a certain source of conflict over property speculation because its supply, unlike other commodities, is not influenced by demand.

Tory housing philosophy is based on an organic view of society, with careful attention to the needs of every member of the social group. While private finance and construction and planning consultants play a significant role, care is taken to secure all a place in an attractive residential environment through subsidies and controls on land speculation. Tories build buffers to provide protection from the dynamism of the real estate market, which, if uncontrolled, prices the poorest members of society out of any home and all but the most affluent out of their cherished dwellings.

The biggest exercise in social planning undertaken by Canadian Red Tories was the Milton Park co-operative conversion, which led to the largest citizen-developer confrontation in Canadian history. Its transformation of a six-block area of downtown Montreal into co-operative and non-profit housing associations represented a vigorous setback to the mechanisms of the real estate market. Elsewhere, the land market replaced the previous mix of income groups with more affluent condominium owners. The Milton Park campaign, by contrast, produced islands of blue-collar workers, roomers, students and artists, political and social activists, and immigrant families in the midst of a sea of gentrification. One of the key promoters of this social solidarity in one neighbourhood was Montreal publisher Dimitrios Roussopoulos, who said of its victory, "It takes so much effort to make so little change."[2] Roussopoulos had indeed captured the paradox whereby a massive social protest achieved the conservative victory of keeping things as they were. Milton Park halted the forces of displacement that create homelessness.

If Roussopoulos is the quintessential Canadian exponent of community, then his polar opposite, promoting anomie, was Jean Drapeau. While Roussopoulos successfully mobilized residents to fight developers and take their neighbourhood out of the real estate market, Drapeau ruthlessly made a virtue out of the massive demolition of low-income housing for commercial redevelopment. Earlier he had also led resistance to Union Nationale efforts to introduce public housing to Montreal in the 1950s. In fighting provincially imposed public housing or activist-generated support for co-operative conversion, Drapeau was a leading Whiggish defender of the real estate market against those who would ban property speculation from large corners of downtown Montreal.

Achieving difficult victories such as Milton Park is symbolic of the trials faced by social-housing activists. The pervasiveness of such tensions in Canadian society underlies George Grant's essential pessimism over the possible survival of a Tory Canada in the shadow of the vigorous empire of American Whiggery. Grant's idiosyncratic insight into the conservative nature of socially based challenges to the real estate market arose from his experience on the pioneering Toronto Citizens' Housing Committee during the 1940s. He posed the alternatives available to cities in the future, of being "worlds in which human beings attempt to lead the good life" or bleak "encampments on the road to economy mastery."[3]

Canadian housing policy will continue to fluctuate between the poles of a compassionate, normative community and rapacious striving for economic mastery. Its contours will be shaped by such factors as the chilly self-interest of the affluent faced with proposals for a neighbouring social-housing project, annual budgetary battles, the zeal of tenant organizers promoting co-operative conversion, the federal-provincial battles at first ministers' conferences, and the success of trade unions and their allies in building up funds for ethical investment. The hearts of religious congregations may be tested over the use of their parking lots for housing, and the majority of voters, who are adequately housed, will have to consider, when entering the polling booth, whether any consideration is due those who are homeless. In such varied ways the values of Canadian society will shape the housing and communities it produces, determining whether the future is in the direction of W.C. Clark's mile-high skyscrapers or Humphrey Carver's greener, warmer, and more compassionate landscape. Clark's sterile wastelands of concrete, such as the Parkchester Apartments he favoured, could be financed "economically." Carver's vision required that society create an attractive residential environment worth paying for. Clark's technocratic Utopia and Carver's call for a socially integrated community in harmony with nature will continue to be conflicting versions of the choices we have in housing policy for the foreseeable future.

Notes

ABBREVIATIONS

ECA	Edmonton City Archives
MUA	McGill University Archives
NA	National Archives of Canada
PANS	Public Archives of Nova Scotia
PAS	Public Archives of Saskatchewan
TCA	Toronto City Archives
VCA	Vancouver City Archives

CHAPTER ONE

1 Catherine Bauer, *Modern Housing* (Boston: Houghton Mifflin 1934), passim.; Peter Marcuse, "The Myth of the Benevolent State: Notes towards a Theory of Housing Conflict," in Michael Harloe, ed., *Urban Change and Conflict*, York University Centre for Environmental Studies, Planning and Development Publications, (Toronto 1977), 40–52.

2 Michael Dennis and Susan Fish, *Programs in Search of a Policy* (Toronto: Hakkert 1972), 173; Ontario Association of Housing Authorities, *Good Housing for Canadians* (Toronto 1964), 79; Humphrey Carver, *Compassionate Landscape* (Toronto: University of Toronto Press 1975), 189–90.

3 In the 1890s Montreal social investigator Sir Herbert Ames remarked at the absence of overcrowding in Montreal in comparison to Europe. While Ames's research was valid for this period, conditions had

changed by 1913. See Herbert Ames, *The City below the Hill* (1897; repr. Toronto: University of Toronto Press 1972), 61, 62; and note rapid rent increases in Canada, *Report of the Board of Inquiry into the Cost of Living*, 2 vols. (Ottawa: King's Printer 1915), 1:483, 511.

4 Helen Searing, "With Red Flags Flying: Housing in Amsterdam, 1915–1923," in Henry Millon and Linda Nochlin, eds., *Art and Architecture in the Service of Politics* (Cambridge: MIT Press 1978), 230–69.

5 Eric Einhorn and John Logue, *Welfare States in Hard Times* (Kent: Kent Popular Press 1980), 35.

6 S.S. Duncan, *Housing Reform, The Capitalist State and Social Democracy* (Sussex: Department of Urban and Regional Studies, University of Sussex, 1978), 28, 29.

7 Ibid., 28–32: For a good account of Swedish land policy, see H. Darin-Drabkin, *Land Policy and Urban Growth* (Toronto: Pergamon Press 1977), 309–46. Darin-Drabkin found that land prices in Stockholm rose at a lower rate than in any of the other industrialized nations he studied (including Germany, France, Italy, Spain, and Israel) despite both high GNP and population growth.

8 Duncan, *Housing Reform*, passim.

9 Alfred J. Kahn and Sheila B. Kamerman, *Not for the Poor Alone* (Philadelphia: Temple University Press 1975), 82–9, 151–71; John Greve, *Voluntary Housing in Scandinavia* (Birmingham: Centre for Urban and Regional Studies 1971), passim. For an interesting examination of how Swedish housing policy even in the 1930s relied on financial incentives rather than coercion in the area of low-income housing, see Elizabeth Denby, *Europe Re-housed* (London: George Allen & Unwin 1937), 48–97. Denby contrasted the payment of shelter subsidies in Sweden with the segregation of problem families in Italian and Dutch housing projects of this period.

10 George R. Nelson, ed., *Freedom and Welfare: Social Patterns in the Northern Countries of Europe* (Ministries of Social Affairs of Denmark, Finland, Iceland, Norway, Sweden, 1953), 290–310.

11 Ibid., 293–302; Ernest Marcussen, *Social Welfare in Denmark* (Copenhagen; Det Danske Selskab 1980), 165, 166.

12 Nelson, *Freedom and Welfare*, 300–5.

13 Richard Scase, *Social Democracy in Capitalist Society* (London: Croom Helm 1977), 48, 86–98; Walter Korpi, *The Working Class in Welfare Capitalism* (New York: International Library of Sociology 1978), 80–9; Gil Burke, *Housing and Social Justice* (London: Longmans 1981), 115.

14 Nathan Straus, *The Seven Myths of Housing* (New York: Alfred A. Knopf 1944), 11.

15 Marcuse, "The Myth of the Benevolent State." Veiller's comments were favourably cited by pioneer Canadian planner Thomas Adams in

his regional plan of New York, which drew a spirited reply from Lewis Mumford, who would later influence Catherine Bauer in post-war European housing achievements. See the *New Republic*, 6 July and 22 June 1932, for Mumford's and Adams' replies, and Thomas Adams and Wayne P. Heydecker, "Housing Conditions in the New York Region," in *Buildings, Their Uses and the Spaces about Them* (New York: Russell Sage Foundation, 1931), 274–92, 297–8.

16 Suzanne Stephens, "Voices of Consequence: Four Architectural Critics," in Susan Torre, ed., *Women in American Architecture: A Historic and Contemporary Perspective* (New York: Watson-Guptill 1977), 136–8.

17 Albert Rose, *Regent Park* (Toronto: University of Toronto Press 1958), passim.

18 Harold Clark Papers, City of Toronto Archives (CTA), passim.; personal communication with Mrs Harold Clark; Carver, *Compassionate Landscape*, 85–9; James Lemon, *Toronto since 1918* (Toronto: James Lorimer 1985), 96–8.

19 Carver, *Compassionate Landscape*, 84.

20 Francis Frisken and Dale Hauser, "Local Autonomy vs Fair Share Intergovernmental Housing Issues in the Toronto Region, 1950–1980," paper presented at Urban and Local Politics section, Canadian Political Science Association annual meeting, 1983, passim.

21 Dennis and Fish, *Programs in Search*, 137, 173, 293.

22 Memorandums by Alexander Laidlaw for CMHC, in National Archives of Canada (NA), Alexander Laidlaw Papers, MG31 B32, file 11.

23 Newspaper clipping files, Housing 1950–1963, Vancouver City Archives (VCA), Housing docket H6, M4289–6.

24 T.C. Douglas to Louis St Laurent, 2 Mar. 1949, Public Archives of Saskatchewan (PAS), T.C. Douglas Papers, file Housing.

25 David Mansur to Robert Winters, 10 Aug. 1949, in NA, Privy Council Papers, RG2 18, vol. 127.

26 Harold Kaplan, *Reform Planning and City Politics: Montreal, Winnipeg, and Toronto* (Toronto: University of Toronto Press 1982), 380–3.

27 Ibid., 506–25.

28 Copy of letter from David Mansur to John Flanders, 14 June 1937, in NA, RG19, vol. 3435.

29 Kaplan, *Reform Planning*, 515–25.

30 "The Fear of Co-op Homes," *St Catharines Standard*, 20 Jan. 1990, 13.

31 Ibid.

32 Personal communication from Albert Rose, who, at the time of these events, was a director of the Ontario Housing Corporation.

33 Albert Rose, *Canadian Housing Policies, 1935–1980* (Toronto: Butterworth 1980), 134.

34 For details of these heated battles, see files regarding housing subsidies in the files of the Metropolitan Toronto chairman's office.
35 Rose, *Canadian Housing Policies*, 87.
36 Ibid., 81–7.
37 N. Lloyd Axworthy, *The Task Force on Housing and Urban Development: A Study of Democratic Decision-Making in Canada*, PhD diss. Princeton 1972, 70.
38 Rose, *Canadian Housing Policies*, 87–91; Dennis and Fish, *Programs in Search*, 173–218.
39 Claire Helman, *The Milton Park Affair* (Montreal: Véhicule Press, 1987), passim.
40 Dennis and Fish, *Programs in Search*, 276–9.
41 Ibid., 266–7.
42 Ibid., 267.
43 Ibid., 266–9.
44 Ibid., 276–9.
45 Personal communication from Mrs Harold Clark.
46 Keith Banting, *The Welfare State and Canadian Federalism* (Kingston: McGill-Queen's University Press 1982), passim.
47 Carver, *Compassionate Landscape*, 55, 56, 89.
48 Frank Underhill, "The Housing Fiasco in Canada," *Canadian Forum*, Feb. 1936.
49 Carver, *Compassionate Landscape*, 162.
50 Ibid., 190.
51 Ibid., 182–5.
52 Axworthy, *Task Force on Housing*, 214–15.
53 Rose, *Canadian Housing Policies*, 47.
54 Paul Hellyer, "Why They Killed Public Land Banking: A Political Memoir," *City Magazine* 3, no. 2 (Dec. 1977): 27.
55 Personal conversation with George Barker, president of the Metropolitan Toronto Housing Company, Metropolitan Housing Company 1982–1988 annual reports, files re non-profit housing and May Robinson House.
56 Personal conversation with Albert Rose during his chairmanship of Metropolitan Toronto Housing Authority.
57 Rose, *Canadian Housing Policies*, 126.
58 Graham Fraser, *Fighting Back* (Toronto: Hakkert 1972), passim.
59 Leo Panitch, ed., *The Canadian State: Political Economy and Political Power* (Toronto: University of Toronto Press 1977), passim.
60 Ibid.
61 Ibid.
62 Rianne Mahon, "Canadian Public Policy: The Unequal Structure of Representation," in Panitch, ed., *Canadian State*, 165–98.

63 Martin Loney, "A Political Economy of Citizen Participation," in Panitch, ed., *Canadian State*, 446–70.
64 Carver, *Compassionate Landscape*, 214.
65 Fraser, *Fighting Back*, 8.
66 Margaret Daly, *The Revolution Game* (Toronto: New Press 1970), passim.; Ian Hamilton, *Children's Crusade* (Toronto: Peter Martin 1970), passim.
67 Robert Bothwell, Ian Drummond, and John English, *Canada since 1945* (Toronto: University of Toronto Press 1981), 135; J.L. Granatstein, *The Ottawa Men* (Toronto: Oxford University Press 1982), 23.

CHAPTER TWO

1 John T. Saywell, *Housing Canadians: Essays on the History of Residential Construction in Canada*, Economic Council of Canada Discussion Paper no. 24 (Ottawa: ECC 1975), 46–50.
2 Canada, Commons, *Debates* (1917), 2:1708–9.
3 Canada, Commons, *Debates* (1929), 1:1084.
4 James Roberts, "The Housing Situation in Hamilton," *Canadian Municipal Journal* 8 (1912): 256–7.
5 Saywell, *Housing Canadians*, 123–4.
6 Herbert Ames, *The City below the Hill* (Toronto: University of Toronto Press 1972), 61–2, 111–12.
7 Canada, Commons, *Debates* (1919), 3:3542–4.
8 Canada, *Report of the Board of Inquiry into the Cost of Living*, 2 vols. (Ottawa: King's Printer 1915), 1:483, 511.
9 Ibid., 1:463–85.
10 Marc H. Choko, *Crises du Logement à Montréal (1860–1939)* (Montréal: Éditions coopératives Albert Saint Martin 1980), 30, 40.
11 Terry Copp, *Anatomy of Poverty* (Toronto: University of Toronto Press 1975), 25, 70–6.
12 Ibid., 23. Using more reliable data, André Cousineau, Montreal sanitary engineer, estimated the density increase to be from 3.7 persons in 1905 to 4.6 in the census of 1911. Fully 10 per cent of Montreal families were sharing their lodgings with one or more additional families.
13 *Report of the Board of Inquiry*, 3, 1:483.
14 John Weaver, *Shaping the Canadian City* (Toronto: Institute of Public Administration 1977), 26.
15 Andrew Eric Jones, *The Beginnings of Canadian Government Housing Policy, 1918–1924* (Ottawa: Centre for Social Welfare Studies, Carleton University, 1978), 3. Skilled workers also felt the pinch; the average share of a carpenter's wage that went to rentals grew from 19.4 per cent in 1905 to 23.3 per cent in 1913.

16 *Report of the Board of Inquiry*, 1:482.
17 *Labour Gazette*, Nov. 1905, 498.
18 Charles J. Hastings, *Report of the Medical Health Officer Dealing with the Recent Investigation of Slum Conditions in Toronto*, 5 July 1911, 8–10.
19 *Annual Report of the Toronto Board of Health* (1913), May, 101; Nov. 247.
20 Oliver Zunz, *The Changing Face of Inequality: Urbanization in Detroit, 1880–1920* (Chicago: University of Chicago Press 1982), 400.
21 Shirley Spragge, *The Provision of Workingmen's Housing: Attempts in Toronto, 1904–1920*, MA thesis, Queen's University 1974, 92–4.
22 Charles A. Hodgetts, *Unsanitary Housing* (Ottawa: Commission of Conservation 1911), 31.
23 *Globe* editorial cited in John T. Saywell, *Housing Canadian: Essays on the History of Residential Construction in Canada*, Economic Council of Canada Discussion Paper no. 24 (Ottawa 1975), 123–4.
24 *Report of the Board of Inquiry*, 1:488.
25 Ibid., 1:487.
26 Ibid., 1:483–4.
27 Ibid., 482.
28 Ibid.
29 Melvin Clarke, "Halifax the Great National Port," *Halifax Herald*, 31 Dec. 1913; *Report of the Board of Inquiry*, 1:472.
30 *Report of the Board of Inquiry*, 1:486–7.
31 *Labour Gazette*, Nov. 1905, 498. Phillips Thompson reported that the employees of the Standard Ideal Sanitary Company "waited upon their manager complaining of their inability to find suitable housing accommodation and asking his assistance," *Labour Gazette*, Oct. 1904, 378.
32 "The Housing Problem in Canada," *Labour Gazette*, Oct. 1904, 378.
33 Alan Artibise, "In Pursuit of Growth," in Gilbert A. Stelter and Alan F.J. Artibise, eds. *Shaping the Urban Landscape* (Ottawa: Carleton University Press 182), 134–5.
34 *Report of the Board of Inquiry*, 1:463, 481; M.C. Urquhart and K.A.H. Buckley, eds., *Historical Statistics in Canada* (Toronto: MacMillan 1965), 87–8.
35 A.J. Dalzell, *Housing in Canada*, 2 vols. (Toronto: Social Service Council of Canada 1927), 1:14–15, 19–20, 23–4, 27.
36 A.J. Dalzell, *Housing in the North American Continent*, NA, Dalzell Papers, MG28, file 1–27, vol. pp 16, 6–13.
37 Don Kerr and Stan Hanson, *Saskatoon: The First Half Century* (Edmonton: New West 1982), 111.
38 Norbert MacDonald, "C.P.R. Town: The City Building Process in Vancouver," in Gilbert A. Stelter and Alan F.J. Artibise eds., *Shaping the Urban Landscape* (Toronto: MacMillan 1982), 404.

39 Testimony of Fred W. Walsh, secretary, Vancouver Metal Trades Committee, to Royal Commission on Industrial Relations, Vancouver evidence, pp 535–7, Canadian Department of Labour Library, Hull Quebec.
40 Warren Caragata, *Alberta Labour: A Heritage Untold* (Toronto: James Lorimer 1979), 45.
41 Alan Artibise, "The Housing Problem in Canada," *Winnipeg: Social History of Urban Growth* (Montreal: McGill-Queen's University Press 1977), 239–44; Walter Van Nus, *The Plan Makers and the City*, PhD diss., University of Toronto 1975, p 27.
42 Caragata, *Alberta Labour*, 43–7.
43 Ibid., testimony of Fred W. Walsh, 535–7.
44 Kerr and Hanson, *Saskatoon*, 125–30.
45 Artibise, "In Pursuit of Growth," 136.
46 Weaver, *Shaping the Canadian City*, 13.
47 Bureau of Municipal Research, *What is "The Ward" Going To Do with Toronto?* (Toronto: Bureau of Municipal Research 1918), 20, 32.
48 Ibid., 19, 23.
49 Ibid., 23–31.
50 Ibid., 38–54.
51 Ibid., 39–54.
52 Wayne Roberts, *Studies in the Toronto Labour Movement, 1896–1914* PhD diss., University of Toronto 1978, pp 367–71.
53 Ibid.; Paul Adolphus Bator, *Slums and Urban Reform in Toronto, 1910–1914* MA thesis, University of Toronto 1973, p 15.
54 Hastings, *Report of the Medical Health Officer*, 8; *Report of the Board of Health*, (Feb. – Apr. 1914) 64. Hastings saw landlords as catering to the demands of foreign workers. His opinion that the housing problems of foreign workers were rooted in their own depraved tastes was shared by the Toronto elite. Even the populist *Toronto Star* on 15 June 1912 wrote that foreign workers were "disposed to be content with lodgings that seem miserably cramped to those who are accustomed to the freedom and space of America." The more conservative *Mail and Empire*, on 5 July 1911, condemned the "refusal of the foreign element to move away from the central districts." Both cited in Bator, *Slums and Urban Reform*, 4, 22–3.
55 Paul Adolphus Bator, *Saving Lives on the Wholesale Plan: Public Health Reform in the City of Toronto, 1900–1930*, PhD diss., University of Toronto 1979, p 69; Dr James Roberts, "Insanitary Areas," *Public Health Journal* (Apr. 1912): 177–82.
56 Ibid.
57 Bator, *Saving Lives*, 193–7; John Weaver, "Tomorrow's Metropolis Revisited: A Critical Assessment of Urban Reform in Canada,

1890–1920," in Gilbert Stelter and Alan F. Artibise, eds., *The Canadian City* (Toronto: McClelland and Stewart 1977), 407.

58 Lt.-Col. Laurie, Port Arthur, Ont., "Sanitary Work among the Foreign Population," *Public Health Journal* (1913): 455–7.

59 *Annual Report of the Toronto Board of Health* (Nov. 1913), 247; (May 1913), 101.

60 J.S. Schoales, "The Importance of Housing Inspection," *Canadian Practitioner and Review* 3 (1917): 103–11.

61 *Annual Report of the Toronto Board of Health* (1914), 64.

62 Ibid. (May 1914), 112.

63 Ibid. (Mar. 1913), 113.

64 Ibid. (Jan. 1913), 70.

65 Ibid. (Nov. 1913), 249.

66 Ibid. (12 Aug. 1913), 247; Paul Bator, "Health and Poverty in Toronto, 1910 to 1921," *Journal of Canadian Studies* (Spring 1979): 43–8.

67 Copp, *Anatomy of Poverty*, 86.

68 *Annual Report of the Toronto Board of Health* (November 1913), 251.

69 Percy Nobbs, "The Statistics of Housing and Co-partnership Schemes," *Public Health Journal* (Aug. 1912): 451–3; MacKay Fripp, "Speculations on the Problem of Housing the Working Classes in Vancouver," *Engineering and Contract Record* 28 (1914): 1278. Fripp's article was the only call for subsidized shelter in Canada from any expert until 1932.

70 Hastings, *Report of the Medical Health Officer*, 8.

71 Artibise, *Winnipeg*, 278; "The Housing Problem in St. John, N.B.," *Labour Gazette* 13 (1913): 1171–2.

72 Shirley Spragge, "The Toronto Housing Company: A Canadian Experiment," paper presented at 1975 Canadian Historical Association Conference, Edmonton, 28; John C. Weaver, *Hamilton: An Illustrated History* (Toronto: James Lorimer 1983), 107.

73 Spragge, *Provision of Workingmen's Housing*, 114–15.

74 Bator, *Slums and Urban Reform*, 42–3; Gene Howard Homel, *James Simpson and the Origins of Canadian Social Democracy*, PhD diss., University of Toronto 1979, pp 661–7.

75 Parliamentary Committee on Housing, *Minutes of Proceedings and Evidence of Special Committee on Housing* (Ottawa: King's Printer 1935), 181–200.

76 Spragge, "The Toronto Housing Company," 22–4.

77 "Sanitary Conditions Essential to Healthy Homes," *Conservation of Life* 1 (1914): 19–20.

78 Dr A.K. Chalmers, "How to Deal with Slums," *Conservation of Life* 1 (1914): 19–20.

79 Walter Van Nus, "Fate of the City: Beautiful Thought in Canada," in Stelter and Artibise, eds., *The Canadian City*, 225.

80 Thomas Adams, "Housing Conditions in Canada," *Conservation of Life* 2 (1916): 10.
81 Thomas Adams, *Rural Planning and Development* (Ottawa: Commission of Conservation 1917), 102–18; Thomas Adams, "The Use of Land for Building Purposes," *Conservation of Life* 3 (1917): 29–32.
82 Adams, *Rural Planning*, 102–18.
83 Ibid., 117–18.
84 Copps, *Anatomy of Povery*, 33.
85 Louis Simpson, "Economic Housing of Industrial Works," *Industrial Canada* 19 (1918): 61–2.
86 Thomas Adams, "Housing Shortage in Canada," *Canadian Municipal Journal* 13 (1918): 238.
87 Wallace, *Housing Problem in Nova Scotia*, 6–7.
88 Thomas Adams, "Government Housing during War," *Conservation of Life* 4 (1918): 25–33.
89 Saywell, *Housing Canadians*, 150–1; C.B. Sissons, "Housing," *Social Welfare* 1 (1918): 128; A.J. Dalzell, *Housing in Canada*, 1:22; Michael Piva, *The Condition of the Working Class in Toronto, 1900–1921*, PhD diss., Concordia 1975, p 214.
90 Thomas Adams, "The Housing Problem and Production," *Conservation of Life* 4 (1918): 49–57.
91 Resolution of Toronto Branch of CMA re Industrial Housing in Ontario, PAO, William Hearst Papers, pp 35–7.
92 Notes of a Deputation That Waited on the Prime Minister with Regard to the Housing Shortage on May 28, 1918, Hearst Papers, p 39, 1–4.
93 Saywell, *Housing Canadians*, 153–4.
94 Ibid.
95 Letter from G. Frank Beer to Sir William Hearst, 4 Sept. 1918, Hearst Papers. The lack of realism in Beer's views is illustrated by the estimates made by the Sault Ste Marie and Brantford municipalities for homes worth from $3,000 to $4,000.
96 Ontario Housing Committee, "Memorial to the Dominion Government," Hearst Papers.
97 Saywell, *Housing Canadians*, 154–5.
98 Fred Jones, *The Federal Housing Program of 1919* (Ottawa: Carleton University School of Social Work 1978), 12.
99 Ibid., 9–14; James Struthers, *No Fault of Their Own: Unemployment and the Canadian Welfare State, 1914–1941* (Toronto: University of Toronto Press 1983), 16–30. Borden's government seriously contemplated the introduction of unemployment insurance.
100 For the influence of similar pressures in Great Britain, see Mark Swenarton, *Homes Fit for Heroes: The Politics and Architecture of Early State*

Housing in Britain (London: Heinemann Educational Books 1981). Jones, *Federal Housing Program*, 13.

101 Letters to J.A. Ellis and Sir John Willison, 23 Nov. 1918, 28 Nov. 1918, Hearst Papers.

102 *Report of the Housing Committee of the Dominion Cabinet*, in *Labour Gazette* 19 (1919): 447–50.

103 Michael Simpson, "Thomas Adams in Canada, 1914–1930," *Urban History Review* 11 (1982): 6, 7.

104 Letter from Thomas Adams to W. Burditt, 13 Mar. 1919, NA, Burditt Papers, MG28, I-275, vol. 16, file 16–15.

105 *Report of the Housing Committee*, 447–50.

106 Ibid.

107 Thomas Adams, "Canada's Post War Housing Progress," *Conservation of Life* 5 (1919): 3:2542–4.

108 Jones, *Federal Housing Program*, 24.

109 Letter of 11 Feb. 1919 from Ansell and Ansell to J.A. Ellis; letter, 6 Mar. 1919 from J.A. Ellis to Sir William Hearst; letter, 13 Feb. 1919 from J.A. Ellis to Sir John William; memo, 17 Apr. 1919, re proposed amendments of Ontario Housing Act, all in Hearst Papers.

110 Jones, *Federal Housing Program*, 26–7.

111 Parliamentary Committee on Housing, *Minutes of Proceedings and Evidence of Special Committee on Housing* (Ottawa: King's Printer 1935), 174.

112 A.E. Grauer, *Housing* (Ottawa: Royal Commission on Dominion Provincial Relations 1939), 37, 81. Quebec municipalities experiencing heavy losses included Chicoutimi, La Tuque, and Quebec; some, like St Lambert, by careful management avoided significant losses.

113 Grauer, *Housing*, 37–8; Jones, *Federal Housing Program*, 26–30.

114 Parliamentary Committee on Housing, *Minutes of Proceedings* (1935), 174.

115 Canada, Commons, *Debates* (1921), 2:1142–6.

116 Harold F. Greenway, *Housing in Canada* (Ottawa: Dominion Bureau of Statistics 1941), 56–60.

117 Ibid., 57–8.

118 Committee of Research, Social Service Council of Canada, *Summary of Findings in Regards to Housing in Canada* (Toronto: Social Service Council of Canada 1929), in Cauchon Papers, NA, MG30, B93, vol. 10, p 5.

119 Ibid., 5–7.

120 Ibid., 7–8.

121 Edith Elmer Wood, "Is Government Aid Necessary in House Financing?" *Social Welfare* 9 (1929): 169–71.

122 Dalzell, *Housing in Canada*, 218.

123 William McCloy, "The Social Worker's Attitude toward Housing," *Social Welfare* (1929): 178–80.
124 Dr Grant Fleming, "Health and Housing," *Municipal Review of Canada* 22 (1926): 159–61.
125 Thomas Adams, *State, Regional and City Planning in America*, 7, cited in Simpson, *Urban History Review*, 11.

CHAPTER THREE

1 Citizens' Committee on Housing, *Housing in Halifax* (Halifax: City Board of Health 1932), Department of Labour Papers, RG27, vol. 3357, file 11, 3–6, passim.
2 Citizens' Committee, *Housing in Halifax*, 16–18, 29–47.
3 Ibid., 3, 52–75.
4 ibid., 23–5
5 Ibid., 25.
6 Ibid., 26, 27.
7 Percy Nobbs, President's Message, *Journal of the Royal Architectural Institute of Canada* (JRAIC) 9 (1932): 3; Percy Nobbs, "Speculative Builders, Loan Companies and the Architects," *Journal of the Royal Architectural Institute of Canada* 9 (1932): 4; Gordon M. West, "There Is Work To Be Done," JRAIC 9 (1932): 239.
8 "Plan to Revive the Construction Industry," JRAIC 9 (1932): 256; "Component Parts of Construction Council Organize Permanent Council," JRAIC 10 (1933): 96; James H. Craig, "What Can Be Learned from State Housing in Great Britain and the United States?" JRAIC 11 (1935): 79–81; "National Survey of Housing Needs Planned," *Canadian Engineer* (1935): 13; "A Plan to Revive the Construction Industry," JRAIC 10 (1933): 25–7; Walter Van Nus, *The Plan Makers and the City: Architects, Engineers, Surveyors and Urban Planning in Canada, 1890–1939*, PhD diss., University of Toronto 1975, pp 125, 126.
9 Parliamentary Committee on Housing, *Minutes of Proceedings and Evidence of Special Committee on Housing* (Ottawa: King's Printer 1935), 103, 104, 105.
10 Bessie E. Touzel, "Housing Conditions in the Capital," *Canadian Welfare* (1935): 49, 50.
11 Montreal Board of Trade, Civic Improvement League, *A Report on Housing and Slum Clearance for Montreal* (Montreal: Montreal Board of Trade 1935), NA, Department of Finance Papers, RG19, vol. 706, file 203–1A, 5, 21, 22.
12 Ibid.
13 Ibid., 13.
14 Ibid., 7, 29–34.

15 Ibid., 23, 24.
16 Ibid., 34, 35.
17 Harold F. Greenway, *Housing in Canada* (Ottawa: Dominion Bureau of Statistics 1941), 35, 36.
18 James I. Craig, "Proposed Low Cost Housing Development for the City of Winnipeg," *JRAIC* 11 (1934): 109–12; Harold E. Greenway, *Housing in Canada* (Ottawa: Dominion Bureau of Statistics 1941), 35, 36.
19 Van Nus, *The Plan Makers*, 130–1; press clipping, *Winnipeg Free Press*, 19 Mar. 1937, NA, RG27, vol. 3380, file 17.
20 *Report of the Lieutenant-Governor's Committee on Housing Conditions in Toronto* (Toronto: Toronto Board of Control 1934), 13–22, 37–44.
21 Ibid., 73–6, 94–112.
22 Ibid., 121.
23 Letter from R.B. Bennett to Herbert A. Bruce, 2 Feb. 1935, NA, Bennett Papers, Housing, no. 399, 499837; Alvin Finkel, *Business and Social Reform in the Thirties* (Toronto: James Lorimer 1979), 78, 135, 141; J.R.H. Wilbur, *The Bennett New Deal: Fraud or Portent*? (Toronto: Copp Clark 1968), 80–90.
24 Wilbur, *Bennett New Deal*, 113–48; Finkel, *Business and Social Reform*, 100–3.
25 Stevens protested repressive measures against the unemployed and the relief camps at the time of this break with Bennett. This was coupled with a call for an extensive public-works program and an expansive housing policy. Writing to Vancouver mayor G.S. McGeer at the time of his break with Bennett, Stevens stressed, in regards to housing, the need for "loosening up of credit by those private organizations which control the credit of our country today." He also urged McGeer to undertake a housing survey of Vancouver in anticipation of a low-rental housing project. Stevens to Sir George Perley, 30 Apr. 1935; Stevens to G.S. McGeer, 23 May 1935, in NA, King Papers, MG26, J1, vol. 207.
26 Canada, Commons, *Debates* (1935), 1:153–62.
27 Ibid., 1:168–71.
28 Canada, Commons, *Debates* (1935), 4:3930.
29 Parliamentary Committee on Housing, *Minutes of Proceedings* (1935), 170–81, 364.
30 Ibid., 94–103.
31 Ibid., 101–32.
32 Ibid., 135.
33 Ibid., 292–325.
34 Ibid.
35 Ibid., 326, 327.
36 Ibid., 364.

37 Ibid., 374.
38 Ibid., 24, 25.
39 Collin McKay, "What's Holding Housing Back?" *Canadian Unionist* 9 (1935): 11, 12.
40 Canada, Commons, *Debates* (1935), 4:3949.
41 Parliamentary Committee on Housing, *Minutes of Proceedings* (1935) 334, 335.
42 Ibid., 347, 348.
43 Ibid., 349.
44 Ibid., 351.
45 Ibid., 354.
46 Ibid., 354, 357.
47 Letter of 9 Apr. 1935 from F.W. Nicolls to W.H. Yates; letter, 13 Apr. 1935, W.C. Clark to F.W. Nicolls; letter, 16 Apr. 1935, Nicolls to Clark, all in NA, Finance Department Papers, RG19, vol. 710, file 203–1A to 25.
48 Parliamentary Committee on Housing, *Minutes of Proceedings* (1935), 354, 357.
49 Ibid., 374.
50 Letter from T. D'Arcy Leonard to W.C. Clark, 8 May 1935, NA, RG19, vol. 705, file 203–1A.
51 Noulan Cauchon, "Urbanism – Whither?" *JRAIC*, 11 (1935): 99.
52 W.C. Clark and J.L. Kingston, *The Skyscraper: A Study in the Economic Height of Modern Office Buildings* (New York: American Institute of Steel Construction 1930), 4, 31.
53 Ibid., 32–151.
54 Ibid., 81–143.
55 McKay, "What's Holding Housing Back?" 11, 12. McKay suggested that one "might think the ruling classes were insane" for refusing to finance the "war on poverty" while being able to find "ways to finance preparation for a bigger, if not better war of destruction of human life."
56 W.C. Clark, "Regularization of National Demand for Labour by Government Employment," paper given to International Association of Public Employment Services, NA, RG19, vol. 3993, passim. These views are particularly significant in view of the conventional presentation of Clark as the introducer of Keynesianism: e.g. John Porter, *The Vertical Mosaic* (Toronto: University of Toronto Press 1955), 452–8.
57 Finkel, *Business and Social Reform*, 101; W.C. Clark "What's Wrong with Us?" address to the Professional Institute of the Civil Service of Canada, Dec. 1931, NA, RG19, vol. 3993.
58 Letter from W.C. Clark to F.W. Nicolls, 13 Apr. 1935, NA, RG19, vol. 710, file 203–1A-1 to 25.

59 Canada, Commons, *Debates* (1935), 4:3770.
60 Letter from T. D'Arcy Leonard to W.C. Clark, 2 June 1935, NA, RG19, vol. 705, file 203–1A, 4:3947–8.
61 Cable from D'Arcy Leonard to W.C. Clark, 25 June 1935, and cable from W.C. Clark to D'Arcy Leonard, 26 June 1935, in ibid., file 203–1A; Canada, Commons, *Debates* (1935), 4:3771–3.
62 H. Woodward, *Canadian Mortgages* (Don Mills: Collins 1959), 8–11.
63 Copy of letter to J. Clark Reilly, 27 June 1935, NA, RG19, vol. 711, file 203–1L.
64 "Committee on Housing," NA, Bennett Papers, Finance, 525484.
65 Collin McKay, "The Housing Shortage," *Canadian Unionist* 9 (1935): 139; Canada, Commons, *Debates* (1935), 4:3928–9.
66 Ibid., 3771–3, 3948.
67 Ibid., 3947.
68 Ibid.
69 Richard Wilbur, *H.H. Stevens* (Toronto: University of Toronto Press 1977), 184–5.
70 Press release, NA, RG19, vol. 710.
71 Letter of 6 July 1937, D.B. Mansur to W.C. Clark, NA, RG19, vol. 710, file 203–1A, 51–100, 4.
72 Memo on Dominion Housing Act Booth, NA, RG19, vol. 706, file 203–1A.
73 Ibid.
74 Memo from David B. Mansur to Arthur Purvis re Dominion Housing Act 1935, NA, RG19, vol. 711, file 203–2L. Mansur's memo to Purvis is highly significant for he was writing in candid manner, drawing on his enormous experience in dealing with both the federal government and the mortgage market. Mansur clearly described how the DHA benefited a wealthy suburban elite that amounted to only 20 per cent of households in major metropolitan centres. He had little faith in cost-saving architectural designs to alter this basic situation, as he felt DHA's own minimum standards were such to prohibit low-cost housing, a point that Clark refused to concede. Mansur's recommendation in this memo – that experienced mortgage men be released by their companies to examine the faults in DHA – was not followed up; the firms were simply not sufficiently enthused about the joint-loan scheme to make such a constructive effort. He also vividly grasped the lack of any social purpose in the DHA, noting that "although the act has a reputation of being benevolent legislation, it is not the intention of the lending company nor the Department of Finance to lose any money on loans under the Act."
75 Ibid.
76 Letter from W.C. Clark to D.B. Mansur, 10 Aug. 1936, NA, RG19, vol. 711, file 203–2L.

CHAPTER FOUR

1 John Rockefeller Jr to W.L.M. King, 27 Nov. 1935, NA, King Papers, MG26 JI, vol. 210, file R-4, 181448–50.
2 King Diary, 4 Apr. 1936, 139, 140, NA, MG26, J13, vol. 78.
3 King Papers, memos and notes 1933–39, NA, MG26, J4, vol. 197, 137452, 137454, 137494.
4 Ibid.
5 King Diary, 3 Sept. 1936, NA, MG26, J13, vol. 718, 311. King's comments are all the more significant in that he seems to have misinterpreted Purvis's recommendations in these remarks to suit his own inclinations. Purvis in this 2 Sept. 1936 Cabinet meeting outlined plans requiring moderate spending increases. The NEC outlined a national employment service, and a national volunteer conservation service, and a national apprenticeship job-training program. King ignored these products of the NEC's efforts, adopting for almost a year only the HIP plan, which required no government expenditure. See James Struthers, *No Fault of Their Own: Unemployment and the Canadian Welfare State, 1914–41* (Toronto: University of Toronto Press 1983), 156–8.
6 Struthers, *No Fault of Their Own*, 175–207.
7 Letter from Gordon West to Arthur Purvis, 1 May 1936, NA, RG27, vol. 335, file 13.
8 W.A. Mackintosh, "Preliminary Report on Housing," NA, RG27, vol. 3358, file 4.
9 Memo from Norman McLarty, NA, RG27, vol. 3357, file 11.
10 Ibid. Letter of W.D. Flanders, deputy administrator, Federal Housing Administration, 8 Oct. 1936, NA, RG27, vol. 3357, file 11.
11 Norman McLarty, "How To Modernize Your Home," FHA form no. 792, NA, RG27, vol. 3357, file 11.
12 W.A. Mackintosh, "House Renovation and Construction as a Source of Employment," NA, RG27, vol. 3347, file 1.
13 "Renovation and Modernization of Houses as a Source of Employment," NA, RG27, vol. 3347, file 1.
14 "Government To Act on Purvis Report," *Toronto Mail and Empire*, 11 Sept. 1939, clipping in NA, RG27, vol. 3347, file 1. This file contains numerous newspaper reports praising the HIP scheme: 1 Oct. 1936, King to Dunning, NA, RG10; 28 Sept. 1936, president of Canadian Bankers Association to Dunning; 14 Oct. 1936, S.H. Logan to C.A. Dunning, all in NA, RG19, vol. 714. Some of King's pleasure with the scheme can also be gleaned from his appointment of McLarty as minister of labour to replace Rodgers when the latter was elevated to the post of minister of national defence upon the outbreak of the Second World War.

15 "Local Advisory Committees: Some Suggestions for Their Organization," NA, RG27, vol. 3347, file 1.
16 Ibid.
17 "Outline of Activities for the Home Improvement Plan under the Organization provided by the National Employment Commission," 26 Oct. 1936, NA, RG27, vol. 3355, file 3.
18 Ibid.
19 Ibid.
20 "Media recommended for plan for promoting and popularizing Home Improvement Plan," NA, RG27, vol. 3355, file 3.
21 NA, RG27, vol. 3355, file 3.
22 Ray Brown, director of publicity, National Employment Commission, bulletin no. 9, 12 Mar. 1937, NA, RG27, vol. 3376, file 3.
23 Arthur B. Purvis, Home Improvement Plan speech, delivered to the 1936 annual dinner of the Canadian Institute of Plumbing and Heating, NA, RG27, vol. 3373, file 36; letter, 12 Nov. 1936, Arthur Purvis to Charles Dunning, NA, RG19, vol. 714.
24 Canadian Construction Association members' newsletter no. 11, "Sell your services, sign up on the job," NA, RG27, vol. 3373, file 35.
25 Prof. W.W. Goforth, "Co-operative action needed for HIP success: Efforts of plumbing and heating industry on behalf of measure outlined," in NA, RG19, vol. 706
26 Minutes of a meeting held at the office of Cockfield, Brown and Company Limited, Tuesday, 8 Dec. 1936, NA, RG27, vol. 3355, file 3.
27 Memo re progress of home-improvement plan, NA, RG27, vol. 3357, file 3.
28 Letter, 21 Nov. 1936, Ray Brown to W.L.M. King, NA, King Papers, MG26, J1, vol. 8, 184371–6.
29 *Profit for You: From the Home Improvement Plan*, booklet in NA, RG19, vol. 712, file 203–6A.
30 Brown, bulletin no. 9, 12 Mar. 1937, NA, RG27, vol. 3376, file 3.
31 Press release for HIP promotion, in NA, RG27, vol. 3354, files 1, 5, 10.
32 Purvis, Home Improvement Plan speech, 4.
33 Urban workers complained of similar discrimination. A foreman of twenty-six years' experience was told by bank lending officers that they were not interested in loans "on old houses." Letter, E.W. Hartle to C.A. Dunning, 24 July 1937, in NA, RG19, vol. 714.
34 A.E. Grauer, *Housing* (Ottawa: Royal Commission on Dominion Provincial Relations 1939), 42, 43.
35 HIP in Alberta, report by provincial chairmen, 8 Nov. 1937, in NA, RG19, vol. 706, file 203–1A.
36 Memo from Nova Scotia Provincial Home Improvement Plan Committee to Arthur B. Purvis, chairman, National Employment Commission, in NA, RG19, vol. 707, file 203–1A.

37 Canada, Commons, *Debates* (1937), 1:469–70; letter from A. MacG. Young, MD, to C.A. Dunning, 17 Apr. 1939, in NA, RG19, vol. 706, file 203–1A.
38 NA, RG19, vol. 706.
39 Minutes of a meeting of the Economic Committee, NA, RG19, vol. 3890.
40 Letter, 22 Mar. 1950, from W.C. Clark to D.H. McKewe of the advertising department of the Canadian Imperial Bank of Commerce, NA, RG19, vol. 714, file 203–9A. Clark told McKewe that since "the loss to any one lending institution was not in excess of 15% of the aggregate loan ... the banks and other lending institutions did not incur any loss in connection with the Government Home Improvement Plan."
41 Low Cost Housing Diary, P. Shephard, Housing Consultant, NA, RG27, vol. 3356, file 12.
42 NEC Low Rental Housing Proposal, NA, RG27, vol. 3374.
43 Memo to C.A. Dunning re Low Rental Housing Program, 7 Mar. 1937; "A Low Rental Housing Program: Summary of Criticisms," nd, both in NA, RG19, vol. 3888.
44 James Struthers, *No Fault of Their Own*, 169–71.
45 Confidential report of A.N. McLean on slum conditions, NA, RG27, vol. 3375, file 16.
46 Report from Charlotte Whitton on relief standards, in NA, RG27, vol. 337, file 16.
47 Montreal Board of Trade, Civic Improvement League, *Report*, 5, 21, 22, in NA, RG19, vol. 706, file 203–1A. "Flood of Evictions; Flood of Overcrowding," press clipping, 8 Oct. 1937, in VCA Housing docket 1, Nov. 1918–Jan. 1939, M4289–1. For a description of the "amateur" or "informal" economy that produces most of the housing stock afforded by low-income tenants, see Beger S. Krohm, Berkeley Fleming, and Marilyn Manzer, *The Other Economy: The Internal Logic of Local Rental Housing* (Toronto: Peter Martin 1977), passim. For an excellent overview of Whitton's desire to reduce relief expenses, see Struthers, *No Fault of Their Own*, 77–9, 153–4, 170–1.
48 H.M. Cassidy, *Unemployment and Relief in Ontario: 1929–32* (Toronto: J.M. Dent & Sons 1932), 211–17.
49 Whitton, report on relief standards.
50 W.C. Clark to A.B. Purvis, 6 Oct. 1937, in NA, RG27, vol. 337, file 16.
51 Copy of letter from David Mansur to John A. Flanders, 14 June 1937, in NA, RG19, vol. 3435.
52 National Employment Commission Low Rental Housing Proposal, 3 Feb. 1937, in NA, RG27, vol. 3374, file 18; W.C. Clark, "A Low Rental Housing Program: Summary of Criticisms"; National Housing Act, Statutes of Canada, 1939, 358–64.

53 "Civic Plebiscite Looms This Year on Housing Plan," 6 Jan. 1939, newspaper clippings, VCA. Civic Improvement League of Montreal to J.L. Ralston, 26 Mar. 1940; a critique of the National Housing Act 1944 by Percy Nobbs; file commentary (likely written by W.C. Clark) on Nobbs's housing efforts, both in NA, RG19, vol. 7094, file: 203–1A-8015. Letter from W.C. Clark to S.H. Prince, 20 Mar. 1940; letter from Prince to Clark, 25 Apr. 1940, both in NA, RG19, vol. 706, file 203–1A.
54 Program for national conference on housing, NA, Horace Seymour Papers, MG30, B93, vol. 4, 35.
55 Letter from W.C. Clark to H.M. Carver, 19 Aug. 1939; Carver to Clark, 17 Aug. 1934, in NA, RG19, vol. 709.
56 Cassidy, *Unemployment*, 203, 204.
57 Ibid., 207–27, 272–4; Canada, Commons, *Debates* (1939), 2:2874.
58 Winnipeg Board of Trade to W.C. Clark, 9 June 1937, in NA, RG19, vol. 3435.
59 Memo to W.C. Clark from F.W. Nicolls re meeting with Dominion Mortgage Investment Association, 4 Dec. 1935, NA, RG19, vol. 705.
60 NA, RG19, vol. 710, file 2031A, 1 to 23; NA, RG19, vol. 397.
61 A.E. Grauer, *Housing* (Ottawa: Royal Commission on Dominion-Provincial Relations 1939), 39, 42.
62 Letter from Ian MacKenzie to C.A. Dunning, 27 Apr. 1936, NA, RG19, vol. 3979; extract from the *Vancouver Sun*, 25 Sept. 1936, NA, RG19, vol. 711, file 203-1L.
63 Letter, 20 Jan. 1936, from W.C. Clark to D.M. Allan, private secretary of James Gardener, minister of agriculture, NA, RG19, vol. 711.
64 Report of operations under The Dominion Housing Act, to W.C. Clark from Western Loan and Savings Association, NA, RG19, E2 (1), vol. 3455; letter from H.P. McManus, 28 Nov. 1936, to the Western Savings and Loan Association, NA, RG19, E2 (1), vol. 3455. McManus found that in Prince Albert there were "no vacant apartments, vacant houses, or even vacant rooms."
65 Letter from A.J. Brown, president, Edmonton Builders' Exchange, to J. Clark Reilly, manager, CCA, 29 July 1936; "Do you want to own your home?" three radio address on the Dominion Housing Act, in NA, RG19, vol. 3979, file H-1–3.
66 Telegram, 10 Aug. 1936, from C.A. Campbell to W.L.M. King, NA, MG26, JT, vol. 214. Letter from Charles E. Campbell to W.L.M. King, 27 Sept. 1940, in NA, RG19, vol. 710.
67 Telegram from Charles E. Campbell to W.L.M. King, 13 Mar. 1937; letter from Charles E. Campbell to C.A. Dunning, 10 Mar. 1937; letter, Charles E. Campbell to W.L.M. King; 25 Aug. 1937, in NA, MG26 (1), vol. 706, file E-EDG, King Papers, MG26, JT, vol. 214.

68 Newspaper clipping scrapbook, ECA, Tuesday, "Decent Low-cost Homes Needed," 4 Oct. 1938; "Many Dwellings Unfit for Use Health Body Told by M.H.O."; "Housing Conditions in City Attacked at Jobless Meet," 22 Apr. 1938.
69 Letter from Charles E. Campbell to W.L.M. King, 27 Sept. 1940, in NA, RG19, vol. 710.
70 Memo to Ralston from W.C. Clark, 12 June 1940, NA, RG19, vol. 2679.
71 Canada, Commons, *Debates* (1939), 4:3665–70, 4389–99; Senate, *Debates* (1939), 1:575–80, 577–9.
72 Ibid.
73 Alvin Finkel, *Business and Social Reform in the Thirties* (Toronto: James Lorimer 1979), 109–12.
74 Thomas E. Bailey to W.L.M. King, 2 Mar. 1936, King Papers, NA, MG26, J1, vol. 213, 183639; J.C. Nunn to W.C. Clark, 3 July 1937, NA, RG19, vol. 710, file 203–1A-51 to 100.
75 Memo from W.C. Clark to H.C. Dunning re Dominion Housing Act – Port Arthur, NA, RG19, vol. 706, file 203–1A.
76 W.C. Clark, memo re trip to Maritimes, NA, RG19, vol. 711, file 203–12; Grauer, *Housing*, 39–42; Canada, Commons, *Debates* (1938), 4:33773–5.
77 Clark, memo re trip to Maritimes; letter from W.C. Clark to A.L. Wickwire, NA, RG19, vol. 711, file 203; Journals of the House of Assembly of Nova Scotia, 1940, Appendix No. 32, *Report of the Housing Commission for year ending 30 November 1939*, 13.
78 Address by W.C. Clark to Dalhousie University, NA, RG27, vol. 3337, file 29.
79 Grauer, *Housing*, 40, 41.
80 *Statutes of Canada*, 1939, chap. 49, An Act To Assist the Construction of Houses, 353–67.
81 Horace Seymour, "A Housing Achievement," *Canadian Unionist* (June 1939): 16–18.
82 Press release, 30 Sept. 1939, NA, RG19, vol. 705, file 203–1A; Canada, Commons, *Debates* (1939), 3:2750; Grauer, *Housing*, 44.
83 Press release, 30 Sept. 1939; Canada, Commons, *Debates* (1939), 3:2750.
84 Copy of letter from S.E. Briand, loan inspector, Sun Life, to D.B. Mansur, inspector of mortgages, Sun Life Assurance Company of Canada, 17 Jan. 1939; copy of letter from J.A. Gray, loan manager of Sun Life, Winnipeg, to S.E. Briand; both in NA, RG19, vol. 711, file 203–2L.
85 O.J. Firestone, *Residential Real Estate in Canada* (Toronto: University of Toronto Press 1951), 483; NA, RG19, vol. 710.

CHAPTER FIVE

1 Memo to the minister from W.C. Clark, 11 Oct. 1939, NA, RG19, vol. 2679; Summary of Housing Programs, in NA, RG19, vol. 706, file 203–1A; memo to Ralston re National Housing Act, from W.C. Clark, 5 Dec. 1939, in NA, RG19, vol. 704, file 203–1A.
2 Minutes of a meeting of the Economic Advisory Committee, 15 Oct. 1940, 2, NA, RG19, vol. 3890.
3 W.C. Clark, Report of the Economic Advisory Committee, 13 Nov. 1940, 7, NA, RG19, vol. 3890.
4 Minutes, Economic Advisory Committee, 15 Oct. 1940, 3.
5 Memo to W.C. Clark from F.W. Nicolls, 19 Aug. 1940; memo for the Economic Advisory Committee on Wartime Housing Policy, both in NA, RG19, vol. 3890.
6 W.C. Clark, Report of the Economic Advisory Committee on Housing Policy, 4, 5.
7 Confidential letter from A.P. Heeney to W.C. Clark, 28 Nov. 1940, NA, RG19, vol. 3890.
8 Letter, 30 Dec. 1940, from W.C. Clark to Angus Macdonald; minutes of a meeting of the Economic Advisory Committee, both in NA, RG19, vol. 3890.
9 Ibid.
10 Memo for the Economic Advisory Committee on Housing Policy, 4.
11 "Efficiency, Economy and Morale Need NHA," *Daily Commercial News and Building Record*, 21 Apr. 1942, in NA, RG19, vol. 709. "Wire Your Member Today," *Daily Commercial News*, in RG19, vol. 706, file 203–1A. H.M. Moore, assistant treasurer, Sun Life, to J.L. Ilsley, 9 Jan. 1942; M.J. Smith, general manager, Equitable Life, to J.L. Ilsley, 6 July 1942; both in NA, RG19, vol. 706, file 203–1A. *Building in Canada*, 31 Mar. 1942, in NA, RG19, vol. 704, file 203–1A.
12 C. Blake Jackson, controller of construction, brief on housing with special reference to the National Housing Act, NA, RG19, vol. 704, file 203–1A.
13 Letter from Victor A. Goggin, general manager, Wartime Housing, to C.D. Howe, 2 May 1942; memo from W.L. Sommerville, vice-president, Wartime Housing, both in NA, RG19, vol. 704, file 203–1A.
14 Letter, 8 May 1942, from F.W. Nicolls to J.L. Ilsley; memo to J.L. Ilsley and W.C. Clark from F.W. Nicolls, 24 Jan. 1942, both in NA, RG19, vol. 704, file 203–1A.
15 Summary of government housing programs in NA, RG19, vol. 706, file 203–1A.
16 Ibid., 4.
17 Ibid., 4, 5; David Bettison, *The Politics of Canadian Urban Development*

(Edmonton: University of Alberta Press 1975), 81. Memo from D.B. Mansur to A.E. Purvis, NA, RG19, vol. 711, file 203–2L.
18 Department of Munitions and Supply, NA, RG28, vol. 2226.
19 Wartime Prices and Trade Board Papers, NA, RG64, vol. 706.
20 Leslie R. Thomson, Preliminary Report on the Housing Situation in Canada and Suggestions for Its Improvement, NA, Wartime Prices and Trade Board Papers, vol. 89.
21 Advisory Committee on Reconstruction, *Report of the Advisory Committee on Reconstruction, 4, Housing and Community Planning* (Ottawa; King's Printer 1944), 37, 38.
22 Owen Lobley, "The Course of Rentals in Canada," NA, RG19, vol. 3582.
23 Ibid.
24 Ibid.
25 *Ottawa Citizen*, 30 Oct. 1939; 3 May 1940.
26 Letter, 20 July 1940, from Norman McLarty to W.C. Clark, in NA, RG19, vol. 3993.
27 Report of the Advisory Committee on the Control of Rentals, NA, RG19, vol. 4663.
28 Ibid.
29 Memo for the board re rent control, NA, RG19, vol. 4663.
30 *Report of the Advisory Committee on Reconstruction*, 37, 38.
31 Second draft, "A Roof over Your Head," speech by Donald Gordon, NA, RG19, vol. 3961.
32 *Report of the Advisory Committee on Reconstruction*, 37.
33 Memo on prices and wages policy, NA, Privy Council Records, RG2, ser. 18, vol. 8.
34 Ibid.
35 Minutes of a meeting of the Economic Advisory Committee, NA, RG19, vol. 3890. Memo to J.L. Ilsley and W.C. Clark, from F.W. Nicolls, 8 Feb. 1941, NA, RG19, vol. 3890, file H-1–15.
36 Ibid.
37 Ibid.
38 Memo from F.W. Nicolls to W.C. Clark, 19 Feb. 1940, NA, RG19, vol. 3890.
39 Ibid.
40 Ibid.
41 Chairman's report of Halifax Project, in Wartime Housing minute book, located in NA, Defence Construction Limited Papers, RG83, vol. 70, bk 1.
42 Memo from J.D. Forbes to F.W. Nicolls, NA, RG19, vol. 3540; memo to J.L. Ilsley and W.C. Clark from F.W. Nicolls, NA, RG19, vol. 3890.
43 Minutes of Wartime Housing Executive Committee, 24 June 1941, NA, RG83, vol. 70, bk 1.

44 Ibid.
45 Thomson, Preliminary Report, Wartime Prices and Trade Board Papers, NA, RG64, vol. 89, 229–30.
46 Ibid., 230–1.
47 Ibid., 243–4.
48 Ibid., 251.
49 Ibid., 252.
50 Ibid., 254.
51 Ibid., 248.
52 Ibid., 257. Pigott at one point told Howe that the only remaining problem was "finding space for women workers"; Pigott to Howe, 14 Oct. 1942, NA, RG19, vol. 709.
53 Minutes of Wartime Housing Board of Directors, NA, RG83, vol. 70, bk 1.
54 Ibid.
55 Ibid.
56 Ibid.
57 Ibid.
58 Thomson, Preliminary Report, 241.
59 Pamphlet, *Wartime Control Established by Canadian Construction Industry*, report of National Conference of Employers and Employees in Construction Industry, in McGill University Archives (MUA), James Papers, RG2, C-1file 4187.
60 Ibid.
61 Ibid.
62 Minutes of Wartime Housing Executive Board, NA, RG83, vol. 70, bk 1.
63 Ibid.
64 Memo to Miss H.M. McKenna from F.W. Nicolls, NA, RG19, vol. 715, file 203C-11.
65 Letter from W.C. Clark to A.K. Shiels, 30 Aug. 1941, NA, RG19, vol. 6425.
66 Minutes of Wartime Housing Executive Committee; *Journals and Proceedings of the House of Assembly of Nova Scotia*, app. no. 33; *Report of the Nova Scotia Housing Commission for the Year ending 30 Nov. 1942*.
67 Minutes of Wartime Housing Executive Committee.
68 Ibid.
69 Inter-office correspondence from DeMara to Carr; letter from Gordon to Ilsley, 18 Feb. 1943, NA, RG19, vol. 699, file 25–1–1.
70 Ibid.
71 Letter from W.C. Clark to Donald Gordon, 9 Nov. 1942, NA, RG19, vol. 3980; letter from J.L. Ilsley to C.D. Howe, 30 Oct. 1942, NA, RG19, vol. 3539.

72 Minutes of Wartime Housing Executive Board; *Report of the Nova Scotia Housing Commission for the Year ending 30 Nov. 1942*, Nova Scotia Legislative Assembly, *Journals and Proceedings* (1945), app. 31, 12.
73 Letter of the Greater Toronto Permanent Housing Committee to W.L.M. King, 23 Sept. 1942, NA, RG19, vol. 645, file 2–2–0.
74 Ibid.; brief "Respecting the Construction of Housing Facilities in Wartime," from the Dominion Mortgage and Investments Association to C.D. Howe, NA, RG19, vol. 3540.
75 Letter from W.C. Clark to Donald Gordon, NA, RG19, vol. 3539.
76 Letter from C.D. Howe to J.L. Ilsley, 2 Nov. 1942, NA, RG19, vol. 3539.
77 Ibid.; letter 7 Dec. 1942, C.D. Howe to J.L. Ilsley, in NA, RG19, vol. 645.
78 Minutes of Wartime Housing Executive Board; O.V. Firestone, *Residential Real Estate in Canada* (Toronto: University of Toronto Press 1951), 125.
79 Confidential memo re housing, Russel Smart to Donald Gordon, NA, RG64, vol. 699, file 25–1.
80 Letter from C.B. Jackson to E.M. Little, director of National Selective Service, NA, RG64, vol. 699, file 25–1.
81 Firestone, *Residential Real Estate*, 498.
82 Letter, 11 Sept. 1943, from H.L. Robson, assistant secretary, Canadian Bankers' Association, to F.W. Nicolls, NA, RG56, vol. 11.
83 Firestone, *Residential Real Estate*, 491.
84 Letter from Henry Borden, chairman, Housing Co-ordinating Committee, to J.L. Ilsley, 15 Nov. 1943, NA, Munitions and Supply Papers, RG28-A, vol. 141.
85 Letter, 13 June 1943, Borden to Ilsley, NA, RG28-A, vol. 141.
86 Memo and recommendation re Real Property Administration housing surveys and registries, by deputy real property administrator Norman W. Long, in NA, RG28-A, vol. 141; letter, Donald Gordon to J.L. Ilsley, 18 Feb. 1943, NA, RG64, vol. 699, file 25–1–1.
87 Letter, 14 Oct. 1942, J.M. Pigott to C.D. Howe, NA, RG19, vol. 709; newspaper clipping, 22 Apr. 1943, "Start Monday on City Check," VCA, Housing docket 2, M4289–2.
88 Taken from monthly reports of housing registries, Mar. to Aug. 1943, in NA, RG64, vol. 215.
89 Memo from Russel Smart to Donald Gordon, 24 Sept. 1943, NA, RG19, vol. 714; memo, Smart to Gordon, re second meeting of Housing Co-ordination Committee, NA, RG64, vol. 699, file 25–1–1.
90 Memo on the housing situation in Canada as of 24 May 1943, from Russel Smart to Donald Gordon, in NA, RG64, vol. 699.
91 Letter, 13 Apr. 1943, C.D. Howe to J.L. Ilsley; letter, 10 Apr. 1943, Victor T. Goggin to C.D. Howe; letter, 2 Apr. 1943, Joseph M. Pigott to

Russel Smart; letter, 6 Apr. 1943, Russel Smart to J.M. Pigott, all in NA, RG19, vol. 3540.

92 Submission to R.S. Smart, national real-property administrator, on the housing problem in Toronto, from a deputation of the city council, 13 May 1943, NA, RG28-A, vol. 141.

93 Letter, 5 May 1943, from J.L. Ilsley to C.D. Howe; memo from Mitchell Sharp re Edmonton housing project, 30 Apr. 1943, both in NA, RG19, vol. 645, file 183–2–1.

94 Thomson, Preliminary Report, passim.

95 Ibid., 10, 11, 303.

96 Ibid., 91, 92, 122.

97 Ibid., 125.

98 Ibid., 172–4.

99 Ibid., 202, 129.

100 Ibid., 305.

101 Ibid., 319–30.

102 Memo, 19 Nov. 1943, Russel Smart to Donald Gordon, NA, RG19, vol. 714, file 203-C, vol. 1; letter, Borden to Ilsley, 15 Nov. 1943, NA, RG28-A, vol. 141.

103 Memo from Russel Smart to Donald Gordon, 24 Sept. 1943, NA, RG19, vol. 714.

104 Report on housing conditions in Montreal, by MacKay Fripp, 10 Sept. 1943, NA, RG64, vol. 700.

105 Fripp appended a long series of correspondence to his memo calling for action on Montreal's housing problems, to show how the city had been getting the proverbial "run-around" from federal authorities. See letters of 14 June, Joseph Pigott to J.O. Asselin; 12 June 1943, C.D. Howe to J.O. Asselin; 17 June 1943, H.R.L. Henry to J.A. Mongeau; 18 June 1943, Victor T. Goggin to J.A. Mongeau. These and additional letters, all in NA, RG64, vol. 700.

106 Memo to W.C. Clark re Montreal housing scheme, from M.W. Sharp, NA, RG19, vol. 715, file 203C-12.

107 Letter, 20 Nov. 1943, J.W. McConnell to J.L. Ilsley; letter 25 Nov. 1943, J.L. Ilsley to J.M. McConnell, both in NA, RG19, vol. 3980.

108 Letter, 25 Feb. 1944, Donald Gordon to J.L. Ilsley; memo, 25 Feb. 1944, Russel Smart to W.C. Clark, both in NA, RG19, vol. 3980.

109 George W. Spinny, private and confidential memo for J.L. Ilsley, 9 Sept. 1944, NA, RG19, vol. 3980.

110 Memo to J.L. Ilsley re Montreal housing project, from W.C. Clark, NA, RG19, vol. 3980.

111 Letter from Russel Smart to Donald Gordon, 21 Nov. 1942, re congested housing in Halifax, NA, RG19, vol. 715.

112 Letter, C.D. Howe to J.L. Ilsley, 7 Dec. 1942, NA, RG19, vol. 645.

113 Letter to W.L.M. King from Betty Paice, 15 Dec. 1945, Privy Council Records, NA, RG2, ser. 18, vol. 9.
114 Letter from P.B. Carswell to C.D. Howe, 8 May 1943, NA, RG2, ser. 18, vol. 9.
115 Memo, 19 July 1943, re Halifax housing, from H.D. Fripp, executive assistant, real-property administrator, RG2, ser. 18, vol. 9.
116 Minutes of a meeting, 17 Sept. 1943, to deal with the housing situation in Halifax in accordance with Cabinet War Committee minutes dated 8 Sept. 1945; memo from F. Kent Hamilton, 18 Sept. 1943, both in NA, RG2, ser. 18, vol. 9.
117 Letter from P.B. Carswell to C.D. Howe, 8 May 1943, NA, RG2, ser. 18, vol. 9.
118 Letter, 17 July 1944, re census returns for Halifax, from C.E. Cousins to C.D. Howe, NA, RG2, ser. 18, vol. 9, file H-13. *Report of the Nova Scotia Housing Commission for the Year ending 30 November 1941*, Nova Scotia Legislative Assembly, *Journals and Proceedings* (1945), app. 31.
119 Letter, Cousins to Howe, 17 July 1944; secret memo to Cabinet War Committee, doc. no. 738, 21 May 1944, Cabinet War Committee Records, NA, reel C-4876.
120 *Rental Control and Evictions in Canada World War II and the Post War Years*, in NA, RG64, vol. 29, 38–41.
121 Ibid., 41, 42.
122 Ibid., 42.
123 "Rent Control Housing Accommodation Maximum Rentals," NA, RG19, vol. 3961. Letter, Donald Gordon to A.L. Wickwire, 29 May 1945; report on the state of rent control in the Halifax division, both in NA, RG64, vol. 706, file 25–9.
124 Ibid.
125 *Rental Control and Evictions*, 29–31.
126 What makes both Lobley's and construction controller C.B. Jackson's observations so peculiar in this regard is that doubling up actually increased with the advent of war. Lobley seems to have been grasping at a rationale to leave the private real estate market undisturbed by further government intervention. Memo to Ilsley, 1 May 1942, NA, RG19, vol. 3961.
127 *Rental Control and Evictions*, 43.
128 Ibid., 44, 45.
129 Ibid., 44, 45.
130 "You Can Build Again," *Financial Post*, 11 Mar. 1944.
131 Letter, 12 May 1944, from Donald Gordon to J.G. Godsoe, chairman, Wartime Industries Control Board, NA, RG28, A, vol. 226.
132 Report of meeting to discuss housing situation, G.W. Withwell to J.G. Godsoe, 31 May 1944; housing survey, report on recent changes

in organization and operation policy in construction control, both in NA, RG28-A, vol. 226.

133 Minutes of 17th meeting, Housing Co-ordinating Committee, NA, RG64, vol. 699, file 25–1–1.

134 Letter, J.G. Godsoe to Donald Gordon, 12 Sept. 1944; letter, J.G. Godsoe to Donald Gordon, 21 Aug. 1944, both in NA, RG64, vol. 337. Letter, J.M. Fogo, to Donald Gordon, 31 May 1944, NA, RG64, vol. 699, file 25–1–1.

135 Letter from Donald Gordon to M.W. Sharp, 14 June 1944, NA, RG19, vol. 714, file 203-C-O. Gordon complained to Sharp that the WPTB was "handicapped without having any knowledge of what constitutes government housing policy."

136 Firestone, *Residential Real Estate*, 215, 485.

137 Ibid., 394.

CHAPTER SIX

1 For an account of how Rogers protected the NEC from the conservative suspicions of Mackenzie King, see James Struthers, *No Fault of Their Own*, 175–84.

2 Jack Granatstein, *Canada's War* (Toronto: Oxford University Press 1975), 247–57.

3 Malcom G. Taylor, *Health Insurance and Canadian Public Health Policy* (Montreal: Queen's University Press 1979), 17–20.

4 Ibid., 35.

5 Memo to the prime minister from Brooke Claxton, 2 July 1943, in Privy Council Papers, NA, RG2, ser. 18, vol. 12.

6 Granatstein, *Canada's War*, 257.

7 See the long exchange between Curtis and Ilsley between 2 June 1942 and 25 Aug. 1942 over housing policy, in NA, RG19, vol. 706. This exchange would be a rehearsal for the disputes over housing policy before the NHA of 1944.

8 George Mooney, "A Post-War Housing Program for Canada," *Canadian Welfare* (July 1943): 3–9.

9 First meeting of the Sub-committee on Housing and Community Planning, 12 Mar. 1943, NA, James Papers, RG2, C-191, file 4186.

10 Advisory Committee on Reconstruction, *Final Report of the Sub-committee on Housing and Community Planning* (Ottawa: King's Printer 1944), 162–9.

11 Ibid., 163–9.

12 Ibid., 163–76.

13 Ibid., 177–82; 9 and 10 Apr. 1943, Curtis subcommittee minutes, McGill University Archives (MUA), James Papers, RG2, C-191, file 4186.

14 Advisory Committee on Reconstruction, *Final Report of the Subcommittee*, 205–32.
15 Curtis subcommittee minutes, 14 May 1943, MUA, James Papers, RG2, C-191, file 4186.
16 Ibid.
17 Advisory Committee on Reconstruction, *Final Report of the Sub-committee*, 92–100, 110–22, 193–204.
18 Ibid., 266–71.
19 Ibid., 222–4.
20 Ibid., 222–34.
21 Leonard Marsh, *Report on Social Security for Canada*, 2nd ed. (Toronto: University of Toronto Press 1975), xxix–xxxvi.
22 Letter, Norman A. White, assistant secretary, Dominion Mortgage Investments Association, to W.C. Clark, in NA, RG19, vol. 3540.
23 Dominion Mortgage and Investment Association brief, Housing in Relation to Post-War Reconstruction, 24 Feb. 1944, in NA, RG19, vol. 3980, file "Housing"; Canada,*Statutes* 1944, NHA.
24 B.H. Higgins, "Appraisals of the Canadian Housing Act," *Public Affairs* (Spring 1945): 169–72; Canada, Commons, *Debates* (1944), III:3607–70; IV:5972–6011; Canada, *Statutes* 1944, Farm Loans Act, 347–53.
25 Higgins, *Public Affairs*, 70, 71.
26 Ibid.; Canada, *Statutes*, 1944, NHA; Canada, Commons, *Debates* (1944), IV:5985–8, 5900–93, 6179–81.
27 *Toronto Globe and Mail*, 28 Oct. 1947, 1.
28 Granatstein, *Canada's War*, 281.
29 Letter of J.L. Ilsley to Mayor Luciene Bourne, 6 Feb. 1945; Bourne to Ilsley, 31 Jan. 1945, both in NA, RG19, vol. 4018.
30 W.C. Clark to H.H. Stevens, 4 May 1945, NA, RG19, vol. 4018.
31 Letter from C.D. Howe to J.L. Ilsley, 24 Feb. 1945, NA, RG19, vol. 709.
32 Letter from Ilsley to Howe, 12 Mar. 1945, NA, RG19, vol. 709; newspaper clippings from Clark's personal files, NA, RG19, vol. 3975.
33 F.N. Anderson, memo re meeting with insurance companies, 3 July 1945; operation of rental projects by life insurance companies, proposal for emergency operation by life insurance companies, 5 July 1945, NA, RG21-A, vol. 343.
34 Ibid.
35 *Rental Control and Evictions*, 52–6.
36 Eric Gold, memo re freezing of tenants, NA, RG64, vol. 700.
37 Ibid.
38 Ibid.; *Vancouver Daily Province*, July 1945, VCA, Housing docket 4, M4289–4.

39 Memo to the minister of finance re notices to vacate given by landlords of self-contained dwellings on the ground that the landlord desires to live in the dwelling himself, 23 July 1945, NA, RG64, vol. 700.
40 *Rental Control and Evictions*, 56.
41 Memo to the minister of finance re notices to vacate.
42 Telegram from C.D. Howe to B.K. Boulton, 24 July 1945; telegram from B.K. Boulton to C.D. Howe, 24 July 1945, both in NA, RG28-A, vol. 343.
43 Letter, C.D. Howe from B.K. Boulton, 25 July 1945, NA, RG28-A, vol. 343; minutes of seventh meeting of interdepartmental Housing Committee, 8 Aug. 1945, NA, RG19, vol. 714, file IHC.
44 Letter from Eric Gold to Mitchell Sharp, 5 Sept. 1945, NA, RG64, vol. 700.
45 Letter from David Mansur to W.A. Mackintosh, 30 July 1945, NA, RG64, vol. 700.
46 Ibid.
47 Emergency Shelter bulletin, 12 Sept. 1945, NA, RG64, vol. 708.
48 Emergency Shelter memo, 24 Nov. 1945, NA, RG64, vol. 708.
49 Emergency Shelter bulletin, 28 Nov. 1945, NA, RG64, vol. 708.
50 Letter from J.G. Godsoe, Donald Gordon, David Mansur, and Mitchell Sharp to I.L. Ilsley and C.D. Howe, NA, RG19, vol. 4018.
51 A.J.P. Heeney, memo to the prime minister re government housing policy and ministerial responsibility, 24 Nov. 1944, NA, RG2, ser. 18, vol. 12.
52 Central Mortgage and Housing Corporation, *Annual Report* (1949), 3.
53 Carver, *Compassionate Landscape*, 112.

CHAPTER SEVEN

1 Carver, *Compassionate Landscape*, 106–8.
2 Ibid., 107, 108; *A National Housing Policy for Canada* (Ottawa: Canadian Welfare Council 1947), 21; Canadian Social Welfare Council, "A National Housing Policy for Canada," *Canadian Welfare* (1 Sept. 1947): 33–7.
3 Memo to C.D. Howe from David Mansur, 30 Oct. 1945, in NA, RG19, vol. 727.
4 Ibid.
5 Ibid.
6 Ontario Association of Public Housing Authorities, *Good Housing for Canadians* (Toronto 1964), 1–99; Dennis and Fish, *Programs in Search*, 173; Ontario Social Welfare Council, *A Study of Housing Policies in Ontario* (Toronto 1973), 9.
7 Memo re housing, 3 June 1949; minutes of CMHC board of directors, 25 June 1949, both in NA, RG19, vol. 726, file 203 MHC-1.

8 Memo re housing, T.C. Douglas to Louis St Laurent, 2 Mar. 1949, PAS, T.C. Douglas Papers, file Housing.
9 David Mansur, memo re housing discussions with Ontario, 17 Aug. 1949, NA, RG19, vol. 3439; Mansur to Winters, 18 Aug. 1949, both in NA, Privy Council Papers, RG2, 18, vol. 127.
10 Minutes of CMHC Executive Committee, 23 May 1950, in NA, RG19, vol. 723; CMHC, *Annual Report* (1951), 16.
11 Minutes of a meeting of board of directors, 20 Feb. 1950, NA, RG19, vol. 724. As Mansur had predicted earlier, at its 17 Apr. 1950 meeting CMHC's board of directors observed that Ontario and Quebec were only interested in using the legislation for land-assembly projects. Manitoba quickly discouraged construction of any public housing by making its municipalities responsible for 100 per cent of the provincial share of costs. See NA, RG19, vol. 723.
12 Newspaper clipping, Housing 1950–1963, VCA, Housing docket 6, M4289–6.
13 Vancouver Housing Association bulletin, 1 Dec. 1954, VCA, RG3, vol. 98, file 35-C-1.
14 Vancouver Housing Association bulletin, extracts of study by W.A. Wilson re housing conditions among social-assistance families, VCA, RG3, vol. 115, file 35-F4. *"We Too, Need Housing": A Survey of the Housing Needs of Single Girls Employed in Downtown Vancouver* (Vancouver Housing Association, May 1951), in VCA, RG3, vol. 83, 34–57.
15 Vancouver Housing Authority, *Annual Report* (1955–57), VCA, RG3, vol. 121, file 35-F-3.
16 Letter from R.R. Hamlin, president, Bayers Road Housing Association, to A.A. DeBard, city manager, 28 Jan. 1954; letter, A.E. Smith, regional supervisor CMHC, to A.A. DeBard, 8 Nov. 1954; letter, James A. Cody, general secretary, Halifax and District Trades and Labour Council, to Mayor R.E. Donahue, 11 July 1953; press clippings, PANS, RG35–102, 4.A6; Rose, *Regent Park*, 123.
17 Harriet Parsons, "Where Will the People Go?" *Canadian Welfare*, 1 Feb. 1957. Parsons noted that most shelters were former barracks, which were "long wooden buildings, mostly covered with tar paper" and divided into "tiny three room apartments by flimsy fibre board partitions." Some of the families were rehoused in homes acquired by the city and awaiting demolition for urban-renewal projects and the future Regent Park South public-housing development. See confidential memo to Mayor R.A. Donahue from A.A. DeBard re emergency shelters, 25 Feb. 1954; letter to A.A. DeBard from Kenneth A. Ross, secretary, board of Trade; press clippings from city manager's files re emergency housing; memo, 2 Mar. 1955, A.A. DeBard to city council, all in PANS, RG35–102, 4A, 4, 28. Carver, *Compassionate Landscape*, 108.

18 CMHC *Annual Report*, (1950) 4, 5; Carver, *Compassionate Landscape*, 109.
19 Memo to W.C. Clark, 19 Jan. 1948, in NA, RG19, vol. 3980.
20 Rental-housing proposals, 24 Nov. 1947, in NA, RG19, vol. 727; 1 Dec. 1947, meeting of CMHC board of directors, NA, RG19, vol. 727.
21 Ibid.; memo to W.C. Clark re housing proposals, 26 Nov. 1947, NA, RG19, vol. 727; 1 Dec. 1947, meeting of CMHC board of directors.
22 Rental-housing proposals, 24 Nov. 1947, in NA, RG19, vol. 727.
23 Ibid.
24 Memo to D.B. Mansur from H. Woodward re Survey of rental situation in Montreal, in NA, RG19, vol. 726, file 203 CM2d-1. Meeting of board of directors, 25 June 1949, NA, RG19, vol. 726, file 203, CM2d-1.
25 Some idea of the severity of the post-war housing shortage can be gained from the fact that only 5 to 10 per cent of Montreal's residents at this time who were seeking rental accommodation could afford serviced units with hot water. In NA, RG19, vol. 726, file CM2d-1, meeting of board of directors, 25 June 1949. See also minutes of CMHC board, 10 May 1948. The Executive Committee of CMHC decided to curtail rental insurance rather than have certain projects built that would charge such high rentals as to prove a political embarrassment to the government. NA, RG19, vol. 727.
26 Memo from Mansur to G.J. McIlraith, parliamentary assistant to the minister for reconstruction and supply, NA, RG2, ser. 18, vol. 80.
27 Memo from Humphrey Carver re housing co-operatives in Nova Scotia, 18 Nov. 1949, in NA, RG19, vol. 726, file 203 CM2d-1. Memo from Humphrey Carver to David Mansur re housing co-operatives in Quebec, 5 Dec. 1949, in NA, RG19, vol. 726.
28 CMHC, *Annual Report* (1950), 4, 5; Carver, *Compassionate Landscape*, 109.
29 *Rental Control and Evictions*, 61.
30 Ibid., 62.
31 Ibid., 62–5.
32 Letter, 9 May 1947, Owen Lobley to K.W. Taylor, in NA, RG19, vol. 361, file 101–102–1.
33 *Rental Control and Evictions*, 67–9.
34 Memo, 13 May 1947, from D.B. Mansur re Rent Controls, in NA, RG19, vol. 361, file 101–102–1.
35 Letter to D.C. Abbott from Owen Lobley, 22 May 1947, re decontrol of rentals, in NA, RG19, vol. 361, file 101–102–1.
36 A Plan and Program for Decontrol of Rents, 27 Oct. 1949, in NA, RG19, vol. 361, file 101–102–1.
37 Letter, K.W. Taylor to D.C. Abbott, re Newfoundland price and rent controls, NA, RG19, vol. 361, file 101–102–1. The Supreme Court did uphold federal rent controls in the interest of "peace, order and

good government." Mansur's strategy, however, had allowed the government to overcome its constitutional powers to delegate a responsibility it wished to see delegated to the provinces.

38 Memo re the present state of decontrol, in NA, RG19, vol. 361, file 101–102.
39 Ibid.
40 Memo to the minister re rent control, from Mitchell Sharp, 20 Oct. 1944, in NA, RG19, vol. 361, file 101–102.
41 Memo to the Cabinet re rent control, from D.C. Abbott, 1 Nov. 1949, in NA, Privy Council Papers, RG2, ser. 18, vol. 119, file W-32-1 (1943–49).
42 Ibid.
43 Statement by the chairman of the Wartime Prices and Trade Board re the establishment of wartime controls with particular reference to rent controls, 3 Nov. 1949, in NA, RG19, vol. 723.
44 Memo to the Cabinet re rent control.
45 Letter, 14 Dec. 1949, Douglas Campbell to D.C. Abbott; letter, 27 Jan. 1950, Abbott to Campbell, both in NA, RG19, vol. 362, file 101–102–1, sec. 1.
46 Translations of telegrams of 16 Dec. 1949 to and from D.C. Abbott and M.L. Duplessis, in NA, RG19, vol. 362, file 101–102–1, sec. 1.
47 Letter from T.C. Douglas to D.C. Abbott, 6 Dec. 1949, in NA, RG2, ser. 18, vol. 119, file W-32-1 (1943–49).
48 Cross-reference sheet regarding Cabinet conclusions, 7 and 8 Dec. 1949, of interest to Treasury Board, including rent controls, Supreme Court reference, and correspondence with the premier of Saskatchewan, 10 Dec. 1949, in NA, RG2, ser. 18, vol. 119, file W-32-1 (1943–49).
49 Department of Finance, memo to the Cabinet re rent controls in Saskatchewan, 21 Nov. 1949, in NA, RG2/18, vol. 119, file W-32-1.
50 For an example of the hostility within the CCF to the rent increases of the federal government, see editorial, "Abbott in Wonderland," in the *Canadian Forum*, Dec. 1949. Douglas was also influenced by an account by the *Financial Counsel*, 21 Nov. 1949, which maintained that landlords' return on investment had "risen considerably during the past few years as an inevitable outcome of the housing situation." The *Counsel* pointed out that landlords benefited from 100 per cent occupancy, which also allowed them to make fewer repairs. The inequities of the federal control program led Saskatchewan attorney general J.W. Corman to propose a separate provincial system as early as 11 Oct. 1949, before the federal increase announced on 3 Nov. 1949. Corman believed that the provincial mediation board had exercised a "restraining influence on grasping landlords" and that this would also happen in the domestic field. See correspondence, telegrams, and

memos in PAS, Douglas Papers, Controls-rentals, 98 (234 and 2–34–1).

51 Press statement, 25 Nov. 1949; press release on increase in rents, 4 Nov. 1949, both in PAS, Douglas Papers, RG31, H98 (2–34–1).

52 Vancouver Housing Association, "Evictions in Toronto," *Housing Bulletin*, Feb. 1949; Albert Rose, "Rental Problems of 1,000 Canadian Families," *Canadian Welfare*, Mar. 1950.

53 Notes respecting rent decontrol, 2 Mar. 1950; newspaper clipping, "Two Foes of Rent Controls Disagree on Ottawa Rights," *Montreal Gazette*, in NA, RG19, vol. 361.

54 Vancouver Housing Association bulletin, Mar. 1955.

55 Halifax council minutes, 30 Apr. 1951, 330–5; 26 Apr. 1951, 317–29; 19 Apr. 1951, 302–10; rental by-law, 20 June 1951, 1–4: letter, 7 Jan. 1953, George H. Ferguson, rental-control officer to mayor, memo re rent control from H. Ferguson, 19 Sept. 1956, all in PANS, RG35–102.

56 Bettison, *Politics of Canadian Urban Development*, 65–80; confidential memo re housing, Aug. 1950, in NA, RG62, ser. 18, vol. 163.

57 Letter from R.H. Winters to Louis St Laurent, 12 Jan. 1951, in NA, RG2, ser. 18, vol. 163.

58 "Mortgage Lending by CMHC," by James E. Coyne, 27 Dec. 1950, in NA, RG19, vol. 3439.

59 Letter from R.H. Winters to Louis St Laurent, 11 Jan. 1951; memo for N.A. Robertson, 13 Jan. 1951, in NA, RG2, ser. 18, vol. 163.

60 Minutes of CMHC board of directors, 9 Apr. 1951, in NA, RG19, vol. 725.

61 Memo of a meeting with housing committee of Dominion Mortgage and Investments Association in Toronto, 2 May 1950, from Denys H. Block, regional loans manager, to J.A. Jones, regional supervisor, in NA, RG56, vol. 52.

62 Ibid.

63 Memo from J. Allan Jones, BC regional office, to D.B. Mansur re direct-loan procedures, 30 June 1950, in NA, RG56, vol. 52.

64 Memo from T.K. Shoyama to T.C. Douglas, 4 June 1951; letter, 9 Mar. 1951, Robert Winters to John H. Sturdy; letter, 19 Mar. 1951, John H. Sturdy to R.H. Winters; telegram, 19 Mar., 1951, John H. Sturdy to R.H. Winters, all in PAS, Douglas Papers, Construction file, 433 (30–1). It is of interest that no documents relating to the federal government's desire to curtail public housing during the Korean War have been transferred to the government records section of the National Archives despite numerous references in the Douglas Papers to this plan. The federal government also refused to release supplies for the building of a community centre for Regent Park; see Albert Rose, *Regent Park*, 167, 168.

65 Memo re housing for defence workers, 11 Apr. 1951, in NA, RG19, vol. 725.
66 Ibid.
67 Ibid.
68 CMHC, *Annual Report* (1951), 11.
69 Ibid.
70 Ibid., 16–18.
71 Memo from Loans Department to Angus McClaskey, supervisor, Ontario Regional Office, re availability of mortgage funds, 12 June 1952, in NA, RG56, vol. 53.
72 Memo from H. Woodward to D.B. Mansur, 22 Jan. 1952, re direct-loan policy; memo from D.B. Mansur re direct-loan and agency policy, both in NA, RG56, vol. 56, file 6.
73 Canada, Commons, *Debates* (1952), 4:2116; see Clark obituary tributes in NA, RG19, vol. 3993.
74 *Ottawa Journal*, 29 Dec. 1952, *Ottawa Evening Citizen*, 29 Dec., 1952, clippings in NA, RG19, vol. 3393.
75 W.E. McLaughlin "Mortgage Lending by Canadian Banks," *Canadian Bankers* (Spring 1955): 54.
76 Ibid., 55.
77 Ibid., 55, 57.
78 Ibid., 56.
79 Ibid.
80 Ibid., 56, 57.
81 Ibid.
82 Evidence of the Standing Committee on Banking and Commerce, re Bill 102, Thursday, 25 Feb. 1954: testimony of R.K. Fraser, chairman of the Legislative Committee, National House Builders Association, 415–48; testimony of R. Brunet, president, Canadian Construction Association, V.L. Leigh, chairman, Housing Committee, CCA, S.D.C. Chutter, assistant manager, CCA, 448–74.
83 Ibid.
84 Ibid.
85 Canada, Commons, *Debates* (1954), 3:2735.
86 Ibid., 2:1425–39. Conservative MP Gordon Churchill also noted that Winnipeg still had 199 families living in emergency shelters and that, in 1953, 132 families had applied for such accommodation (1412), Evidence of the Standing Committee, 491–516.
87 Testimony of George Mooney to Standing Committee on Banking and Commerce.
88 *Report of the First Joint Conference on Co-operative Housing*, PAS, Douglas Papers, XV:20T-Co-Op Housing, 1932–58.

89 Memo from R.S. Staples to national directors, provincial secretaries, delegates to housing conference, in PAS, Douglas Papers, XV:20T-Co-Op Housing.
90 Submission to the Standing Committee on Banking and Commerce by the Co-Operative Union of Canada and Le Conseil canadien de la co-opération, in PAS, Douglas Papers, XV:20T-Co-Op Housing, 1932–58.
91 Staples, memo to national directors, provincial secretaries, delegates to housing conference.
92 Memo to Robert Winters from P.S. Secord, 2 Dec. 1954; letter from Anne Rivkin, 11 Jan. 1954, to Allan Armstrong, both in NA, RG56, vol. 55.
93 Submission to Standing Committee on Banking and Commerce.

CHAPTER EIGHT

1 Carver, *Compassionate Landscape*, 125–40; Dennis and Fish, *Programs in Search*, 137, 173, 293.
2 Carver, *Compassionate Landscape*. "Mayor Says NHA Change Would Be Godsend to City," *Halifax Mail-Star*, 23 Dec. 1955, PANS, RG35–102, 4A, 10, 57, file re Slum Clearance, 1951–59, press clippings scrapbook.
3 Memo, Stewart Bates to Robert Winters, 1 June 1956; letter to Stewart Bates, 12 Feb. 1957, cited in Dennis and Fish, *Programs in Search*, 137, 293.
4 "Residents, Owners Reported Opposing Development," 31 Dec. 1955, *Halifax Mail-Star*, PANS, RG35–102, 4A, 10, 57, file re Slum Clearance, 1951–59, press clippings scrapbook.
5 George Stephenson, *A Redevelopment Study of Halifax* (Halifax: City of Halifax and CMHC 1957), 54–9.
6 Ibid., 23–8.
7 Ibid.
8 "Redevelopment, Halifax, Nova Scotia," 14 Nov. 1957, 6, in PANS, RG35-M (42 unsorted).
9 Ibid.
10 Minutes of the Housing Committee, 16 Apr. 1958, 10–12, PANS, RG35-M (42 unsorted).
11 Institute of Public Affairs, Dalhousie, *The Conditions of the Negros of Halifax City, Nova Scotia* (Halifax 1962), 13–15; minutes of Redevelopment Committee, 15 Apr. 1959, 4, in PANS, RG35-M (42 unsorted).
12 "Qualification of Tenants – Mulgrave Park Housing Project," 5–6, PANS, RG35–102 (42 unsorted); Allocation Policy Mulgrave Park, 25 May 1961, PANS, RG35–102, 39 (c.1); "Mulgrave Park Tenants Face 12 Percent Rent Increase," *Mail-Star*, 25 May 1962; "Evictions Start at

Mulgrave," *Mail-Star*, 30 May 1962; "Halifax to Oust Mulgrave Park Big Pay Tenants," *Mail-Star*, 27 Feb. 1962; "Mulgrave Park to Evict Families,"*Mail-Star*, 9 May 1962, all in PANS, Dept. of Public Works newspaper scrapbook on housing, MG9, vol. 461.

13 Housing Policy Review Committee minutes, 11 July 1961, 1–3, PANS, RG35–102, 34 (6.2).

14 Ibid., 6.

15 Report of the Housing Policy Review Committee, 12 May 1961, 2, PANS, RG35–102, 34 (6.3).

16 T. Penfold, "An Analysis of the Causes of Deteriorating Housing Conditions in Selected Areas of Halifax," PANS, MG100, vol. 154, #43; H.S. Coblentz, "Relocation," background paper for Halifax housing study, 28 June 1961, in Library of Nova Scotia Archives, WA-9130, H17, no. 6.

17 Lembi Buchanan, "The Men Who Built Halifax," *Halifax Magazine*, June 1981, 40–4; Eleanor O'Donnell, "Leading the Way: An Unauthorized Guide to the Lobley Empire," in Gary Burwell and Ian McKay, eds., *People, Resources and Power* (Fredericton: Acadiensis Press 1987), 48.

18 "Re-development Plans Breed Suspicion among City's Chinese," 4 Feb. 1961, VCA, press clipping docket, in Housing Developments, M4292; "It May Be a Slum but It's Home for Many Folk," *Vancouver Sun*, VCA, Clippings, Civic Slum Clearance, MS12, 293.

19 Carver, "Community Renewal Programming," 8; "Rebuilding Vancouver," Vancouver Housing Association bulletin, Mar. 1958, VCA, 36-A-3; "Urban Renewal," Vancouver Housing Association bulletin, Aug. 1965, VCA, 36–5–7.

20 Carver, "Community Renewal Planning," 9; Donna Sabeline, Dane Zaken, Gordon Pape, *Montreal at the Crossroads* (Montreal: Harvest House 1982), 65–81; Ron Haggert and Aubrey Golden, *Rumours of War* (Toronto: New Press 1971), 149, 150; "Table One – Housing Activities under Federal Provincial Partnership," in letter from John R. Nicholson to Lester B. Pearson, in NA, Alexander Laidlaw Papers, MG31, B32, vol. 29, file 29–14.

21 H. Spence Sales, "Urban Redevelopment of Moncton," *Atlantic Advocate*, Apr. 1965, 19–23; M.M. Sommerville, "Urban Renewal and Public Relations,"*Urban Renewal and Public Housing in Canada* 2, no. 1 (1906).

22 Albert Potvin, "Neglect and Urban Decay," *Habitat* 1, no. 7: 20–1

23 Amherst urban-renewal study, in 1965, in PANS, Library Division, F107, AM5, UR1; urban-renewal study of Pictou County, 1966, PANS, Library Division, F106, 58.

24 Norman Pearson, *Town of Glace Bay, Urban Renewal Study* (1965), in Library, Institute of Public Affairs, Dalhousie University, 35–8.

25 Ibid., 36–9.
26 Ibid., 40–2.
27 Ibid., 42, 43.
28 Ibid., 42.
29 Ibid., 70.
30 Ibid., 77–81.
31 Ibid., 73, 79.
32 Peter G. Burns "Enforcement of Minimum Standards for Existing Housing," *Urban Renewal and Public Housing in Canada* (1965): 5–9.
33 Victor K. Copps, "The Urban Renewal Program in Hamilton, Ontario," *Urban Renewal and Low Income Housing* 8, no. 3 (1972): 6–10.
34 Ibid., 12.
35 Dennis and Fish, *Programs in Search*; Franklin J. Henry, *The Consequences of Relocation:A Study of Hamilton's North End* (Hamilton 1974), passim.
36 Henry, *The Consequences of Relocation*, 30–61.
37 Ibid., 98–129.
38 Ibid., 130–6.
39 Ibid., 100–29.
40 Straus, *Seven Myths*, 69–93; Carver, *Compassionate Landscape*, 122, 190; Carver, "Community Renewal Programming," 6. In his 1965 article on urban renewal, Carver indicated that he conceived of the name of urban renewal at a New York City conference, "in a taxi-cab between the Pennsylvania Station and the Grand Central."
41 Fraser, *Fighting Back*, 60, 61, 108, 109.
42 Ibid., 150, 151.
43 Paul Hellyer, "Why They Killed Public Land Banking: A Political Memoir," *City Magazine* 3, no. 2: 27.
44 Metropolitan Toronto Housing Authority, *Regent Park South: A Study* (1962), passim.
45 Dennis and Fish, *Programs in Search*, 317.
46 Ibid., 197–206.
47 Annual reports, Metropolitan Toronto Housing Company, 1982–87.
48 Personal communication, George Barker, president, Metropolitan Toronto Housing Company; articles on R.J. Smith, *Ontario Housing* (1965) 23–7.
49 Michael Audain, *Beyond Shelter* (Ottawa: Canadian Council on Social Development 1973), 235–45.
50 Ibid., 297–381.
51 Ibid., 405–17.
52 Canadian Welfare Council, *Who Should Manage Public Housing*? (Ottawa 1970), passim.
53 Dennis and Fish, *Programs in Search*, 250.

54 Ibid.

55 *History of Willow Park Co-operative Community* (Winnipeg 1973), 1.

56 Ibid., 2–8.

57 Clipping from *Enterprise*, May 1978, in NA, MG31, B31, vol. 23, file 23–3; Co-operative Housing Foundation, "Co-operative Housing: What Credit Unions Can Contribute," "Co-operative Housing: A Solution to the Housing Problems of Labour Members," in NA, MG31, B32, vol. 23, file 23–6.

58 Co-operative Housing Federation of Canada, *The First Ten Years of the Co-operative Housing Foundation*, 1, 2; Dennis and Fish, *Programs in Search*, 218, 248–62.

59 Dennis and Fish, *Programs in Search*, 225–62.

60 Carver, *Compassionate Landscape*, 164.

61 Hellyer, "Why They Killed Public Land Banking," 37.

62 Ibid.; Peter Spurr, *Land and Urban Development* (Toronto: James Lorimer 1976), 316–24.

63 Hellyer, "Why They Killed Public Land Banking," 34.

64 Spurr, *Land and Urban Development*, 81–182, 196–293; James Lorimer, *The Developers* (Toronto: James Lorimer 1978), 89–94; Dennis and Fish, *Programs in Search*, 323, 324.

65 Graham Barker, Jennifer Penner, Wally Seccombe, *Highrise and Superprofits* (Kitchener: Dumont Press Graphix 1973), 16–23, 62–72; Spurr, *Land and Urban Development*, 18.

66 Canada, Commons, *Debates* (1969), 79–93.

67 Records of Runneymeade filed with Companies Branch of the Ontario Ministry of Consumer and Corporate Relations.

68 Canada, Commons, *Debates* (1969), 7979, 7980; Hellyer, "Why They Killed Public Land Banking," 35.

69 Ibid.

70 Dennis and Fish, *Programs in Search*, 37–8.

71 Albert Rose, *Citizen Participation in Urban Renewal* (Toronto: Centre for Urban and Community Studies 1974), 204, 205, 210, 224, 225.

72 Donald Gutstein, Jack Long, Dorothy McIntosh, and the editors, "Neighbourhood Improvement: What It Means in Calgary, Vancouver and Toronto," *City Magazine* (Aug. – Sept. 1975): 23–7.

73 Ibid., 15, 16.

74 Karen Hill and Janet McClain, "Redefining the Inner City," and Julia Weston, "Gentrification," both in *Habitat* 15, no. 4 (1982).

75 David Greenspan, *Down to Earth* (Toronto: Federal-Provincial task force on the Supply and Price of Serviced Residential Land 1978), 7.

76 Ibid., 14–31; Lorimer, *The Developers*, 109–12.

77 Lorimer, *The Developers*, 229–36.

78 Ibid., 233–6.

79 Greenspan, *Down to Earth*, 56.
80 Spurr, *Land and Urban Development*, 149, 150; Lorimer, *The Developers*, 104–9.
81 James Lorimer, "Avoiding Real Issues of Developers' Profits," *Globe & Mail*, 26 Sept. 1978.
82 Greenspan, *Down to Earth*, 39.
83 Lorimer, *The Developers*, 112–21; Niagara Regional Planning Department, *Residential Lot Inventory* (1977); Central Mortgage and Housing, *Land Infrastructure Report: Hamilton Region* (1978); Land Banking and New Communities Division, Central Mortgage and Housing, *Servicing and Infrastructure Analysis* (1978), passim.
84 Ibid.
85 Greenspan, *Down to Earth*, 43; Land Banking and New Communities Division, CMHC.
86 Lorimer, *The Developers*, 117–18.
87 Carver, *Compassionate Landscape*, 189.
88 Memo, A.F. Laidlaw, The Organization of CMHC for Third Sector Housing, in NA MG31, B-32, file 11–26; Alexander Laidlaw, comments on ten advantages of condominiums, 7 July 1972, in NA, MG31,B-32,
vol. 11, file 11–3.
89 Dennis and Fish, *Programs in Search*, 200–80.
90 Ibid.
91 Alexander Laidlaw, Current Activity in Co-operative Housing, mid-Feb. 1974, 20 Feb. 1974, in NA, MG31, B-32, vol. 11, file 11–30.
92 Alexander Laidlaw, notes on prospects for co-operative housing in Saint John, Moncton, Halifax, and Sydney, based on visit by Laidlaw and Hadrell, 29 Nov. 1974, in NA, MG31, B-32, file 11–58.
93 City of Toronto Housing Department, *No Vacancy: Will the New Federal Housing Programs Work in Toronto*? (Toronto 1979), 17–99.
94 City of Toronto Housing Department, *Vanishing Options: The Impending Rental Crisis in Toronto* (Toronto 1980), 37, 38.
95 Social Planning Council of Metropolitan Toronto, *A New Housing Agenda For Metropolitan Toronto* (Toronto 1984), 36.
96 Ibid., 37.
97 Zuhair Kashmeri, "AHOP Debacle Leaves CMHC a Giant Landlord – but Broke," *Toronto Globe and Mail*, 11 Nov. 1980, clipping in NA, MG31, B-32, vol. 23, file 23–3.
98 Ibid.; Glen Warner, "The Great Condo Con Job," in NA, MG31, B-32, vol. 23, file 23–3.
99 Canada Mortgage and Housing, *Annual Report* (1983), 34.
100 Ibid., 4–66.

101 Barbara Baker, "Latest of 35 Metro Co-ops First To Offer Single Houses," *Globe & Mail*, 13 Aug. 1979, clipping in NA, MG31, B-36, vol. 23, file 23–3.

102 Paul Weinberg, "DelZotto Dealt in for 'Non-profit' Development," *Toronto Clarion*, 27 Nov. 1979. The development company in question was DelZotto Enterprises. One of its executives, Angelo DelZotto, was charged by an Ontario royal commission with introducing "a sinister array of characters" (i.e., organized-crime figures) into the Ontario construction industry to serve as labour racketeers and to intimidate the competitors of a subsidiary lathing company. See Justice Harry Waisberg, *The Report of Certain Sectors of the Building Industry* (Toronto: Queen's Printer 1974), 41–8, 143, 199–209, 229. Angelo DelZotto also bribed an Ontario Housing Corporation official: see Kozer Worth, "Chemalloy Confusion Continues," *Financial Post*, May 1970, 40. The Cimpello Charitable Foundation worked closely with the DelZotto companies in all its non-profit projects, including its property-management firm, Del Realty. This was seen as an asset by CMHC Toronto director Tom Hughes. He maintained that large apartment buildings cannot "be run by tenants"; see Weinberg.

103 Ottawa Co-operative Council, "Co-op Housing: A Solution to the Housing Problems of Labour Union Members," *Directory of Co-operatives in the Ottawa District* (Ottawa 1983), passim.; *Ottawa Citizen*, 17 Nov. 1977, "Labor Housing Co-op One of the Largest," clipping in NA, MG31, B-32, vol. 29, file 29–14.

104 Douglas Mulhall, "CMHC's Bill Teron Tells Housing Co-ops to: Get Your Act Together," *Credit Lines*, 15 Feb. 1978, in NA, MG32, vol. 23, file 23–3.

105 Co-operative Housing Foundation, "Response to CMHC's Non-Profit Housing Program, 1978," in NA, MG31, B-32, vol. 23, file 23–6.

106 Ibid.

107 Mark Goldblatt, "Saving a Movement," in NA, MG31, B-32, vol. 23, file 23–6.

108 Ibid.

109 "Co-op Housing at the Cross Roads," "Suggestions to Regional Co-ordinators," "CHT Annual Meeting Resolutions re Unit Allocations," in NA, MG31, B-32, vol. 23, file 23–6.

110 Gabriella Golager, "How the People of Milton Park Saved Their Neighbourhood," *Habitat* 25, no. 3 (1982): 2–9; *Milton Park: A Co-operative Neighbourhood* (Montreal 1983), passim.

111 Toronto Housing Department, *Vanishing Options*, 37, 38; Co-operative Housing Foundation, *Positive Findings From the 56.1 Evaluation* (Ottawa 1983), 1–6.

112 Personal communication with staff of Metropolitan Toronto Housing Company.

113 Mark Goldblatt, "Launching a Worker Co-op Sector: Lessons from Co-op Housing," *Worker Co-ops* (Spring 1988).

114 Sidney Kling and Adam Fuerstenberg, *Senior Citizen Development in Metropolitan Toronto: The Role of Individual Religious Congregations* (Toronto: Ryerson Polytechnical Institute 1983), 55, 56.

115 Tom Carter and Deborah Lyon, *Housing in Alberta*. Institute of Urban Studies research and working paper no. 29 (1987), 33.

116 Report of Ontario Housing Corporation to the Social Services and Housing Committee of the Council of the Municipality of Metropolitan Toronto (1972), 7.

117 Tom Carter and Ann Mcaffe, "The Municipal Role in Housing the Homeless and Poor," in Alex Murray and George Fallis, eds., *Housing the Homeless and Poor* (Toronto: University of Toronto Press 1990), 231–50.

118 Keith G. Banting, "Social Housing in a Divided State," in Murray and Fallis, eds., *Housing the Homeless and Poor*, 133, 134.

119 Carter and Mcaffe, "The Municipal Role," 115.

120 Banting, "Social Housing," 147, 152.

121 Social Planning Council of Metropolitan Toronto, *People without Homes: A Permanent Emergency* (Toronto 1983), passim.

122 Ibid.

123 Social Planning Council of Metro Toronto, "Innovations in Regulations and Legislation Affecting the Rights of Persons without Secure Housing," in Heather Lang Rutz and Doyne C. Ahern, eds., *New Partnerships – Building for the Future*, Proceedings of the Canadian Conference to Observe the International Year of Shelter for the Homeless (Ottawa: Canadian Association of Housing and Renewal Officials 1988), 96; Department of Community Services, Metropolitan Toronto, *No Place To Go: A Study of Homelessness in Metropolitan Toronto* (1983), passim.

124 Task Force on Housing for Low Income Single People of Metropolitan Toronto, *Final Report* (1984).

125 John Sewell, "Trapped by Inaction," *Now Magazine*, 29 Sept. – 5 Oct. 1988, 11.

126 Alex Murray, "Homelessness, the People," in Murray and Fallis, eds., *Housing the Homeless and Poor*, 41.

127 Peter Smith, introduction to Lang Rutz and Ahern, *New Partnerships*, 3.

128 "Canadian Native Housing Issues," in Lang Rutz and Ahern, *New Partnerships*, 44–6.

129 Ibid.

130 Social Assistance Review Committee, *Transitions* (Toronto: Queen's Printer for Ontario 1988), 69.
131 Ibid., 60.
132 Ibid., 583.
133 Ibid., 8.
134 Sewell, "Trapped," 10, 11.
135 David Lewis Stein, "Is Ontario Set to Privatize Public Housing?" *Toronto Star*, 23 Sept. 1988.
136 "Funding Appropriate Shelter," in Lang Rutz, Ahern, *New Partnerships*, 71, 72.

CHAPTER NINE

1 Alex Murray, "Homelessness: The People," in Alex Murray and George Fallis, eds., *Housing the Homeless and the Poor* (Toronto: University of Toronto Press 1990), 42–5.
2 Claire Helman, *The Milton Park Affair* (Montreal: Véhicule Press 1987), 170–1.
3 George Grant, "An Ethic of Community," in Michael Oliver, ed., *Social Purpose for Canada* (Toronto: University of Toronto Press 1961), 12.

Index